CURRENT INJECTION IN SOLIDS

ELECTRICAL SCIENCE
A Series of Monographs and Texts

Editors: Henry G. Booker and Nicholas DeClaris
A complete list of titles in this series appears at the end of this volume

CURRENT INJECTION IN SOLIDS

MURRAY A. LAMPERT
DEPARTMENT OF ELECTRICAL ENGINEERING
PRINCETON UNIVERSITY
PRINCETON, NEW JERSEY

AND

PETER MARK
DEPARTMENT OF ELECTRICAL ENGINEERING
PRINCETON UNIVERSITY
PRINCETON, NEW JERSEY

1970

ACADEMIC PRESS New York and London

ACADEMIC PRESS, INC.
111 Fifth Avenue, New York, New York 10003

United Kingdom Edition published by
ACADEMIC PRESS, INC. (LONDON) LTD.
Berkeley Square House, London W1X 6BA

Library of Congress Catalog Card Number: 75-117115

PRINTED IN THE UNITED STATES OF AMERICA

To ALBERT ROSE

who chose simplicity over precision
and thereby gave to all of us insight.

CONTENTS

3. ANALYTIC FORMULATION AND GEOMETRIC PROPERTIES

4. ANALYTIC AND APPROXIMATE SOLUTIONS

5. RESEARCH APPLICATIONS AND EXPERIMENTAL RESULTS

6. TRANSIENT INJECTION CURRENTS

7. SPECIAL TOPICS

8. INJECTION FROM A POINT AND OTHER GEOMETRIES

PREFACE

Following the spectacular success of the bipolar transistor in solid-state electronics there has been a continually growing interest in the properties of currents injected into more insulating materials. This interest is fed from many sources, both scientific and technological. On the scientific side, such currents have proven capable of yielding valuable information about the defect states in the forbidden gap of insulators and also about the transport coefficients in these materials. On the technological side, interest is fed from the growing successes of the unipolar, field-effect devices and the injection, electroluminescent, and lasing devices.

The intent of this book is to provide a thorough exploration of the bulk-controlled behavior of injection currents in solids, with emphasis on the underlying physics. It is hoped that the book will prove useful both to those working in research or development in this general area and to those wishing to learn something about the field. The subject matter is covered at several levels, ranging from the purely phenomenological to the purely analytical. In the latter approach, the characteristic differential equations are studied both approximately and, where possible, rigorously. One of us (M.A.L.) has, for several years, given a one-semester graduate course based on the subject matter of the book. The only background required for comprehending the subject matter is a previous exposure to the basic concepts of modern solid-state physics, such as provided in an introductory course in solid state for electrical engineers or physicists. These concepts, namely energy bands and energy gap, effective masses for electrons and holes, localized electronic states (defect states) in the gap, Fermi–Dirac statistics, and Fermi level, provide an exceedingly convenient language in which to discuss the problems of current injection.

The subject matter divides naturally into two parts: one-carrier injection currents which are necessarily space-charge-limited, and two-carrier injection currents in which recombination, as well as space charge, can play an important role. The discussion is largely confined to steady-state, planar-flow current injection. Two exceptions are the chapter on the single-injection, planar-flow transient and the chapter on the single-injection, steady-state spherical flow. Representative experimental data are presented, where available.

Since the central subject matter of the book is the bulk control of injected currents, as compared to their control by the contacts, contact constraints are entirely neglected in the first go-around via the assumption of idealized, injecting contacts. In this respect, the book is complementary to a whole host of well-known books on semiconductor physics and electronics. Since *p–n* junction phenomena are already adequately covered in most of these texts, we have not gone over this ground once again. On the other hand, we have paid attention to the corrections imposed on the pure bulk theory of current injection by the constraints associated with 'real' contacts, including the *p–n* junction. These corrections are of particular interest in the case of double injection into highly purified semiconductors (the injected plasma problem).

This book is a considerably expanded and updated version of a comprehensive, but highly condensed, review of the subject by one of us (M.A.L.), "Volume Controlled Current Injection in Insulators," published in *Reports on Progress in Physics*, Volume 27, 1964.

The book owes much, and in different ways, to the RCA Laboratories in Princeton, New Jersey. They have been, for many years, a prominent sponsor of research in the field of the electronic activity of insulators. The area of injection currents in insulators was nurtured to maturity there, largely by Dr. A. Rose and his collaborators. Much of the work of the authors in this area was done at these Laboratories. The Drafting Department of RCA Laboratories is responsible for drawing or redrawing virtually all of the figures in the text. The authors would also like to express their gratitude to Princeton University for continued support of their work in this field. One of us (P.M.) would also like to express appreciation to the Office of Naval Research for their support of his work during the writing of this book. Finally, special thanks go to Mrs. Pearl Bazel for her careful typing of the manuscript.

CURRENT INJECTION IN SOLIDS

PART I

ONE-CARRIER CURRENTS

1

CONCEPTUAL BASIS

Historically an insulator was simply a nonconducting material, and its function in electrical technology was to furnish electrical insulation. Under an applied voltage any insulator carried relatively little, or negligible, current, at least up to electrical breakdown at very high field strengths. Based on such a definition, a program for the study of current injection into insulators would hardly have appeared promising. This completely empirical viewpoint grew out of, and hence was necessarily in accord with, the then available information; further, there was no theory available with which to criticize such a position. Under the impact of the modern quantum theory, a newer, quite different point of view has emerged, which is the subject of this book.

Very soon after the appearance of the theory of quantum mechanics, Bloch (1928) applied the new theory to the electronic structure of solids. The energy-band picture which emerged from this study provided, for the first time, a sound theoretical framework for the understanding of electrical conduction in nonmetallic solids. In 1940, Mott and Gurney, in their classic book, "Electronic Processes in Ionic Crystals," made the important observation that, based on the energy-band viewpoint, it should be possible to obtain injection of electrons from a suitable contact into an insulator, in a manner closely analogous to their injection from a thermionic cathode into vacuum.

1.1 Energy-Band Basis for Injection

The energy-band basis for the injection of electrons into an insulator is easily understood by simply inspecting the relevant energy-band diagrams. Figure 1.1a shows a metal-vacuum contact and Fig. 1.1b an

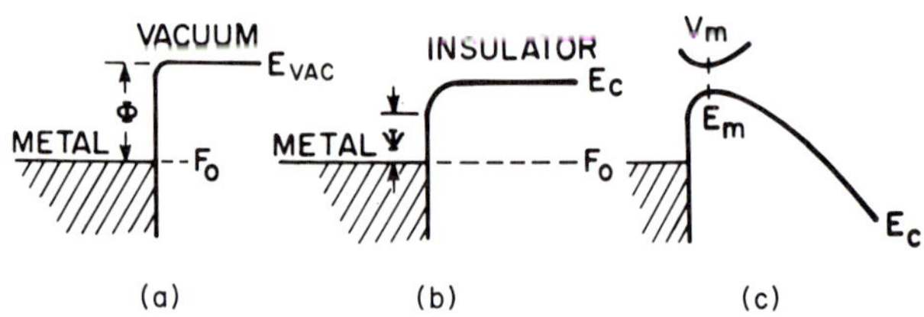

FIG. 1.1. Schematic energy-band diagrams for: (a) the contact of a metal to the vacuum; (b) the contact of a metal to an insulator. With application of voltage (c), a potential minimum (electron energy maximum) appears near the interface for case (b) as well as for case (a).

electron-injecting, metal-insulator contact, both contacts being in thermal equilibrium (no applied voltage). F_0 is the Fermi level, E_{vac} the lowest vacuum level, and E_c the lowest conduction-band level. The two diagrams are clearly very similar. Electrons will "boil off" from the metal into the conduction band of the insulator just as they do from the heated cathode into vacuum. The electrons in the metal available for emission are those thermally excited to an energy sufficiently large to overcome the energetic barrier at the emission surface. It was noted by Mott and Gurney that the interface dipole energy barrier Ψ at the metal-insulator contact can be substantially smaller than the corresponding work function barrier Φ for the metal-vacuum contact. As a result, even at room temperature or lower, there may be a sufficient number of electrons available at the contact to support space-charge-limited electron flow into the insulator—an advantage of exceeding technological interest. The space-charge-limited current flow into the insulator from the "ohmic" contact schematically illustrated in Fig. 1.1b is characterized by the same hallmark as space-charge-limited current flow in vacuum, namely by a potential minimum near the emission surface, indicated by V_m in Fig. 1.1c, which illustrates schematically the effect of an applied voltage of proper polarity (contact negative). V_m corresponds to the energy maximum E_m in Fig. 1.1c, potential being the negative of energy for electrons.

Electron flow in the conduction band of the insulator is dominated by frequent collisions with the thermal vibrations of the insulator (phonons), and with the chemical impurities and structural imperfections in the insulator. Consequently, the mathematical description of the space-charge-limited current flow will be somewhat different from that of the corresponding flow in vacuum. However this is rather a matter of detail than of principle. A more important effect will be produced by the localized electronic states, associated with impurities and structural imperfections, which are inevitably present in actual materials but which have been omitted from the highly idealized model shown in Figs. 1.1b and c.

1.2 Volume Considerations

In their brief study of the current-injection problem, Mott and Gurney made the passing observation that the presence of localized electron traps, energetically located in the forbidden gap, for example at levels E_{tn1} and E_{tn2} in Fig. 1.2a, could drastically interfere with the passage of injected current, particularly at sufficiently low temperatures where the capture of injected electrons would be "stable." Mott and Gurney's very brief remarks were later elaborated into an incisive theory by Rose (1951, 1955) who gave a detailed description of the form and magnitude of the reduction in injected current arising from localized trapping of the injected carriers. The earlier work of Mott and Gurney and the two papers by Rose laid the foundations for the field of injection currents in insulators as it stands today. The effects of trapping on one-carrier injection currents constitute the central subject of Part I of this book.

In order to present clearly the analogy between current injection into vacuum and into an insulator, we have thus far only discussed electron injection. From the well-known duality between electrons and holes we should, of course, equally well expect the possibility of hole injection into the valence band of an insulator. Figure 1.2b illustrates an "ohmic" metal-insulator contact suitable for this purpose. In this figure E_v is

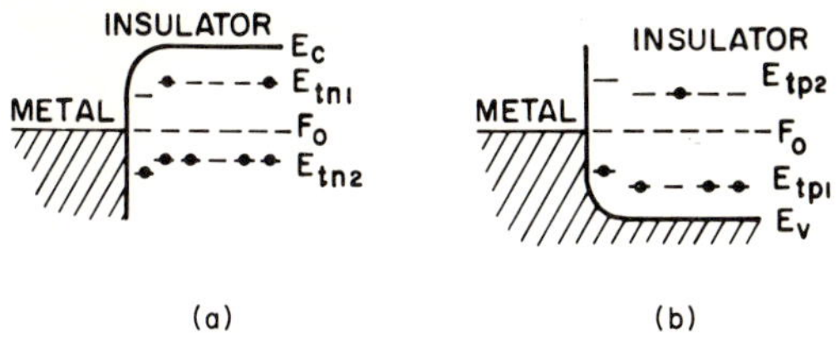

FIG. 1.2. Schematic energy-band diagrams for: (a) a contact of a metal to an insulator, ohmic for electron injection, including electron-trapping states; (b) a contact of a metal to an insulator, ohmic for hole injection, including hole-trapping states.

the highest valence-band level. Included in the diagram are localized hole-trapping states, the analogs of the electron-trapping states, at levels E_{tp1} and E_{tp2}.

1.3 Contact Considerations

Up to this point, the only injecting contacts referred to have been metal-insulator contacts with proper insulator band-bending to promote injection. In practice, other kinds of injecting contacts are possible as illustrated in Figs. 1.3a–d. In Figs. 1.3a and 1.3b, the bulk proper of the insulator lies to the right of the plane C. In Fig. 1.3a, the electron

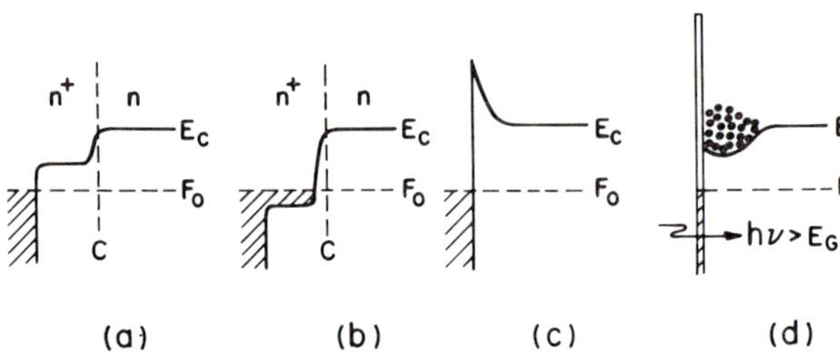

Fig. 1.3. Schematic energy-band contacts for electron injection: (a) nondegenerate n^+–n contact; (b) degenerate n^+–n contact; (c) tunneling contact (blocking at low bias); (d) light-generated reservoir contact.

injection takes place from the n^+ region to the left of C over the n^+–n or n^+–i (intrinsic) junction. The n^+ region must of course be contacted by a suitable electrode. The n^+ region shown in this diagram is nondegenerate and therefore will become useless as an electron-injecting contact at temperatures so low that carrier freeze-out takes place. The contact shown in Fig. 1.3b is very similar to that shown in Fig. 1.3a, the only difference being that the n^+ region is now degenerate. Since carrier freeze-out never takes place in this degenerate region, such a contact can remain electron injecting down to the lowest temperatures. In Fig. 1.3c, the electron-injecting contact is a metal-insulator contact which, at low voltages, is a conventional electron-blocking contact. However, the potential barrier is assumed thin enough that, with a modest voltage across it, copious tunneling of electrons takes place from the metal into the insulator. Beyond this voltage the contact will function as an injecting contact. A metallic point contact which is sufficiently sharp will, in principle, function in this manner, as an injecting contact, independent of energy-band details at the contact. The electric field intensity close to the point will be high enough, at relatively small applied voltage, to promote injection through field emission (tunneling). The electron-injecting contact shown in Fig. 1.3d is simply a reservoir of electrons created through illumination of the insulator surface with an intense beam of strongly absorbed light of frequency ν such that $h\nu > E_G$, the bandgap. Not shown on the diagram is the corresponding reservoir of holes in the valence band simultaneously created by the absorbed light. A transparent conducting film is provided for electrical contact. Instead of light, a moderately high-voltage electron beam might be used to create electron and hole pairs in the insulator near the surface. In this case, the transparent conducting film might no longer be necessary for electrical contact. All five of the injecting contacts shown in Fig. 1.2a and Figs. 1.3a–d have proved useful in injection studies.

The usual sketching of schematic energy-band diagrams might suggest that the contact problem is a relatively simple, straightforward one. In practice the problem of fabricating a suitable injecting contact is more often than not a difficult one. Frequently, a not insignificant portion of a research program on the electronic properties of an insulator is spent solving, or trying to solve, the contact problem. The technology of contacts is in a primitive state and the "science" of contacts likewise. Consequently, where the difficulties are stubborn—for example, in fabricating a good hole-injecting contact to CdS—it is not clear just how fundamental they are. A properly bending band at a metal–insulator contact is, in itself, no guarantee of a useful injecting contact. Obviously, if the energy step at the contact interface (the quantity Ψ in Fig. 1.1b for the electron-injecting contact) is too large, only a small temperature-limited saturation current can be drawn out of the contact, as, for example, with an insufficiently heated, thermionic cathode in a vacuum tube. The possible use of heterojunctions—for example, Cu_2S, grown or evaporated onto CdS, as a hole-injecting contact—has been too little explored for definitive evaluation. In principle, the photon or electron bombardment technique for "generating" the contact circumvents the untidy difficulties of chemistry and metallurgy. And, indeed, this technique has been successfully used with some materials, e.g., iodine and anthracene (see Chapter 5). The recent availability of laser sources of light will likely popularize this technique.

Although the problem of the injecting contact can hardly be avoided by the experimentalist, it generally need not be faced by the theoretician, at least in the study of one-carrier injection. The fact of the matter is that the behavior of one-carrier, space-charge-limited injection currents is, generally, completely dominated by the bulk properties of the sample. An adequate theory of the current injection can be constructed neglecting the diffusion-current contributions to the total current. Concomitantly, the problem of the detailed structure of the injecting contact is entirely circumvented through use of an idealized boundary condition, namely, vanishing of the electric field intensity at the cathode interface. Although this boundary condition yields unphysical properties in the immediate vicinity of the cathode (an injected free-carrier concentration going to infinity at the cathode interface), it yields a correct description of the current-flow properties over the bulk of the sample. The situation here is quite analogous to that of the vacuum diode problem. There too, in the derivation of the famous Child's law, a simplified approach is used: The thermal velocity spread of the emitted electrons is ignored and, with it, the "structure" of the contact (the potential minimum); the boundary condition employed at the cathode is the same as for the solid (vanishing of the electric-field intensity at the cathode) and yields the same unphysical result right at the cathode. In both cases, the vacuum and the solid, an enormous simplification of the theoretical problem is attained at a very small price indeed, namely an unphysical description of the current-flow properties in the immediate vicinity of the cathode. This simplified approach is used throughout most of Part I; however, the "exact" theory is also examined in Chapter 9.

1.4 Role in Solid-State Research and Development

A growing body of experience indicates that injection currents can be a powerful "probe" for the study of insulators. Their widest, but by no means only, contribution is undoubtedly in obtaining information about defect states in the forbidden gap. As remarked above, localized defect states can strongly influence the injected current flowing in response to an applied voltage. Not only is the magnitude of the current-response controlled by the defect states (e.g., reduction of the injected current through trapping in localized states) but the actual form of the current-voltage characteristic is determined by the interaction of the injected carriers with such states. The point is well illustrated by citing in advance some of the more dramatic results of the theory. Figure 1.4 shows, on a log–log plot, a predicted dc current-voltage characteristic for one-carrier, space-charge-limited current injection into an insulator with a single set of carrier traps which are, initially, largely empty. This might correspond to the situation illustrated in Fig. 1.2a if only the states at level E_{tn1} are present in the insulator, or to the situation illustrated in Fig. 1.2b if only the states at level E_{tp1} are present. Figure 1.5

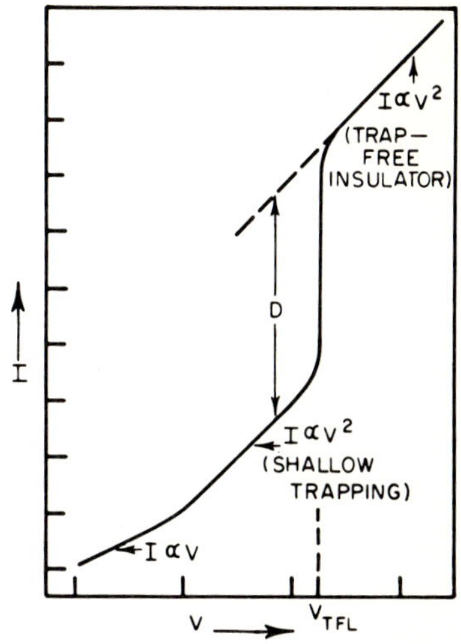

FIG. 1.4. Schematic current-voltage characteristic for one-carrier SCL current injection controlled by a single set of traps; log–log plot, each unit corresponding to a decade.

shows, on a log–log plot, a theoretical, dc current-voltage characteristic for double injection into an insulator with a single set of recombination centers (appropriately located in the forbidden energy gap and with an appropriate ratio of electron and hole capture cross sections). A striking feature of both of these current-voltage characteristics is the considerable

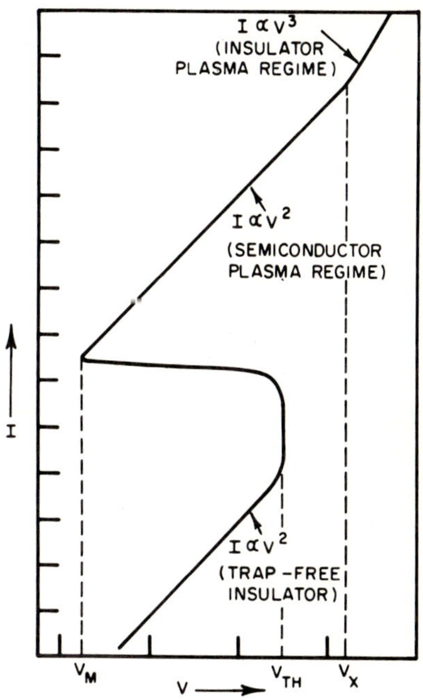

FIG. 1.5. Schematic current-voltage characteristic for double injection controlled by a single set of deep-lying recombination centers (initially filled and with $\sigma_p \gg \sigma_n$, or initially empty and with $\sigma_n \gg \sigma_p$, σ_p and σ_n being the hole and electron capture cross sections, respectively, for the center); log–log plot, each unit corresponding to a decade.

amount of structure contained in them. These curves will be derived and discussed in detail in later chapters. We wish to point out here only that the structure in these curves must obviously contain useful information about the single set of defect states, in each case, which is responsible for that structure. For example, the voltage V_{TFL} at which the current rises nearly vertically in Fig. 1.4 yields directly the concentration of the trapping states and the displacement D of the two square laws yields their energetic location in the forbidden gap. In Fig. 1.5 the voltage ratio V_{TH}/V_M yields the ratio of the two free-carrier capture cross sections of the recombination centers, and the coefficient of the square law in the "semiconductor regime" yields one of the capture cross sections. Thus both capture cross sections are obtained, and at this point V_M or V_{TH} alone yields the concentration of recombination centers, which can also be obtained, independently, from the voltage V_X of crossover from the square law to the cube law.

Prototype, transient, one-carrier space-charge-limited current characteristics are illustrated in Fig. 1.6. Here current is plotted against time, both measured in suitable normalized units, following application of a step function of voltage. Again a striking amount of structure is exhibited in the current characteristic and, accordingly, useful information is obtained about the insulator. There is even richness of structure for the trap-free insulator as well as for the insulator with traps. In either case the location in time of the current cusp yields directly the free-carrier mobility, and the long-time current falloff in the bottom curve yields a trapping time. This transient technique has found important application in the measurement of low mobilities in molecular crystals.

From these several examples it is seen that the kind of information about defect states yielded directly by a study of injection currents includes their concentration, their energy location in the forbidden gap, and their capture cross sections for free carriers. If present in too large concentration, the defect states inhibit injection currents at practical voltages. In effect, the purer the insulator the easier it is to obtain information using injection currents. Electronically dominant, defect-state concentrations as low as 10^{12} cm^{-3}, and much lower, are easily detected by this technique. In this respect the technique may be considered to be complementary to the majority of techniques, chemical, optical, etc., which become increasingly difficult to employ with increasing purification.

On the experimental side, single-injection space-charge-limited current measurements are now a well-established technique for obtaining information about defect states in insulators. Such measurements have

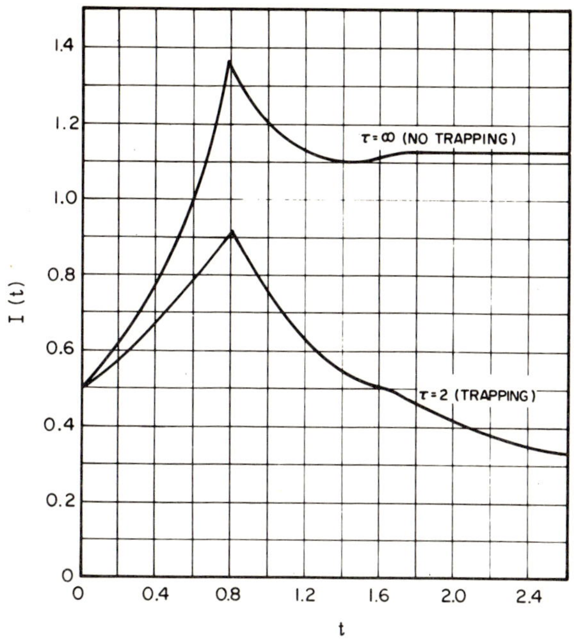

FIG. 1.6. Computed transient current response (current $I(t)$ versus time t) for one-carrier SCL current injection; linear plot with I and t measured in suitable normalized units. (See Chapter 6.) Upper curve is for a trap-free insulator, lower curve includes trapping effects with trapping time $\tau = 2$ in normalized time units. (From Many and Rakavy, 1962, Fig. 5.)

been made on a substantial number of materials, ranging from elemental and compound semiconductors and ionic insulators to organic and inorganic, molecular crystals and vitreous materials. This work is summarized in Chapter 5 (dc measurements) and Chapter 6 (transient measurements). In most cases, for a given material, the one-carrier, space-charge-limited currents have been observed for only one sign of carrier. Their absence for the other sign of carrier is a measure mostly of the contact difficulties, previously mentioned, and to some extent also of purification difficulties.

Double injection studies, both theoretical and experimental, are of generally more recent vintage than the single-injection studies. Injected plasmas are reasonably well understood and the experiments accurately check the theory. Much remains to be done, both theoretically and experimentally, with double injection with trapping. In general, the outstanding experimental problem in observing double injection is that of fabricating the contacts. On one and the same sample are needed a good electron-injecting cathode and a good hole-injecting anode. Sample purification may also present difficulties.

The injected plasma has a particular scientific interest over and above its relationship to the defect states of the solid, namely, it provides a useful vehicle for the experimental study of instabilities in plasmas. Although we shall not be concerned in this book with the instabilities *per se*, in Part II we study in detail the properties of the stable plasma, thereby establishing the necessary framework for analysis of the instabilities.

Since the field of injection currents in insulators grew out of concern with technological problems, it has remained in intimate contact with technological developments. One of the earliest applications to technology was the delineation of upper limits on insulator conversion device (photoconductors, light amplifiers) performance due to the onset of space-charge-limited current injection. The relevant arguments are sketched briefly in Chapter 7. Development work has been done on insulator devices analogous to vacuum devices (rectifier diodes, amplifier triodes) and on insulator transistor structures (thin-film transistor). References to some of this work are given in Chapter 5.

The attempt to generate light efficiently from solids has lent technological interest to double injection phenomena in insulators for many years. The work up to 1960 is reviewed by Henisch (1962); a comprehensive bibliography is also given by Ivey (1961). The thorny problems and prospects for injection electroluminescence in wide-bandgap materials are considered by Fischer (1961). Observations of injection electroluminescence in an organic, anthracene, are reviewed by Helfrich (1968). The search for high-output injection electroluminescence in lower-bandgap materials, namely semiconductors, was crowned with brilliant success with the achievement of coherent light emission (laser action) in GaAs by several groups in 1962. This work, as well as similar work on other semiconductors, is reviewed by Pilkuhn (1968). Progress in nonlasing injection electroluminescence in semiconductors, mainly GaP, is reviewed by Thomas (1968).

1.5 Further Remarks

In the interest of simplicity, throughout the entire book the discussion is confined to homogeneous samples. For the same reason we consider, for the most part, only steady, dc current flow (with the exception of Chapter 6) and planar, one-dimensional flow (with the exception of

Chapter 8). In the same spirit, a simplistic view of localized defect states is taken; they are considered to be single-level states, for the most part, any effects due to excited states being ignored. All breakdown phenomena are excluded from consideration and almost all of the detailed studies presented are based on the low field, field-independent mobility.

In Part I, for the sake of avoiding a stilted, generalized terminology, it has been assumed throughout that the current carriers are electrons. This has made it possible to avoid the use of a carrier-identification subscript "n" or "e." Any result obtained for electron injection currents holds equally well for hole injection currents with the appropriate change of terminology. It has been assumed that the electrons flow from left to right, in the usual, pertinent diagrams. Mks units are employed throughout the book except where specified otherwise. The more significant formulas are also presented in practical units in Appendix B. Likewise, a complete table of notation is presented in Appendix A.

A few words are in order about terminology. There is not at the present time a unique, universal terminology in insulator physics. We do not attempt to distinguish "insulators" from "semiconductors" in any sharply defined sense. At room temperature, by an "insulator" we mean, roughly, a "wide-bandgap material" (with, say $E_G \gtrsim 2$ eV) and by "semiconductor" a "narrow-bandgap material" (with, say $E_G \lesssim 2$ eV). On the other hand, a characterization of insulators as high-resistivity materials and semiconductors as low-resistivity materials is also frequently a useful distinction. We use the terminology, "ohmic contact," "reservoir contact," and "injecting contact" interchangeably. They all denote contacts capable of current injection with only a relatively small fraction of the applied voltage absorbed across the contact. By "trap" we mean, in the context of single injection, *any* localized, defect electronic-state in the material. Where quasi-thermal equilibrium between trapped and free carriers obtains, it is characterized by a "quasi-Fermi level" (called elsewhere in the literature also a "steady-state Fermi level" and "imref"). "Shallow" and "deep" trapping are defined quite precisely with respect to the position of the quasi-Fermi level and so are not absolute concepts but, rather, dependent on the injection level.

2

PHENOMENOLOGICAL ANALYSIS

A considerable understanding of the properties of one-carrier injection currents may be gained through simple phenomenological analysis. This kind of analysis, introduced into the subject by A. Rose, attempts to describe only the gross, overall features of the current flow. By circumscribing the goal of the analysis in this manner, there is achieved a simplicity of argument which lends a great deal of insight into the underlying physics of one-carrier injection.

The quantities that the theory of one-carrier currents must deal with are free and trapped charge concentrations, free-carrier drift velocity, and electric field intensity. The phenomenological analysis ignores the spatial variation of these quantities and instead concerns itself only with their respective average values and the relations connecting them. This procedure works because, for planar current flow, all of the quantities vary relatively gently with position over most of the distance between cathode and anode. Therefore the average value of a quantity is also a good representative value. In keeping with the spirit of this approach, the average values actually used are not always precise mathematical averages, although they are always good approximations to these, as shown in Chapter 4. The final result is that the relations derived by the phenomenological analysis, for example, the current-voltage characteristic, are correct in their functional dependence on all physical quantities and, at worst, contain only an incorrect numerical coefficient. Further, the error in this coefficient is not large; it is generally smaller than a factor of 2, so that a small price indeed is paid for the insights gained by the simple analysis.

We now proceed with the discussion of specific prototype problems, working our way from the simplest model through gradually more complex ones. We assume, in every case, that the cathode is a reservoir contact, that is, a contact that can freely supply whatever current is required by the bulk-determined solution to the problem. Except under very special circumstances, namely very small cathode–anode spacings and low applied voltages, such a contact plays a negligible role in the determination of the current-voltage characteristic and can accordingly be ignored in our discussions.

2.1 The Perfect Insulator

We first discuss the current-voltage characteristic for the simplest possible situation, namely the perfect insulator, free of traps and with a negligible concentration of free carriers in thermal equilibrium. This is the solid-state analog of the thermionic vacuum diode. All injected electrons remain free, i.e., in the conduction band, and all contribute to space charge.

The current density may be written as

$$J = \rho v, \tag{2.1}$$

or, equivalently, as

$$J = Q/t, \tag{2.1a}$$

with ρ the average, injected, free charge concentration, v the average drift velocity of a free electron, Q the *total* injected free charge, per unit area, between cathode and anode, and t the transit time of a free electron between cathode and anode. The quantities v and t, and ρ and Q are related by

$$t = L/v \tag{2.2}$$

and

$$Q = \rho L, \tag{2.2a}$$

where L is the spacing between cathode and anode.

In writing the current as a pure drift current we have, of course, neglected any diffusive contribution to the current. Diffusion currents are sizable only in the immediate neighborhood of the contacts and neglect of such currents is consistent with our neglect of the role of the contacts. These are matters discussed at some length in Chapter 9.

It is apparent from (2.1) or (2.1a) that we need only determine the functional dependence of Q, or ρ, and t, or v, on the applied voltage V

to determine the current-voltage characteristic. We first consider Q. It is known from elementary electrostatics that the total charge Q_0, per unit area, on one plate of a parallel-plate condenser is proportional to the voltage V across the condenser; namely,

$$Q_0 = C_0 V \quad \text{and} \quad C_0 = \epsilon/L, \tag{2.3}$$

where ϵ is the static dielectric constant of the medium between the plates and L is the spacing between plates. From the viewpoint of electrostatics, the current injection problem is very similar to that of the condenser. A given voltage V between cathode and anode can support only a limited total charge Q, per unit area, injected into the insulator from the reservoir contact. Extrapolating the experience with condensers, it is reasonable to suppose that here too, Q is proportional to V: $Q = CV$. If the injected charge were uniformly distributed between cathode and anode, its average distance from the anode would be $L/2$ and so the capacitance C would be twice the "geometric" capacitance C_0 given by (2.3). In fact, since the charge is all injected at the cathode, we would expect that its distribution is nonuniform, there being greater concentration near the cathode. Thus the average distance of the injected charge from the anode is somewhat greater than $L/2$ but clearly smaller than L, the limiting distance corresponding to no injection at all. On physical grounds alone, then, it is plausible that $C_0 < C < 2C_0$, a theorem that is rigorously established in Chapter 3. Since we are not presently interested in great precision, we can tolerate a factor-of-two inaccuracy and write

$$Q \simeq C_0 V = (\epsilon/L)V. \tag{2.4}$$

Combining (2.1a), (2.2), and (2.4) we obtain

$$J \simeq C_0 vV/L = \epsilon vV/L^2. \tag{2.5}$$

The electron drift in the conduction band of the solid is marked by frequent collisions with the thermal vibrations, the impurities, and the structural defects in the solid. At field strengths that are not too high, this leads to an electron drift velocity proportional to the applied field:

$$v = \mu\mathscr{E} = \mu(V/L) \quad \text{and} \quad t = L^2/\mu V, \tag{2.6}$$

with μ the free-electron drift mobility and $\mathscr{E}$ the average electric field intensity. Combining (2.5) and (2.6), we obtain the current-voltage characteristic

$$J \simeq \epsilon\mu(V^2/L^3). \tag{2.7}$$

The analytically derived result (4.9) differs from (2.7) only through the presence in the former of the numerical factor 9/8.

Although a perfect insulator is a highly idealized object, the result (2.7) is still very useful since it is the limiting form of the current-voltage characteristic for an imperfect insulator at applied voltages high enough that the total number of injected electrons substantially exceeds the total number of initially empty electron traps in the material. It represents the highest one-carrier injection current a given insulator with a given cathode–anode spacing can carry.

It is important to note that Eqs. (2.1)–(2.5) are useful under carrier-flow conditions that are much more general than that given by (2.6). Thus, they are still useful at high fields where the drift mobility is field-dependent so that the drift velocity is no longer simply proportional to the applied field. The modifications of the current-voltage characteristics necessitated by such high-field conditions are discussed at the end of this chapter.

Further, Eqs. (2.1)–(2.5) are not restricted to current flow in solids. It is instructive to apply the above arguments to the thermionic vacuum diode. In this case, electrons move freely without collisions so that their drift velocity is related to the potential difference they have moved through:

$$v \simeq (eV/m_0)^{1/2}, \tag{2.8}$$

with e and m_0 the charge and mass, respectively, of the electron. Substituting for v from (2.8) into (2.5) we obtain

$$J \simeq \epsilon_0(e/m_0)^{1/2}(V^{3/2}/L^2), \tag{2.9}$$

where ϵ has been replaced by ϵ_0, the dielectric constant of vacuum. The correct result, the famous Child's law, differs from (2.9) only through the presence in the former of the numerical factor $4\sqrt{2}/9 = 0.63$.

2.2 The Trap-Free Insulator with Thermal-Free Carriers

The next step in complexity is the inclusion in the problem of the thermally generated free electrons, of concentration n_0. A possible source of these electrons might be, for example, a set of donors so shallow, i.e., having such a small binding energy, that they are not effective as electron traps. It is assumed that there are no electron traps deeper in the forbidden gap.

We now expect that Ohm's law,

$$J = en_0\mu(V/L), \tag{2.10}$$

will be observed at low voltages. For, although at any voltage V, however small, there will be excess charge injected into the insulator, given by (2.4), there cannot be significant departures from Ohm's law until the average, injected, excess, free-electron concentration n_i becomes comparable to the thermally generated, hence neutralized, concentration n_0. Thus the crossover from Ohm's law (2.10) to the trap-free square law (2.7), that is, the onset of space-charge-limited current injection, takes place at the voltage V_X determined by

$$en_0L = Q = CV_X \simeq (\epsilon/L)V_X. \tag{2.11}$$

From (2.11), and further simple manipulation, it is seen that the crossover is characterized by

$$V_X \simeq (en_0L^2/\epsilon), \quad \text{or} \quad t_X \simeq t_\Omega, \qquad \text{with} \quad t_X = (L^2/\mu V_X) \quad \text{and} \quad t_\Omega = (\epsilon/en_0\mu), \tag{2.12}$$

where t_X is, approximately, the electron transit time between cathode and anode at voltage V_X, and t_Ω is the well-known ohmic or dielectric relaxation time. Thus (2.12) confirms a logical expectation: The injected excess electrons will dominate the thermally generated electrons when the transit time for the excess electrons is too short for their charge to be relaxed by the thermal carriers.

It is readily checked that the crossover voltage V_X given by (2.21) corresponds simply to the voltage where Ohm's law (2.10) intersects the trap-free square law (2.7).

2.3 The Insulator with Traps

The presence of electron traps in the insulator—an inevitable attribute of real insulators—will generally result in a greatly reduced current at lower injection levels, since those traps initially empty will capture, and thereby immobilize, most of the injected carriers. On the other hand, the amount of excess charge that can be supported in the insulator at applied voltage V, namely $Q = CV$, is the same whether the excess charges are free or trapped. (The amount of charge that can be stored on the plates of a parallel-plate condenser, at a given voltage, is independent of whether the charge is free or immobilized in surface states.) Thus (2.2a) must be rewritten as

$$Q = (\rho + \rho_t)L \simeq C_0V = (\epsilon/L)V, \tag{2.13}$$

where ρ_t is the average, injected, trapped charge concentration, and where (2.4) has also been used. Note that with Q redefined by (2.13) we can no longer write $J = Q/t$, as in (2.1a), although, of course, $J = \rho v$, (2.1), is still valid.

In order to deduce the space-charge-controlled current-voltage characteristic, a third relationship is required in addition to (2.1) and (2.13), namely, that connecting ρ and ρ_t. The appropriate form for this relationship is arrived at by first considering the relationship between free and trapped carriers in thermal equilibrium. This relationship is conveniently expressed via the thermodynamic Fermi level F_0. For nondegenerate semiconductors or insulators, the thermal-equilibrium free electron concentration is given by the familiar expression

$$n_0 = N_c \exp[(F_0 - E_c)/kT], \tag{2.14}$$

with N_c the effective density of states in the conduction band, E_c the energy of the bottom edge of the conduction band, k Boltzmann's constant, and T the temperature in degrees Kelvin. The concentration $n_{t,0}$ of filled electron traps at level E_t, Fig. 2.1, is then given by the familiar Fermi–Dirac expression

$$\begin{aligned} n_{t,0} &= \frac{N_t}{1 + (1/g)\exp[(E_t - F_0)/kT]} = \frac{N_t}{1 + (1/g)(N/n_0)}; \\ N &= N_c \exp[(E_t - E_c)/kT], \end{aligned} \tag{2.15}$$

where N_t is the concentration of traps and g the degeneracy factor (statistical weight) for the traps. Now, clearly this equilibrium trap occupancy results from a balance between capture of electrons into the traps, the downward arrow in Fig. 2.1, and their thermal re-emission into the conduction band, the upward arrow. *The crucial point for the*

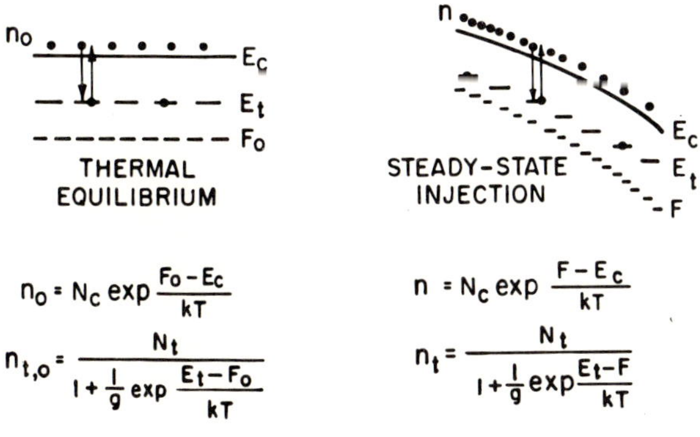

FIG. 2.1. Thermal equilibrium (left) characterized by the thermodynamic Fermi level F_0 and quasi-thermal equilibrium (right) characterized by the quasi-Fermi level F.

development of injection theory is the fact that the presence of a moderate electric field will not affect these elementary microscopic processes of electron capture and thermal re-emission. (For example, the electron capture process can be affected only if the electrons are substantially "heated up" by the applied field *and* the electron capture cross section is strongly dependent on the electron energy.) Thus, in the presence of the applied field, if it is not too strong, the balance between free and trapped electrons is altered *only* through the change in free-electron concentration accompanying injection. In other words, the balance between free and trapped electrons is reached as if the crystal were in thermal equilibrium, only with the free-electron concentration n, achieved under injection, instead of the true thermal-equilibrium concentration n_0. The corresponding Fermi level F is called the electron quasi-Fermi level and is related to n, *by definition*, exactly as n_0 is related to F_0 in (2.14); namely,

$$n = n_i + n_0 = N_c \exp[(F - E_c)/kT], \tag{2.16}$$

where n_i is the average, excess injected free-electron concentration.

The trapped-electron concentration n_t is then given by (2.15), only with F_0 replaced by F:

$$n_t = n_{t,i} + n_{t,0} = \frac{N_t}{1 + (1/g)\exp[(E_t - F)/kT]} = \frac{N_t}{1 + (1/g)(N/n)}, \tag{2.17}$$

with $n_{t,i}$ the average, injected, excess trapped-electron concentration and N given by (2.15).

The limits of validity for the assumption of quasi-thermal equilibrium are examined in somewhat more detail in Chapter 3. We now apply this assumption to the discussion of several trap models.

2.4 Shallow Traps

An electron trap at E_t is said to be shallow if F lies below E_t: $(E_t - F)/kT > 1$. Thus F_0 must lie below E_t, as for example, with states at energy E_{tn1} in Fig. 1.2a. It then follows from (2.17) that for

$(E_t - F)/kT > 1$:

$$\frac{n}{n_t} = \frac{p}{p_t} = \frac{N}{gN_t} = \frac{N_c}{gN_t}\exp\left(\frac{E_t - E_c}{kT}\right) = \theta, \tag{2.18}$$

where θ is a constant, independent of injection level, so long as the trap remains shallow.

The shallow traps of concentration N_t at energy E_t will substantially affect the space-charge-limited injection current if $\theta \ll 1$. Equation (2.13) can then be written as

$$Q \simeq \rho_t L \simeq \rho L/\theta \simeq \epsilon V/L. \tag{2.19}$$

Combining (2.19), (2.1), and (2.6), we obtain

$$J \simeq \theta\epsilon\mu(V^2/L^3). \tag{2.20}$$

(The correct, analytically derived result has the numerical factor 9/8 in front of it, just as for the trap-free case.) Here we have assumed that only a single trap level is effective in influencing the current. If there are several sets of trap levels, all shallow, the set with the smallest θ dominates the current flow and the θ for it appears in (2.20). It must be kept in mind that (2.20) is valid only so long as F lies below E_t. F, of course, keeps rising with injection level, i.e., with applied voltage. What happens when F crosses E_t is discussed shortly.

As already noted in Section 2.2, at sufficiently low voltages, Ohm's law (2.10) will be observed because of the presence in the insulator of the thermal free carriers n_0. (It is, of course, not implied that the n_0 of Section 2.2 is the same as the n_0 in the present discussion. The n_0 will generally vary from crystal to crystal, even for the same material, and will be determined by the total aggregate of defect states in a given crystal. The condition of local charge neutrality determines the position of the thermodynamic Fermi level F_0 in the energy gap and this, in turn, from (2.14), determines n_0. In this chapter, and indeed throughout most of the book, we treat n_0, or equivalently F_0, simply as a given parameter.) It is easily seen that the voltage V_X of crossover from Ohm's law (2.10) to the shallow-trap square law (2.20) will be $1/\theta$ times the crossover voltage (2.12) for the trap-free case; for when the free-electron concentration is doubled through injection, the excess, shallow-trapped electron concentration is given by n_0/θ. Thus,

$$V_X \simeq en_0L^2/\theta\epsilon. \tag{2.21}$$

It is convenient to define the concept of the *effective drift mobility* μ_e. If the free electron concentration is disturbed from its thermal-equilibrium value n_0, say by injection, to a new value n, then on a time scale long compared to the microscopic trapping and detrapping times, the population n_t of the electron traps will be in quasithermal equilibrium with n, as discussed in the preceding Section 2.3. Then the entire body of injected electrons, $n + n_t \simeq n_t$, will have the effective drift mobility

$\mu_e = (n/n_t)\mu$. In terms of μ_e it is readily checked, using (2.18), that (2.21) is equivalent to

$$t_{X,e} \simeq t_\Omega, \qquad \text{with} \quad t_{X,e} = (L^2/\mu_e V_X) \quad \text{and} \quad \mu_e = (n/n_t)\mu = \theta\mu, \tag{2.22}$$

where $t_{X,e}$ is the effective transit time, between cathode and anode, of the entire body of injected charge at voltage V_X, and t_Ω is the ohmic relaxation time given in (2.12).

2.5 Deep Traps

An electron trap at E_t is said to be deep if F lies above E_t: $(F - E_t)/kT > 1$. Consider, for example, the situation where there is a single set of traps of concentration N_t located at an energy E_t below F_0, e.g., the traps at energy E_{tn2} in Fig. 1.2a. In thermal equilibrium, the hole occupancy of the traps, i.e., the concentration of traps not occupied by electrons is, from (2.15),

$$p_{t,0} = N_t - n_{t,0} = \frac{N_t}{1 + g \exp[(F_0 - E_t)/kT]} \simeq \frac{N_t}{g} \exp\left(\frac{E_t - F_0}{kT}\right), \tag{2.23}$$

where the last expression is a valid approximation when $(F_0 - E_t)/kT > 1$.

As in the previous problems, Ohm's law (2.10) will obtain up to that voltage V_X where the injected free-electron concentration n_i becomes comparable to the thermal concentration n_0. From the discussion in Section 2.3, it is evident that this doubling of the thermal concentration corresponds to an upward motion of the quasi-Fermi level F, away from F_0, by an amount kT (more precisely, $0.7kT$). This motion is sufficient to fill, approximately, the deep traps. In this case, then, the voltage V_X for the onset of space-charge-limited current injection coincides with the voltage V_{TFL} required to fill the set of deep traps; namely, from (2.4),

$$V_{TFL} \simeq \frac{Q_{TFL}}{C_0} = \frac{ep_{t,0}L}{C_0} \simeq \frac{ep_{t,0}L^2}{\epsilon}, \tag{2.24}$$

with $p_{t,0}$ given by (2.23) and with subscript TFL used to denote the trap-filled-limit.

We next explore how the current varies with applied voltage beyond V_{TFL}. It is convenient to do this by simply estimating the change in current accompanying a doubling of the voltage, $V = 2V_{TFL}$. Because of the proportionality of total injected charge to voltage, (2.4), the total injected charge is likewise doubled, $Q(2V_{TFL}) = 2Q_{TFL}$. Since the traps were already filled at the voltage V_{TFL}, the additional injected charge, $Q(2V_{\text{TFL}}) - Q_{\text{TFL}} = Q_{\text{FTL}} = ep_{t,0}L$, must all appear in the conduction band. We have then for the ratio of the currents at the two voltages

$$\frac{J(2V_{\text{TFL}})}{J(V_{\text{TFL}})} \simeq \frac{2n(2V_{\text{TFL}})}{n(V_{\text{TFL}})} \simeq \frac{p_{t,0}}{n_0}. \tag{2.25}$$

The additional factor of two appears in the numerator of the middle expression in (2.25) because the average applied field is doubled when the voltage is doubled. Also, we have used $n(V_{\text{TFL}}) \simeq 2n_0$. The ratio $p_{t,0}/n_0$ can easily be many powers of ten in insulators.

A prototype plot of the current-voltage characteristic resulting from this situation is shown in Fig. 2.2a, namely, the curve labeled I. The current rises very steeply, in fact nearly vertically, with voltage at $V \simeq V_{\text{TFL,I}}$. Shortly beyond the voltage $V = 2V_{\text{TFL,I}}$, the characteristic merges with the trap-free square law (2.7), since the injected free charge then dominates the injected trapped charge.

The nearly vertical rise in current accompanying the filling of a

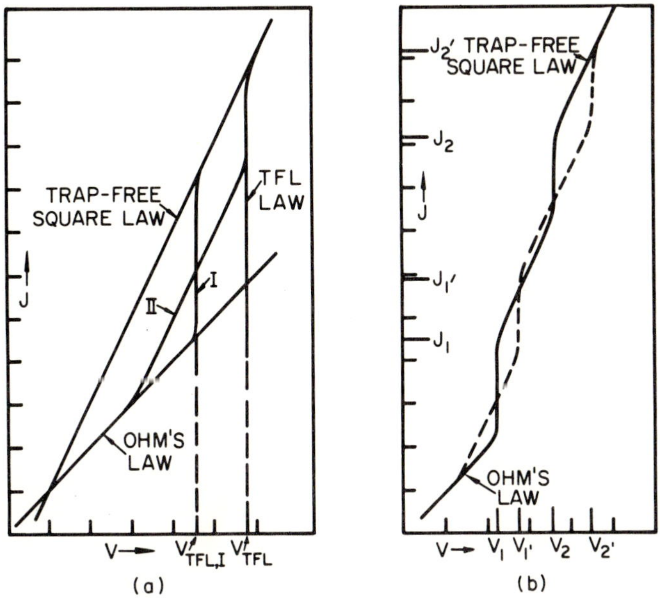

FIG. 2.2. Prototype current-voltage characteristics on log–log plots (each unit corresponding to a decade): (a) prototype curves for a single set of traps; I corresponds to F_0 lying above E_t, II to F_0 lying below E_t; (b) prototype curves for two sets of traps; the solid curve corresponds to F_0 lying above one set of traps and below the other set, the dashed curve to F_0 lying below both sets of traps.

discrete set of traps is one of the most dramatic results of the theory of current injection in insulators. Seen experimentally, this phenomenon would surely be misinterpreted as an electrical breakdown in the insulator in the absence of space-charge-limited current injection theory.

2.6 Complete Current-Voltage Characteristic for a Single Set of Traps

For a single set of traps of concentration N_t, and for a given position of the equilibrium Fermi level F_0, a family of current-voltage characteristics is possible, each member corresponding to a different energy location E_t of the traps. This entire family is, in a manner of speaking, contained in a certain "triangle" in the log J–log V plane, illustrated in Fig. 2.2a. This triangle is bounded by Ohm's law, the trap-free square law, and a "vertical" leg corresponding to the so-called TFL (trap-filled-limit) law. This last is the current-voltage characteristic for the hypothetical (unphysical) situation in which one imagines all traps filled at the outset, before voltage is applied. Since the voltage $V_{TFL} \simeq eN_tL^2/\epsilon$ is necessary to support the total excess charge eN_tL, artificially stuffed into the insulator, no current can flow until the applied voltage V reaches V_{TFL}. Thereafter the current rises nearly vertically with voltage, as discussed above, finally merging with the trap-free square law at $V \simeq 2V_{TFL}$. Returning to the real problem, wherever the trap level E_t is located, the corresponding J–V plot obviously cannot lie above the trap-free square law or below Ohm's law, and finally cannot lie to the right of, i.e., below, the TFL law, since this is obviously the most unfavorable situation vis-a-vis current flow—thus it must lie "inside" the "triangle."

The family of current-voltage characteristics consists essentially of two subfamilies, one for which E_t lies below F_0 and the other for which E_t lies above F_0. The former case has already been analyzed in Section 2.5. A typical member is shown in Fig. 2.2a, the curve labeled I. The vertical section occurs at the voltage $V_{TFL,I} \simeq ep_{t,0}L^2/\epsilon$ with $p_{t,0} = (N_t/g) \exp(E_t - F_0)/kT$. The deeper the trap level, i.e., the larger $(F_0 - E_t)/kT$, the smaller is $V_{TFL,I}$. Finally, when the traps are so deep that $p_{t,0} \simeq n_0$, the vertical portion disappears—the traps no longer affect the current flow.

The latter case has already been analyzed, in part, in Section 2.4. So long as F lies below E_t, the traps are shallow and the current follows the square law (2.20), with θ given by (2.18), after departing from Ohm's law at the crossover voltage V_X given by (2.21). In the neighborhood of

$V = V_{TFL}$, F crosses E_t and the J–V curve merges with the TFL law. A typical curve, labeled II, is shown in Fig. 2.2a. Note that when E_t gets close enough to the conduction band so that $\theta \simeq 1$ (here we are assuming that $gN_t < N_c$), the effect of the traps becomes negligible and only the trap-free square law is observed.

2.7 Current-Voltage Characteristic for Two Sets of Traps

At this point we have the basic building blocks with which to construct the current-voltage characteristic for situations in which there is more than one set of electron traps. We illustrate the procedure for two sets of traps, of concentrations N_{t1}, N_{t2}, respectively, and located at energies E_{t1}, E_{t2}, respectively, where $E_{t2} > E_{t1}$ and $N_{t2} > p_{t1,0}$. The solid curve in Fig. 2.2b corresponds to the situation where $E_{t1} < F_0 < E_{t2}$. The current follows Ohm's law up to the TFL-voltage $V_1 \simeq ep_{t1,0}L^2/\epsilon$. It then rises nearly vertically up to the current J_1 at which the corresponding injection level is such that $n_{t2} \simeq p_{t1,0}$. Thereafter the N_{t2} traps at level E_{t2} dominate and the current follows the shallow-trap square law up to the TFL-voltage $V_2 \simeq eN_{t2}L^2/\epsilon$. Again the current rises nearly vertically up to the current J_2 where $n \simeq N_{t2}$. At this point, the characteristic merges with the trap-free square law.

The dashed curve in Fig. 2.2b corresponds to the situation where $F_0 < E_{t1} < E_{t2}$. Here there is also a shallow-trap square-law segment of the characteristic corresponding to the N_{t1} levels at E_{t1}. Otherwise the characteristic is similar to the previous one (solid curve). Voltages $V_{1'}$, $V_{2'}$ and currents $J_{1'}$, $J_{2'}$ have the same significance as the previously noted voltages and currents, V_1, V_2 and J_1, J_2, respectively.

2.8 Extended Distributions of Traps: The Exponential and Uniform Distributions

In all discussion of electron trapping up to this point, we have regarded the energy level of any given kind of trap as a precisely defined quantity E_t. Thus we have repeatedly made reference to "a set of traps of concentration N_t located at energy E_t." For single-crystal materials of high chemical and structural purity this is undoubtedly a good approximation for the purpose of analyzing current injection. At the other extreme, where one is dealing with vitreous or amorphous insulators, e.g., evaporated thin layers of poorly defined crystallinity, it would not at all be surprising to find that isolated, discrete energy levels for the traps provide a poor working description, even if the material is of high chemical purity. Because of the large amount of structural disorder, a given kind of electron trap, either of chemical or structural origin, does not have a uniquely defined environment. There will be differences, from one trap site to another, in nearest neighbor, next nearest neighbor configurations, and so on. The net result will be a broad smearing out of the level. On the basis of this argument, it would be reasonable to represent a set of traps of a given kind by a Gaussian distribution $\mathcal{N}_t(E) = \mathcal{N}_1 \exp[-(E - E_t)^2/\Gamma^2]$ where $\mathcal{N}_t(E)$ is the concentration of traps per unit energy and Γ provides a measure of the smearing out of the level at E_t. On the other hand, the structural disorder may be expected to give rise to many different sets of traps. One simple possible representation of the many different sets of traps, all smeared out in energy, is the exponential distribution

$$\mathcal{N}_t(E) = \mathcal{N}_0 \exp\left(\frac{E - E_c}{kT_t}\right) = \mathcal{N}_n \exp\left(\frac{E - F_0}{kT_t}\right); \qquad \mathcal{N}_n = \mathcal{N}_0 \exp\left(\frac{F_0 - E_c}{kT_t}\right), \tag{2.26}$$

with T_t a temperature parameter characterizing the trap distribution. In the absence of a detailed theory for the distribution of trapping levels, the physical significance of the parameter T_t is not clear. Possibly, T_t corresponds to that temperature in the final cooling-down stage of the sample preparation at which annealing effectively ceased.

We now assume that $T_t > T$, the lattice temperature. It is not difficult to show that the crossover voltage V_X from the low-voltage Ohm's law (2.10) to the space-charge-controlled current regime is

$$V_X \simeq ekT_t\mathcal{N}_n L^2/\epsilon. \tag{2.27}$$

In the space-charge-limited current regime, the free and trapped injected carrier concentrations are related to F through

$$n = \rho/e = N_c \exp[(F - E_c)/kT]$$

and, from (2.26),

$$n_t = \rho_t/e = kT_t\mathcal{N}_0 \exp[(F - E_c)/kT_t].$$

Eliminating F, the relation between free and trapped charge is

$$\frac{\rho}{e} \simeq N_c \left(\frac{\rho_t}{e\mathcal{N}_0 kT_t}\right)^l \simeq N_c \left(\frac{\epsilon V}{eL^2\mathcal{N}_0 kT_t}\right)^l, \qquad \text{with } l = \frac{T_t}{T}, \tag{2.28}$$

where ρ_t has been replaced by $\epsilon V/L^2$. Finally, (2.1) gives

$$J \simeq e\mu N_c \left(\frac{\epsilon}{e\mathcal{N}_0 kT_t}\right)^l \frac{V^{(l+1)}}{L^{(2l+1)}}. \tag{2.29}$$

It is readily checked that (2.27) corresponds to the voltage of intersection of the space-charge-limited characteristic (2.29) with Ohm's law (2.10).

If $T_t < T$, the empty traps at the top of the distribution, near the conduction band, always dominate those near the quasi-Fermi level, and the situation reduces to the shallow-trapping case previously discussed.

In deriving (2.28) and (2.29) from (2.26), it has been assumed that, although $T_t > T$, kT_t is not so large that the entire excursion of F away from F_0, over the measured range of current, is less than kT_t. If the latter is true, then it is a good approximation simply to regard the traps as uniformly distributed in energy: $\mathcal{N}_t(E) \simeq \mathcal{N}_n$. In this case, $n_t = \rho_t/e = \mathcal{N}_n(F - F_0)$ and $n/n_0 = \exp(F - F_0)/kT$, so that

$$\rho/e \simeq n_0 \exp(\rho_t/e\mathcal{N}_n kT) \simeq n_0 \exp(\epsilon V/e\mathcal{N}_n kTL^2). \tag{2.30}$$

Now (2.1) gives

$$J \simeq e\mu n_0(V/L) \exp(\alpha V/TL^2) \qquad \text{with} \quad \alpha = \epsilon/e\mathcal{N}_n k. \tag{2.31}$$

The correct α [see (4.153)] is larger by a factor of 2. Here the typical error associated with the phenomenological analysis is not negligible because it appears in an exponent.

2.9 The Onset of Space-Charge-Controlled Currents: The General Case

The onset of space-charge-limited currents in the general case is determined by the identical considerations as for the previous special cases: A significant departure from Ohm's law will be observed only when the injected free-electron concentration n_i becomes comparable to the thermal concentration n_0, or, in the language of the quasi-Fermi level, when F moves up in the forbidden gap from F_0 by an amount kT approximately. Let $n_{t,x}^*$ denote the *total* average change in occupancy of all electron traps due to this upward motion of the Fermi level. Then the voltage V_x required to support the total trapped space charge $Q = en_{t,x}^* L$ per unit area, which is also the onset voltage for space-charge-limited current flow, is, from (2.4)

$$V_x \simeq en_{t,x}^* L^2/\epsilon. \tag{2.32}$$

The contributions to $n_{t,X}^*$ are, approximately:

(i) from traps lying above F_0 (shallow traps), the thermal concentrations of electrons in such traps,

(ii) from traps lying within kT of F_0, the total concentrations of such traps, and

(iii) from traps lying below F_0 (deep traps), the thermal concentrations of holes (vacancies) in such traps.

In principle any one of the three contributions can dominate $n_{t,X}^*$. In practice, the dominant contribution is likely to be from (ii) [or a number not much smaller than (ii)] since F_0 is likely to be anchored near a dominant set of traps in order that local charge neutrality be preserved.

In terms of the effective drift mobility μ_e, introduced in Section 2.4, Eq. (2.22), it is easily seen that (2.32) is equivalent to

$$t_{X,e} \simeq t_\Omega, \qquad \text{with} \quad t_{X,e} = (L^2/\mu_e V_X) \quad \text{and} \quad \mu_e = (n_0/n_{t,X}^*)\mu, \tag{2.33}$$

and with t_Ω given by (2.12).

A somewhat more precise calculation of $n_{t,X}^*$ may be made as follows: Let $n_{t,X}$ be the average electron occupancy, at $F = F_0 + kT$, of a given set of traps, of concentration N_t and statistical weight g and located at energy E_t. Define f as the fractional change in occupancy of the traps due to the upward motion of F. Then, from (2.17),

$$\begin{aligned} fN_t &= n_{t,X} - n_{t,0} \\ &= N_t \left\{ \left(1 + \frac{1}{e_0 g} \exp \frac{E_t - F_0}{kT}\right)^{-1} - \left(1 + \frac{1}{g} \exp \frac{E_t - F_0}{kT}\right)^{-1} \right\}, \end{aligned} \tag{2.34}$$

where e_0 is the natural base for logarithms, $e_0 = 2.718\ldots$. The expression (2.34) holds for each set of traps. A plot of f vs $(E_t - F_0)/kT$, for $g = 2$, is given in Fig. 2.3. A plot of f, for $g > 2$, is the same curve displaced to the right by $\ln(g/2)$, and, for $g = 1$, displaced to the left by $\ln 2 = 0.693$. On either side of the maximum, f rapidly approaches a decreasing Boltzmann-type exponential, on the left corresponding to the thermal excitation probability for holes below F_0, and on the right to that for electrons above F_0. Adding up all of the excess trapped carriers, we obtain

$$n_{t,X}^* = \sum_j f_j N_{tj}, \tag{2.35}$$

where subscript j has been added to distinguish the different sets of traps. Where there is a continuous distribution of traps, the discrete summation $\sum_j$ must be replaced by an integration, as for example, in deriving (2.27).

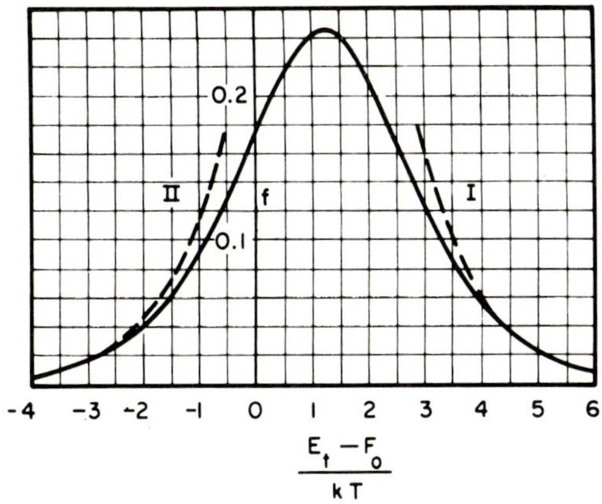

FIG. 2.3. Plot of the fractional change in trap occupancy, $f = (n_{t,x} - n_{t,0})/N_t$, for $F = F_0 + kT$, versus $(E_t - F_0)/kT$. The dashed curves are Boltzmann approximations.

2.10 HIGH-FIELD EFFECTS

It has already been remarked early in this chapter that the simple phenomenological analysis can still be used under high field conditions where the drift mobility is field-dependent. For the perfect insulator problem discussed in Section 2.1, one need only know the functional form of the dependence of drift velocity v on applied voltage V to obtain, using (2.5), the current-voltage characteristic.

For example, if above some critical field strength $\mathscr{E}_1$, the drift velocity varies as the square-root of the applied field (a characteristic of "warm" electrons when acoustic phonon scattering is dominant)

$\mathscr{E} > \mathscr{E}_1$:

$$v = \mu(\mathscr{E}_1\mathscr{E})^{1/2}, \tag{2.36}$$

where μ is the low-field, field-independent mobility, then (2.5) gives

$$J \simeq \epsilon\mu\mathscr{E}_1^{1/2}(V^{3/2}/L^{5/2}). \tag{2.37}$$

If the quasi-thermal relationship between free and trapped injected carriers represented by (2.16) and (2.17) is not strongly impaired by the heating of the carriers (and such impairment requires *both* that the electrons be *strongly* heated and the cross section for their capture by the traps be *strongly* velocity dependent), then the current-voltage characteristic under the high-field condition (2.36) is obtained from the low-field J–V characteristic, at $\mathscr{E} > \mathscr{E}_1$, for *any* given trap model, simply by multiplying the latter by $(\mathscr{E}_1 L/V)^{1/2}$. That is,

$$\mathscr{E} > \mathscr{E}_1, \quad v = \mu(\mathscr{E}\mathscr{E}_1)^{1/2}:$$

$$J_{\mathrm{LF}} = f(V), \qquad J_{\mathrm{HF}} = (\mathscr{E}_1 L/V)^{1/2} f(V), \tag{2.38}$$

where subscripts LF and HF denote "low fields" and "high fields," respectively.

If there is a substantial range of applied fields above a critical field $\mathscr{E}_{\mathrm{s}}$ over which the drift velocity is saturated at the value v_{s} (a situation which may occur when the electrons are heated up by the field to optical phonon energy), then (2.5) gives for the perfect insulator problem

$$\mathscr{E} > \mathscr{E}_{\mathrm{s}}, \quad v = v_{\mathrm{s}}:$$

$$J \simeq \epsilon v_{\mathrm{s}}(V/L^2). \tag{2.39}$$

If the quasi-thermal relationship between free and trapped carriers is still a reasonably good approximation over the range of saturation, then we have analogously to (2.38), for problems involving traps

$$\mathscr{E} > \mathscr{E}_{\mathrm{s}}, \quad v = v_{\mathrm{s}}:$$

$$J_{\mathrm{LF}} = f(V), \qquad J_{\mathrm{HF}} = (v_{\mathrm{s}} L/\mu V) f(V). \tag{2.40}$$

3

ANALYTIC FORMULATION AND GEOMETRIC PROPERTIES

We have seen in the previous chapter that many useful results of one-carrier space-charge-limited (SCL) current theory can be obtained by simple phenomenological arguments, without ever writing down a differential equation. Since those arguments were based on the *average* behavior of the quantities pertinent to current flow, they obviously could not yield information about the *detailed* behavior of these same quantities. Such information can be obtained only by an analytic study of the problem, that is, by writing down the algebraic and differential equations characterizing the current flow and obtaining the solution that satisfies the appropriate boundary conditions. At the same time, there is achieved both a more rigorous derivation of the current-voltage characteristic and a more accurate value of the numerical constant in the characteristic.

The equations that govern the behavior of steady-state, one carrier SCL currents are the current flow equation itself, Poisson's equation, and the steady-state equilibrium equations relating the free-electron concentration to the trapped-electron concentrations in the different traps at the same position. We now consider these equations.

3.1 The Current-Flow Equation

We assume that the electrons are injected at $x = 0$, the cathode, and exit at $x = L$, the anode.

It is convenient to adopt the sign conventions,

$$\mathbf{J} = -J\mathbf{x} \qquad \text{and} \qquad \boldsymbol{\mathscr{E}} = -\mathscr{E}\mathbf{x}, \tag{3.1}$$

where $\mathbf{J}$ is the current density, $\boldsymbol{\mathscr{E}}$ the electric field intensity, and $\mathbf{x}$ a unit vector along the x axis. With these conventions, J is always positive, and likewise $\mathscr{E}$ in the simplified version of the theory that neglects diffusion.

The current-flow equation states simply that the current is the sum of two contributions, a drift current J_{dr} and a diffusion current J_{di}, and is a constant independent of position x:

$$J_{\mathrm{dr}} = e\mu n\mathscr{E}, \tag{3.2}$$

$$J_{\mathrm{di}} = -eD\, dn/dx, \qquad \text{with} \quad D = \mu V_{\mathrm{T}}, \; V_{\mathrm{T}} = kT/e, \tag{3.3}$$

$$J = J_{\mathrm{dr}} + J_{\mathrm{di}} = e\mu(n\mathscr{E} - V_{\mathrm{T}}\, dn/dx) = \text{constant}. \tag{3.4}$$

The constancy of J in the steady state is simply a restatement of particle conservation:

$$\operatorname{div} \mathbf{J} = dJ/dx = 0. \tag{3.5}$$

Throughout most of the book we shall be working with a simplified theory of injection currents in which the diffusion currents are neglected. With this approximation, (3.4) simplifies to:

Simplified theory:

$$J = J_{\mathrm{dr}} = e\mu n\mathscr{E} = \text{constant}. \tag{3.6}$$

Note that we are now using the same symbol n to denote the *position-dependent* concentration that we used in Chapter 2 to denote the *average* concentration, and likewise for other symbols n_{t}, $\mathscr{E}$, F, etc. There should be no confusion on this score because in Chapter 2 *only* averages of quantities entered into the argument. In this, and succeeding chapters of Part I, we shall use a special notation to denote averages.

Equation (3.4) is valid under conditions where the velocity distribution of the free electrons is only slightly perturbed from the thermal-equilibrium Maxwellian distribution by the presence of the applied field. At sufficiently high applied field strengths, the perturbation of the velocity distribution is no longer small and the drift current is then, generally, no longer proportional to the applied field, as already noted at the end of the preceding chapter, Section 2.10, and as discussed, for example, in the review articles by Franz (1956), Gunn (1957), Stratton (1961), Conwell (1967), and Chynoweth (1968). It is not obvious, in such cases, that one can then always regard the current as the sum of a drift current and a diffusion current proportional to dn/dx, and, if one can, what relation $D(\mathscr{E})$ has to $\mu(\mathscr{E})$. This question is not explored further in this book. Fortunately, in most injection problems of interest, at applied fields high enough to produce significant field-dependence of the mobility, the "diffusion current," whatever its functional form, will be relatively insignificant. For this case, then, we can replace (3.4) and (3.6) by

High applied fields:

$$J = env(\mathscr{E}) = en\mu(\mathscr{E})\mathscr{E} = \text{constant}. \tag{3.7}$$

Possible functional forms of $v(\mathscr{E})$ at high fields are cited in Sections 2.10 and 7.4.

The onset field strength $\mathscr{E}_1$ for field-dependence of the mobility depends strongly on the material, its chemical and structural purity, and the temperature. For very pure Ge at 4°K, $\mathscr{E}_1 \approx 1$ V/cm; for Ge and Si at 300°K, $\mathscr{E}_1 \approx 10^3$ V/cm; for CdS and ZnS at 300°K, $\mathscr{E}_1 \gtrsim 10^5$ V/cm.

Note that (3.4) may still be used in high-field regions that are present in thermal equilibrium, such as a region adjacent to a contact interface or a junction region. For, in thermal equilibrium, the Maxwellian distribution holds throughout these regions, and $J = 0$ corresponds to exact cancellation of the opposing drift and diffusion currents, $J > 0$ to only partial cancellation.

3.2 The Poisson Equation

This equation brings in the role of space charge in SCL current flow:

$$(\epsilon/e)(d\mathscr{E}/dx) = (n - n_0) + \sum_j (n_{tj} - n_{tj,0}), \tag{3.8}$$

where j denotes the jth set of traps. Here $n = n(x)$ and $n_{tj} = n_{tj}(x)$ are the actual, spatially varying free electron concentration and trapped electron concentration in the jth set of traps respectively, and n_0, $n_{tj,0}$ are the corresponding, constant, thermal equilibrium values in the bulk, neutral insulator far removed from contacts and surfaces. Equation (3.8) is written for trapping states associated with sharply defined energy levels, E_{tj} being the energy level for the jth set of traps. Trap distributions continuous in energy, such as (2.26), are readily included in (3.8) by regarding $\sum_j$ as a generalized sum which, for a continuous distribution, becomes $\int dE$, the corresponding n_t then being a density in energy as well as a spatial concentration.

3.3 The Equations Relating Free and Trapped Charge

The system of equations characterizing one-carrier, steady-state, SCL current flow is completed by the equations relating free and trapped carrier concentrations. In words, these equations are, for all j:

$$n(x) \text{ and } n_{tj}(x) \text{ are in local, quasi-thermal equilibrium;} \tag{3.9}$$

that is, $n(x)$ and $n_{tj}(x)$ have the ordinary, thermal equilibrium relationship to each other, corresponding to the temperature of the lattice, except that $n(x)$ here has its spatially dependent, steady-state value corresponding to a given current flow. In terms of the quasi-Fermi level F, which is now regarded as position-dependent, $F = F(x)$, (3.9) can be written explicitly as

$$n(x) = N_c \exp\frac{F(x) - E_c(x)}{kT}; \qquad N_j = N_c \exp\frac{E_{tj} - E_c}{kT}, \tag{3.10}$$

$$n_{tj}(x) = \frac{N_{tj}}{1 + (1/g_j)\exp[(E_{tj}(x) - F(x))/kT]} = \frac{N_{tj}}{1 + (1/g_j)(N_j/n(x))}. \tag{3.11}$$

These expressions are simply the position-dependent generalizations of (2.16) and (2.17), respectively. Note that, whereas E_c and E_{tj} both vary with position, $E_c - E_{tj}$ does not depend on position. For a scalar effective mass m, $N_c = 2(2\pi mkT/h^2)^{3/2} = 4.83 \times 10^{15}(m/m_0)^{3/2}T^{3/2}\ \text{cm}^{-3}$, where h is Planck's constant. Where there are s equivalent conduction-band minima and the surfaces of constant energy are ellipsoids of revolution, $N_c = 4.83 \times 10^{15}s(m_l m_t^2/m_0^3)^{1/2}T^{3/2}\ \text{cm}^{-3}$, where m_l, m_t are the longitudinal and transverse effective masses, respectively, in the diagonalized mass tensor. For the warped, constant-energy surfaces near the valence band maximum at $k = 0$ in semiconductors like Ge and Si, the appropriate expressions for the valence-band densities of states are given by Lax and Mavroides (1955).

The occupation probability factor in (3.11) corresponds to the simplest possible situation, namely where each trap can hold only a single electron and there is only one electronic level, of statistical weight g_j, available for that electron. Occupation probability formulas covering multivalent and amphoteric defect states as well as excited states of monovalent defects are given in Blakemore (1962), Chapter 3.

The physical basis for the assumption of quasi-thermal equilibrium was touched upon briefly in Section 2.3. We now examine this assumption more closely. In general, the relation between free and trapped carriers is obtained, for the steady state, by balancing the trapping rate r per unit volume (the downward arrow in Fig. 2.1),

$$r = n\langle v\sigma\rangle(N_t - n_t), \tag{3.12}$$

against the thermal ionization rate g_{th} per unit volume (the upward arrow in Fig. 2.1),

$$g_{th} = n_t\langle e\rangle N_c . \tag{3.13}$$

Thus,

$$r = g_{th}, \qquad \text{giving} \quad n_t = \frac{N_t}{1 + (\langle e\rangle/\langle v\sigma\rangle)(N_c/n)} . \tag{3.14}$$

In the above, v is the *magnitude* of the velocity of the free electron, $\sigma = \sigma(v)$ the cross section for its capture by a trap, $\langle v\sigma\rangle$ the average of $v\sigma$ over the free-electron velocity distribution, and $\langle e\rangle N_c = \sum_k e_k$, where e_k is the probability per unit time of thermal ejection of a trapped electron into the particular conduction band state labeled by k and the sum is taken over all conduction band states. Note that in their presently defined meanings, the symbols v and e are never used except inside the brackets $\langle\ \rangle$; therefore they cannot be confused with v meaning drift velocity (Chapter 2) and e meaning the magnitude of the charge on the electron.

In thermal equilibrium, $n = n_0$ and $n_t = n_{t,0}$. Comparing (3.14) with (2.15), it is seen that

$$\frac{\langle e\rangle_0}{\langle v\sigma\rangle_0} = \frac{N}{gN_c} = \frac{1}{g}\exp\frac{E_t - E_c}{kT}, \tag{3.15}$$

where subscript 0, as usual, denotes thermal equilibrium conditions.

What happens to $\langle e\rangle$ and $\langle v\sigma\rangle$ under an applied voltage, assuming the lattice temperature (hence also N_c) unchanged? We would expect $\langle e\rangle$ to be unchanged until very high fields are reached; for thermal emission probabilities will only be affected by fields high enough to give significant Stark shifts of trap levels or to bring tunneling into play as a factor in ionization of the trap. Such fields are also likely high enough to cause impact ionization and so are, in any case, excluded from our discussion. Taking $\langle e\rangle = \langle e\rangle_0$, Eq. (3.13) can be rewritten, using (3.15), as

$$g_{th} = n_t f^* \exp[(E_t - E_c)/kT], \qquad f^* = (1/g)\langle v\sigma\rangle_0 N_c . \tag{3.16}$$

The quantity f^* is called the frequency-factor or attempt-to-escape frequency. In a classical picture, it represents the frequency with which the trapped electron strikes the potential "wall" of the trap. The Boltzmann factor then represents the probability that the trapped electron will have enough thermal energy to escape when it reaches the wall.

In contrast to $\langle e \rangle$, $\langle v\sigma \rangle$ can change substantially with applied field if the free electrons are significantly heated up by the applied field *and* the capture cross section σ is strongly velocity-dependent, e.g., of Coulomb type. Thus, if σ is Coulomb-attractive, $\langle v\sigma \rangle$ will decrease as the electrons heat up and, from (3.14), the equilibrium will shift towards a higher concentration of free electrons, an effect actually seen in Ge at 4°K in the prebreakdown region by Koenig and Gunther-Mohr (1957). On the other hand, if σ is Coulomb-repulsive, $\langle v\sigma \rangle$ will increase as the electrons heat up and the equilibrium will shift toward a lower concentration of free electrons. This effect can be very large because capture in this case involves activation over the Coulomb-repulsive barrier and therefore will be very sensitive to the detailed structure of the free-electron distribution function. (This argument applies to any activation barrier, whatever its physical origin.) This effect has been strikingly demonstrated, in n-Ge containing Au^{2-} impurities, at 20°K, by Ridley and Pratt (1963, 1965) (see also Pratt and Ridley, 1963, 1965a, b). In this work, the decrease in free electron concentration through capture by the Au^{2-} centers at fields $\mathscr{E} \gtrsim 10$ V/cm was substantial enough to give a negative differential resistance with subsequent electrical instability.

The above remarks lead to the conclusion that the assumption of quasi-thermal equilibrium is certainly a good one, if the free-carrier velocity distribution is not significantly changed by the applied field; and even with moderate changes in this distribution, the assumption will be reasonably good if σ is only weakly dependent on velocity. (If $\sigma \propto 1/v$, $v\sigma$ is independent of velocity and the assumption is good even for large changes in the distribution function.)

3.4 The Simplified Theory

The simplified theory of one-carrier SCL currents is based on two approximations:

(i) Diffusion currents are neglected, that is, Eq. (3.6), instead of Eq. (3.4), is used to describe the current flow, and

(ii) The cathode is taken to be an infinite reservoir of electrons available for injection.

The first approximation simplifies the theory to such an extent that it becomes mathematically tractable to elementary analysis. The second approximation, though not simplifying further the structure of the theory, does make the theory independent of any detailed properties of the contact and thereby makes a universal theory possible.

The two approximations are closely related; the second approximation is not possible without the first, since an infinite concentration of electrons necessarily gives an infinite diffusion current at the contact. The two approximations taken together effectively remove both the injecting and exiting contacts from any role in the theory. The simplified theory built on (i) and (ii), though necessarily in error in the immediate vicinity of the contacts, generally gives a good description of the details of current flow over the bulk of the insulator, away from the contacts, and of the current-voltage characteristic, which is an overall property of the insulator. The approximations (i) and (ii) are considered at greater length in Chapter 9.

The basic equations for the simplified theory are (3.6), (3.8), and the set (3.11). From (3.6) and (3.11),

$$n = n_0(\mathscr{E}_J/\mathscr{E}), \qquad n_{tj} = \frac{N_{tj}}{1 + (\mathscr{E}/\mathscr{E}_{J,j})}; \tag{3.17}$$

$$\mathscr{E}_J = (J/e\mu n_0), \qquad \mathscr{E}_{J,j} = (g_j J/e\mu N_j), \tag{3.18}$$

where $\mathscr{E}_J$ and $\mathscr{E}_{J,j}$ obviously have the dimensions of electric field intensity, hence the notation, and where subscript J has been used to emphasize the (linear) dependence on J. Substituting from (3.17) into (3.8), we obtain the first-order differential equation

$$dx/d\mathscr{E} = f(\mathscr{E}; J), \tag{3.19}$$

$$\frac{1}{f(\mathscr{E}; J)} = \frac{e}{\epsilon}(\mathscr{E}_J - \mathscr{E})\left[\frac{n_0}{\mathscr{E}} + \sum_j \frac{n_{tj,0}}{(\mathscr{E} + \mathscr{E}_{J,j})}\right], \tag{3.20}$$

where $n_{tj,0}$ is given by (3.11) with $n(x)$ replaced by n_0. Note that in $f(\mathscr{E}; J)$, $\mathscr{E}$ is the variable and J is a given (constant) parameter, namely the current density, which does not vary with position.

A single boundary condition is needed to specify a unique solution to (3.19). It is

$$\mathscr{E} = 0 \quad \text{at} \quad x = 0, \tag{3.21}$$

and follows directly from approximation (ii) above. For, clearly, the electric field intensity must vanish at the injecting contact in order that the infinite density of electrons there deliver a finite, i.e., noninfinite, field-driven current. Equation (3.21) is the analytic characterization of an ohmic contact in the context of the simplified theory.

Integrating Eq. (3.19) from the cathode to the anode, and using (3.21), we obtain

$$L = \int_0^{\mathscr{E}_a} f(\mathscr{E}; J)\, d\mathscr{E} = g(\mathscr{E}_a; J), \tag{3.22}$$

where $\mathscr{E}_a = \mathscr{E}(L)$ is the electric field intensity at the anode.

The applied voltage V is calculated simply as the integral of the electric field intensity between cathode and anode:

$$V = \int_0^L \mathscr{E}\, dx = \int_0^{\mathscr{E}_a} \mathscr{E} f(\mathscr{E}; J)\, d\mathscr{E} = h(\mathscr{E}_a; J), \tag{3.23}$$

where we have used (3.19). This procedure ignores the built-in fields present in the insulator in thermal equilibrium, namely the fields associated with band-bending near the contacts. Our neglect of the contacts in this regard is consistent with the general neglect of the role of the contact inherent in the simplified theory.

In many cases of interest, the integrations in (3.22) and (3.23) can be carried out analytically, via expansions of $f(\mathscr{E}; J)$ and $\mathscr{E} f(\mathscr{E}; J)$ in simple partial fractions, yielding explicit analytic expressions for the functions g and h. In these cases, the *purely algebraic* equations

$$L = g(\mathscr{E}_a; J), \qquad V = h(\mathscr{E}_a; J) \tag{3.24}$$

define *implicitly* the relation between J and V, via the intermediate variable $\mathscr{E}_a$. In general it is not possible to eliminate the variable $\mathscr{E}_a$ and so derive an explicit analytic expression relating J to V. However, this frequently can be done, in approximate fashion, over restricted ranges of voltage, yielding different "regimes" of current flow (Ohm's law regime, square-law regime, etc.). Several important simple cases are studied in detail in Chapter 4. In Chapter 9, we consider the modifications of the simplified theory [use of Eq. (3.6)] required by a *finite* reservoir contact [dropping of assumption (ii) above].

A further comment is in order on the simplified theory. Although contact effects are excluded from the theory by its very construction, it is sometimes possible to "graft" them onto the theory in a simple and natural way. Thus, if the cathode is a good, ohmic (reservoir) contact but there is a potential barrier at the anode of height V_t volts, then the simplified theory will, in most cases, give a good description of the current-voltage characteristic if the applied voltage V in (3.23), (3.24) is replaced by $V - V_t$. (The characteristic for $0 < V < V_t$ must then be obtained from other considerations, say, for example, from forward-bias semiconductor-rectifier theory.)

3.5 Miscellaneous Relations and Approximations

Integration of the Poisson equation (3.8), subject to boundary equation (3.21), gives for Q, the total (free plus trapped) injected, excess charge in the insulator

$$Q = \epsilon\mathscr{E}_{\mathrm{a}} . \tag{3.25}$$

This relation is true even under Ohm's law conditions; in this case the injected charge is localized in the immediate neighborhood of the cathode. Taking $Q = CV$ to *define* the capacitance C, (3.25) gives

$$C = \epsilon\mathscr{E}_{\mathrm{a}}/V = (\mathscr{E}_{\mathrm{a}}/\mathscr{E}_{\Omega})C_0 , \tag{3.26}$$

where $\mathscr{E}_{\Omega}$ is the ohmic field: $\mathscr{E}_{\Omega} = V/L$, and C_0 is the geometric capacitance: $C_0 = \epsilon/L$, which we have already used in Chapter 2. The relation (3.26) is used to good advantage in the following section, where we establish close bounds on $\mathscr{E}_{\mathrm{a}}/\mathscr{E}_{\Omega}$, and therefore on C/C_0.

The drift transit time t of the injected free carriers is given by

$$t = \int_0^L \frac{dx}{v(x)} = \frac{1}{\mu}\int_0^L \frac{dx}{\mathscr{E}(x)} = \frac{1}{\mu}\int_0^{\mathscr{E}_{\mathrm{a}}} \frac{1}{\mathscr{E}}\,\frac{dx}{d\mathscr{E}}\,d\mathscr{E} = \frac{1}{\mu}\int_0^{\mathscr{E}_{\mathrm{a}}} \frac{f(\mathscr{E};J)}{\mathscr{E}}\,d\mathscr{E}, \tag{3.27}$$

where $v(x)$ is the local drift velocity given by $v(x) = \mu\mathscr{E}(x)$, and where we have used (3.19). Here t is defined so as to exclude specifically trapping effects, that is, it refers to the transit of free carriers. In a Haynes–Shockley (1949) type of experiment, in which a pulse of excess minority carriers is drifted across a semiconductor sample, a large fraction of the excess minority carriers may be trapped in a quasithermal equilibrium with the free carriers. In this case the observed transit time of the pulse includes the time the excess minority carriers spend in traps and is greater than the trap-free time t by the ratio of trap to free injected carrier concentrations, as already noted in the discussion at the end of Section 2.4.

The average free carrier concentration $\bar{n}$ (injected plus thermal) is given by

$$\bar{n} = \frac{1}{L}\int_0^L n\,dx = \frac{J}{e\mu L}\int_0^L \frac{dx}{\mathscr{E}} = \frac{J}{eL}\,t, \tag{3.28}$$

where we have used (3.6) and (3.27). Since all of the free carriers have precisely the same transit time t, Eq. (3.28) is simply the expected result, $J = e\bar{n}L/t$, which we already used in (2.1a).

From the relationship (3.11) between free and trapped carriers, rewritten here, for convenience, omitting the subscript j and variable x, as

$$n_t = \frac{N_t}{1 + (1/g)\exp[(E_t - F)/kT]} = \frac{N_t}{1 + (1/g)(N/n)};$$
$$N = N_c \exp\left(\frac{E_t - E_c}{kT}\right), \tag{3.29}$$

it follows that

$$p_t = N_t - n_t = \frac{N_t}{1 + g\exp[(F - E_t)/kT]} = \frac{N_t}{1 + g(n/N)}. \tag{3.30}$$

Further, the precise relation between the differential increments of n_t and n is given by

$$\delta n_t = -\delta p_t = (n_t p_t / n N_t)\,\delta n. \tag{3.31}$$

Useful approximations to (3.29) and (3.30) are, for

Shallow traps:

$$(1/g)\exp[(E_t - F)/kT] \gg 1 \quad \text{or} \quad (1/g)(N/n) \gg 1,$$
$$n_t \simeq (1/\theta)n, \quad \theta = (N/gN_t); \quad p_t \simeq N_t. \tag{3.32}$$

Deep traps:

$$(1/g)\exp[(E_t - F)/kT] \ll 1 \quad \text{or} \quad (1/g)(N/n) \ll 1,$$
$$n_t \simeq N_t; \quad p_t \simeq (NN_t/g)(1/n); \quad \delta n_t/\delta n \simeq p_t/n. \tag{3.33}$$

The concepts of "shallow" and "deep" traps have already been discussed in Sections 2.4 and 2.5.

3.6 Geometric Results and the $Q = CV$ Relation

The simplified theory characterized by Eqs. (3.6), (3.8), (3.10), (3.11), and the boundary condition (3.21) predicts certain universal properties of the current flow, namely, properties that are totally independent of the detailed nature of the traps (their number and energy distribution). It is convenient to describe these properties in terms of current flow into a semi-infinite insulator. For a finite insulator, the solution for the corresponding semi-infinite insulator is used out as far as the anode. These universal properties are:

(i) $n(x)$ is monotonic decreasing and approaches n_0 as $x \to \infty$.

(ii) $\mathscr{E}(x)$ is monotonic increasing and convex, that is, $d\mathscr{E}/dx$ is monotonic decreasing.

At the cathode, which we take at $x = 0$, the boundary condition $\mathscr{E} = 0$ requires, from (3.6), that $n = \infty$. Then, from (3.8), $d\mathscr{E}/dx = \infty$ at the cathode so that $\mathscr{E}$ becomes positive moving away from the cathode. From (3.6), n accordingly decreases moving away from the cathode. We first show that the curve $n(x)$ vs x can never cross the horizontal line $n = n_0$. The proof follows readily by contradiction, that is, by assuming it can cross and showing that this leads to a contradiction. If $n(x)$ crossed $n = n_0$ first at some point x_1, as with curve $\mathscr{C}_1$ in Fig. 3.1, then imme-

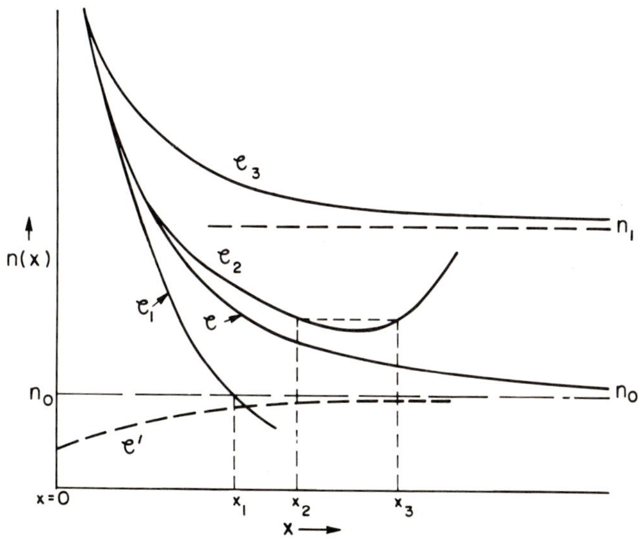

FIG. 3.1. Schematic plot of n versus x for a semi-infinite crystal with an electron-injecting contact. $\mathscr{C}_1$, $\mathscr{C}_2$, and $\mathscr{C}_3$ are impossible situations. An allowed situation is $\mathscr{C}$: $n(x)$ monotonically decreasing toward n_0 as x approaches infinity. For an electron-blocking contact, an allowed variation of $n(x)$ with x is $\mathscr{C}'$.

diately beyond x_1, $n < n_0$ and from (3.11), $n_{tj} < n_{tj,0}$. From (3.8), $\mathscr{E}$, as well as n, is decreasing immediately beyond x_1. But from (3.6), it is impossible to have n and $\mathscr{E}$ decreasing simultaneously. Hence behavior such as curve $\mathscr{C}_1$ is impossible.

We have so far shown that $n(x)$ lies everywhere above n_0. From (3.11), $n_{tj}(x)$ lies everywhere above $n_{tj,0}$. Therefore, from (3.8), $d\mathscr{E}/dx$ is everywhere positive, so that $\mathscr{E}$ is monotonic increasing with x. Clearly, behavior such as curve $\mathscr{C}_2$ is ruled out; that is, there cannot be two points x_2 and x_3 with $n(x_2) = n(x_3)$ because $\mathscr{E}(x_3) > \mathscr{E}(x_2)$ and so (3.6) would be violated. Thus $n(x)$ must be monotonic decreasing and it remains to establish that it approaches n_0 as x approaches ∞. If $n(x)$

decreased monotonically toward $n_1 > n_0$, such as curve $\mathscr{C}_3$, then everywhere $n(x) - n_0 > n_1 - n_0$ and $n_{tj}(x) - n_{tj,0} > n_{tj,1} - n_{tj,0}$, where $n_{tj,1}$ is the value of $n_{tj}(x)$ corresponding to $n(x) = n_1$ in (3.11). Thus, $(\epsilon/e)(d\mathscr{E}/dx)$ everywhere exceeds $(n_1 - n_0) + \sum_j (n_{tj,1} - n_{tj,0})$ and so $\mathscr{E}$ has a minimum, *finite* rate of growth for all x, no matter how large. Thus, for large x, $J \simeq e\mu n_1 \mathscr{E}(x) =$ constant is impossible, violating (3.6). Thus, we conclude that $n(x)$ decreases monotonically toward n_0, as curve $\mathscr{C}$ in Fig. (3.1). From (3.11), $n_{tj}(x)$ decreases monotonically toward $n_{tj,0}$, and from (3.8), $d\mathscr{E}/dx$ decreases monotonically as x increases. Hence, $\mathscr{E}(x)$ increases monotonically with x but with ever-decreasing slope; that is, the curve $\mathscr{E}(x)$ vs x is convex as in Fig. 3.2. The universal properties (i) and (ii) have now been established.

The properties (i) and (ii) of injection currents are valid under considerably more general conditions than (3.6), and (3.10), (3.11). Thus:

(a) If instead of (3.6), we use (3.7) and we require only that $v(\mathscr{E}_2) \geqslant v(\mathscr{E}_1)$ for $\mathscr{E}_2 > \mathscr{E}_1$ (this rules out differential negative resistance, but permits saturation of the drift velocity); and

(b) if instead of (3.10), (3.11), we require only that $n_{tj}(x_2) \geqslant n_{tj}(x_1)$ for $n(x_2) > n(x_1)$ (this rules out impact or field ionization of traps),

then (i) and (ii) are still valid except for one change. In the case of saturation of the drift velocity above a critical field, $n(x)$ will, in general, decrease monotonically to, and level off at, $n_1(J) > n_0$. This is obvious, since $n(x)$ can no longer change with x once the drift velocity has saturated. The proofs for the more general conditions are quite similar to the proofs supplied above.

The convex character of $\mathscr{E}$ has a very useful consequence, namely the inequality

$$\mathscr{E}_\Omega < \mathscr{E}_a < 2\mathscr{E}_\Omega . \tag{3.34}$$

This follows from a comparison of areas, which is given in Fig. 3.2. The inequality sequence (3.34) is clearly equivalent to that written at the bottom of Fig. 3.2. Using (3.26), we see that (3.34) is equivalent to the inequality sequence

$$C_0 < C < 2C_0 , \tag{3.35}$$

which was argued on grounds of physical plausibility in Section 2.1.

In all of the discussion of this section, the boundary condition (3.21) has played a very important role in that it enforces a regime of injection in the solid. The property (i) then asserts that there cannot be anywhere else in the solid a region of depletion. A generalization of this property is valid which does not depend on the particular boundary condition at the cathode, namely,

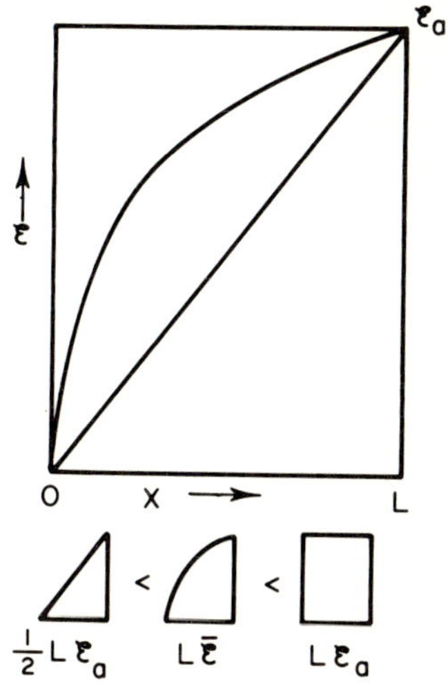

FIG. 3.2. Convexity of the electric field intensity $\mathcal{E}$ for one-carrier SCL current flow (universal property) and consequent inequalities, in the simplified theory. The quantities under the geometric figures are the areas of the corresponding figures.

(iii) there cannot be simultaneously, in a solid, regions of accumulation and depletion.

The proof is essentially that given above in establishing that curve $\mathscr{C}_1$ is not a possible variation of $n(x)$ with x. Whether the solid is entirely in accumulation or entirely in depletion depends on the nature of the cathode. With an injecting cathode, characterized in the simplified theory by boundary condition (3.21), the solid is everywhere in accumulation, and the variation of $n(x)$ with x follows a curve like $\mathscr{C}$ in Fig. 3.1. With a neutral or blocking cathode, the solid is everywhere in depletion and the variation of $n(x)$ with x follows a curve like the dashed curve $\mathscr{C}'$ in Fig. 3.1, namely with $n(x)$ monotonic increasing and approaching n_0 as x approaches ∞.

It must be kept in mind that (iii) is a consequence of the simplified theory, neglecting diffusion currents and high-field phenomena. In practice, (iii) can easily be violated precisely through the agency of one of the neglected factors. Thus, in a solid with an injecting cathode but blocking anode, (iii) is certainly violated at low currents. The current flow near the anode can, in this case, be described adequately only by inclusion of the diffusion current in the analysis. Another example is the electrical instability induced by a differential negative mobility, leading to the formation of dipole domains in the solid, in violation of (iii).

4

ANALYTIC AND APPROXIMATE SOLUTIONS

In this chapter we shall obtain exact analytic and approximate solutions, within the framework of the simplified theory, for the various prototype problems of physical interest, which were studied phenomenologically in Chapter 2. In addition to the current-voltage characteristic, we shall obtain the spatial variation of the injected free-carrier concentration and such other useful information as the space-charge capacitance and the free-carrier transit time.

The exact analytic solutions to injection problems tend to be excessively unwieldy. Therefore, even where such solutions are available, it is still useful to employ simplifying approximations. A very powerful approximation scheme, simple to employ and based on obvious physical considerations, is the Method of Regional Approximations. This method generally gives good accuracy. For example, the current-voltage characteristic obtained from it differs by less than a factor of two from the characteristic obtained from the exact analytic solution for the planar flow, one-carrier injection problems. In the present chapter the use of the Regional Approximation Method is supplemental; it serves to simplify and clarify. In later chapters, it is used extensively to obtain solutions of problems that are beyond the reach of exact analysis.

We rewrite here, for convenience, the equations characterizing the simplified theory:

$$J = e\mu n\mathscr{E} = \text{constant}, \tag{4.1}$$

$$(\epsilon/e)(d\mathscr{E}/dx) = n - n_0 + \sum_j (n_{tj} - n_{tj,0}), \tag{4.2}$$

$$n(x) = N_c \exp \frac{F(x) - E_c(x)}{kT}; \qquad n_{tj}(x) = \frac{N_{tj}}{1 + (1/g_j)[N_j/n(x)]} \tag{4.3}$$

with $N_j = N_c \exp(E_{tj} - E_c)/kT$. These equations are solved subject to the boundary condition

$$\mathscr{E}(0) = 0. \tag{4.4}$$

They are just the Eqs. (3.6), (3.8), (3.10), and (3.11), and (3.21), respectively. The remainder of this chapter is devoted to obtaining solutions of this set of defining equations for various cases of physical interest.

4.1 The Perfect Trap-Free Insulator

This is the direct, solid-state analog of the thermionic vacuum diode. There are neither thermal free carriers nor trapping states in the solid, that is n_0 and all the n_{tj} are identically zero in the above equations. Therefore the defining equations for the problem are (4.1) and

$$(\epsilon/e)(d\mathscr{E}/dx) = n. \tag{4.5}$$

These two equations combine to give

$$\mathscr{E}(d\mathscr{E}/dx) = J/\epsilon\mu. \tag{4.6}$$

This differential equation, subject to the boundary condition (4.4), is readily integrated to give

$$\mathscr{E}(x) = (2J/\epsilon\mu)^{1/2} x^{1/2} \tag{4.7}$$

and

$$V(x) = \int_0^x \mathscr{E}(x)\, dx = (8J/9\epsilon\mu)^{1/2} x^{3/2}. \tag{4.8}$$

Taking $x = L$ and $V = V(L)$ in the latter relation, we obtain the current-voltage characteristic

$$J = (9/8)\,\epsilon\mu(V^2/L^3). \tag{4.9}$$

Equation (4.9) is variously referred to as the trap-free square law, the Mott–Gurney square law, and Child's law for solids, the latter in analogy with the famous Child's law for the vacuum diode.

Further relations of interest are

$$\frac{\mathscr{E}(x)}{\mathscr{E}_\Omega} = \frac{3}{2}\left(\frac{x}{L}\right)^{1/2}; \qquad \frac{V(x)}{V} = \left(\frac{x}{L}\right)^{3/2}; \qquad \frac{n(x)}{\bar{n}} = \frac{1}{2}\left(\frac{L}{x}\right)^{1/2} \tag{4.10}$$

where $\mathscr{E}_\Omega$ is the ohmic (average) field, $\mathscr{E}_\Omega = V/L$, and $\bar{n}$ is the spatial average of $n(x)$:

$$\bar{n} = (1/L) \int_0^L n(x)\, dx = \tfrac{3}{2}(\epsilon V/eL^2), \tag{4.11}$$

where $n(x)$ is obtained from (4.7) and (4.1). From (4.10) and (3.26), it follows that the capacitance per unit area is given by

$$C = \tfrac{3}{2} C_0\,, \tag{4.12}$$

in accord with (3.35). The interelectrode transit time is given by

$$t = \int_0^L \frac{dx}{\mu \mathscr{E}(x)} = \frac{4}{3} t_0\,, \qquad t_0 = \frac{L^2}{\mu V}\,, \tag{4.13}$$

a result which can also be obtained directly from (3.28), (4.9), and (4.11). The above results are summarized in Table 4.1 and are compared with the corresponding quantities of the thermionic vacuum diode. Order-of-magnitude comparisons of currents through either type of device are

TABLE 4.1

PROPERTIES OF SCL CURRENT FLOW

Vacuum diode	Trap-free insulator diode
$J = \dfrac{4\sqrt{2}}{9}\left(\dfrac{e}{m}\right)^{1/2} \epsilon_0 \dfrac{V^{3/2}}{L^2}$	$J = \dfrac{9}{8} \epsilon\mu \dfrac{V^2}{L^3}$
$= 2.3 \times 10^{-6} \dfrac{V^{3/2}}{L^2}$ A/cm²	$= 10^{-13} \kappa\mu \dfrac{V^2}{L^3}$ A/cm²
$\mathscr{E} \propto x^{1/3}$	$\mathscr{E} \propto x^{1/2}$
$\mathscr{E}_a/\mathscr{E}_\Omega = 4/3$	$\mathscr{E}_a/\mathscr{E}_\Omega = 3/2$
$V(x) \propto x^{4/3}$	$V(x) \propto x^{3/2}$
$n \propto x^{-2/3}$	$n \propto x^{-1/2}$
$\bar{n} = 3n_a$	$\bar{n} = 2n_a$
$t = 3L\left(\dfrac{m}{2eV}\right)^{1/2}$	$t = \dfrac{4}{3}\dfrac{L^2}{\mu V}$ sec
$= 5 \times 10^{-8} \dfrac{L}{\sqrt{V}}$ sec	
For $V = 1000$ V, $L = 10^{-2}$ cm and area $A = 0.1$ cm²	For same V, L, area and $\mu = 10$ cm²/V sec, $\kappa = 10$
$I = JA \simeq 100$ A	$I \simeq 1$ A
$t \simeq 10^{-11}$ sec	$t \simeq 10^{-8}$ sec

also given at the bottom of the table. It is seen that the vacuum diode carries a much larger current than does a typical insulating solid of comparable dimensions under the same voltage. The reason lies with the magnitude of μ. By choosing a material with larger μ, a correspondingly larger insulator diode current could be achieved, at least at voltages below which high field effects become prominent. In practice, of course, trapping normally restricts steady-state one-carrier SCL currents in solids to values very much lower than the SCL currents in vacuum diodes of comparable dimensions.

4.2 The Trap-Free Insulator with Thermal Free Carriers

This represents the next step in complication in the hierarchy of one-carrier, SCL injection problems. Here we consider the solid free of traps but characterized by a finite conductivity $\sigma_0 = en_0\mu$. Thus all the n_{tj} are identically zero, and the defining equations are (4.1) and

$$(\epsilon/e)(d\mathscr{E}/dx) = n(x) - n_0 . \tag{4.14}$$

The thermally generated electrons, of concentration n_0, contribute to the current but not to the space-charge. Here we expect a transition from a low voltage regime where Ohm's law obtains so long as $\bar{n} \simeq n_0$. We first present the exact analytic solution to this problem.

4.2.1 *Exact Analytic Solution*

It is convenient to employ dimensionless variables as follows:

$$u = \frac{n_0}{n(x)} = \frac{en_0\mu\mathscr{E}(x)}{J}, \qquad w = \frac{e^2{n_0}^2\mu x}{\epsilon J}, \qquad v = \frac{e^3{n_0}^3\mu^2 V(x)}{\epsilon J^2}. \tag{4.15}$$

The differential equation in $\mathscr{E}$, obtained from (4.1) and (4.14), becomes in the dimensionless variables,

$$u\,du/(1-u) = dw, \tag{4.16}$$

with solution

$$w = -u - \ln(1-u), \tag{4.17}$$

where $u = 0$ at $w = 0$, corresponding to the boundary condition (4.4). Note also that $w \to \infty$ as $u \to 1$. This means that for all finite w, $u < 1$ always, so that in the exact solution, there will always be a finite amount of injected space-charge in a solid of finite dimension irrespective of the injection level. The voltage integral, in the dimensionless notation, is

$$v = \int_0^w u\, dw = \int_0^u u(dw/du)\, du = -\tfrac{1}{2}u^2 - u - \ln(1-u). \tag{4.18}$$

The current J and the voltage V are given, in terms of the dimensionless variables, by

$$J = \frac{e^2 n_0^2 \mu L}{\epsilon} \frac{1}{w_a}, \qquad V = \frac{e n_0 L^2}{\epsilon} \frac{v_a}{w_a^2}, \tag{4.19}$$

where subscript a corresponds to taking $x = L$ in (4.15).

It is clearly impossible to eliminate u analytically from (4.17) and (4.18) and thereby obtain an explicit analytical relation between $1/w_a$ and v_a/w_a^2. However, this elimination of u can be achieved through numerical tabulation of w and v vs u. The final result is given as the solid line of the log–log plot of $1/w_a$ vs v_a/w_a^2 in Fig. 4.1. The dotted line is the solution according to the Regional Approximation discussed in Section 4.2.2. There are two limiting regimes.

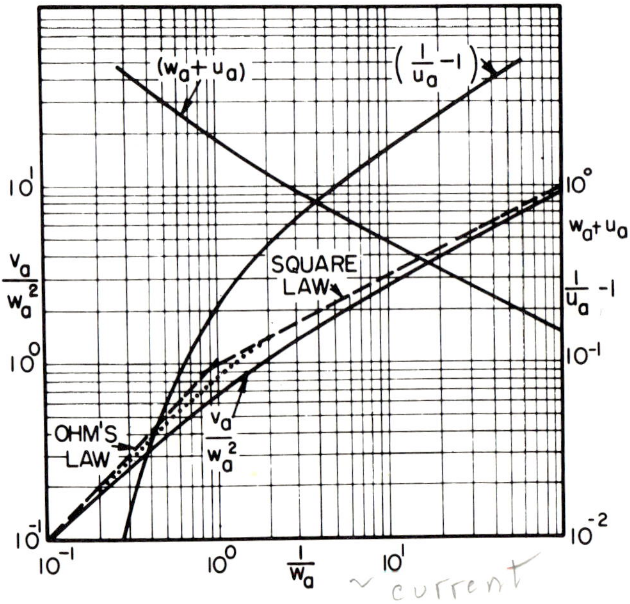

FIG. 4.1. Universal curves for SCL current injection into a trap-free insulator with thermal free carriers. $J \propto 1/w_a$ and $V \propto v_a/w_a^2$; $(1/u_a) - 1 = (n_a - n_0)/n_0$ and $w_a + u_a = t/t_\Omega$. (The curves are equally applicable to the problem of the injected plasma in semiconductors—see Chapter 12.)

Ohm's-Law Regime ($u_a \simeq 1$)

$$w_a \simeq -1 - \ln(1 - u_a), \quad v_a \simeq -\tfrac{3}{2} - \ln(1 - u_a) \simeq w_a, \quad \text{or} \quad v_a/w_a^2 \simeq 1/w_a,$$

which is just Ohm's law, plotted as the dashed line of slope unity in Fig. 4.1.

Trap-Free, Square-Law Regime ($u_a \ll 1$)

$$w_a \simeq u_a^2/2, \qquad v_a \simeq w_a^3/3, \qquad v_a/w_a^2 \simeq (2^{3/2}/3)(1/w_a)^{1/2},$$

which is just the trap-free square law (4.9) plotted as the dashed line of the slope two in Fig. 4.1. Also plotted in Fig. 4.1 are the quantities $(n_a - n_0)/n_0 = (1/u_a) - 1$, which is a measure of the injected electron concentration at the anode, and $t/t_\Omega = w_a + u_a$, which is the ratio of the free-carrier transit time t, defined by (3.27), to the ohmic relaxation time, $t_\Omega = \epsilon/en_0\mu$. Other quantities of interest are: $\mathscr{E}_a/\mathscr{E}_\Omega = u_a w_a/v_a$, $(n_a - n_0)/(\bar{n} - n_0) = w_a(1 - u_a)/u_a^2$, and $t/t_0 = e\bar{n}\mu\mathscr{E}_\Omega/J = v_a(w_a + u_a)/w_a^2$. Some values of these quantities are listed in Table 4.2. In particular,

TABLE 4.2

NUMERICAL VALUES FOR SOME QUANTITIES OF INTEREST IN THE PROBLEM OF SCL CURRENT INJECTION IN A TRAP-FREE INSULATOR WITH THERMAL FREE CARRIERS

w_a	$\to 0$	0.1	1	10	$\to \infty$
$u_a w_a/v_a = \mathscr{E}_a/\mathscr{E}_\Omega$	3/2	1.44	1.30	1.05	1
$\dfrac{w_a(1 - u_a)}{u_a^2} = \dfrac{n_a - n_0}{\bar{n} - n_0}$	1/2	0.42	0.23	2.0×10^{-4}	0
$\dfrac{v_a(w_a + u_a)}{w_a^2} = \dfrac{t}{t_0} = \dfrac{e\bar{n}\mu\mathscr{E}_\Omega}{J}$	4/3	1.29	1.19	1.05	1

$w_a \to 0$ corresponds to the trap-free square law while the limit $w_a \to \infty$ corresponds to Ohm's law.

Also of interest is the spatial distribution of injected carriers $n - n_0$. This is plotted in Fig. 4.2, where the normalized quantity $(n - n_0)/(n_a - n_0)$ is plotted against x/L at three different injection levels: $w_a = 0.20$ which is in the trap-free square-law regime, $w_a = 1.05$ which is in the transition region, and $w_a = 5.0$ which is in the Ohm's law regime. At $w_a = 0.20$, the distribution follows fairly well that for the perfect insulator, $n(x)/n_a = (L/x)^{1/2}$, shown plotted as a dashed curve. At $w_a = 5.0$, the distribution, for $x/L > 0.02$, follows closely the simple exponential $(n - n_0)/(n_a - n_0) \simeq \exp w_a(1 - x/L)$, obtained from

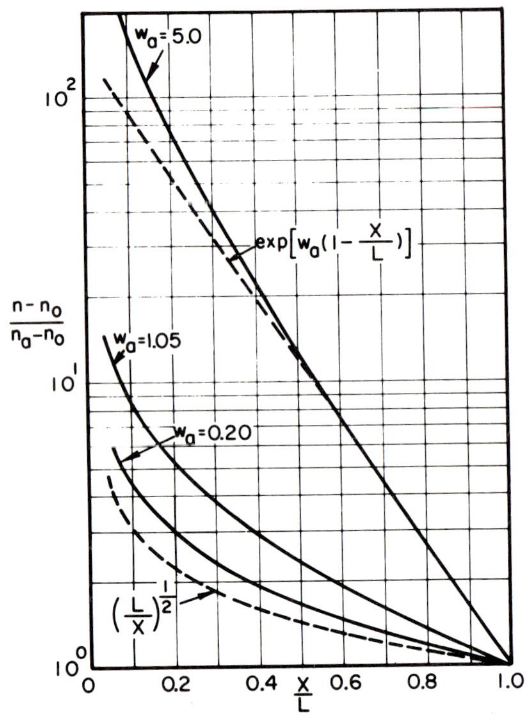

FIG. 4.2. Normalized spatial distribution of injected carriers for SCL current injection into a trap-free insulator with thermal free carriers. The quantities $w_a = 5.0$, 1.05, and 0.20 are representative values for the Ohm's-law regime, the transition regime, and the square-law regime, respectively. (The curves are equally applicable to the problem of the injected plasma in semiconductors—see Chapter 12.)

Eq. (4.17) in the limit $u \to 1$. This corresponds to a scale distance for falloff of injected free carrier concentration given by L/w_a.

4.2.2 *Regional Approximation*

Even though an exact solution to this simple problem has been presented above, it is instructive to study the problem by the Method of Regional Approximation. The method is based on the observation that two components contribute *additively* to the total current flow, the thermal free carriers n_0 and the injected carriers $n_i = n - n_0$:

$$J = e\mu n\mathscr{E}, \qquad n = n_0 + n_i . \tag{4.20}$$

A schematic plot of n_i vs x is shown as the dashed curve in Fig. 4.3. As established with great generality in Chapter 3, and in particular by

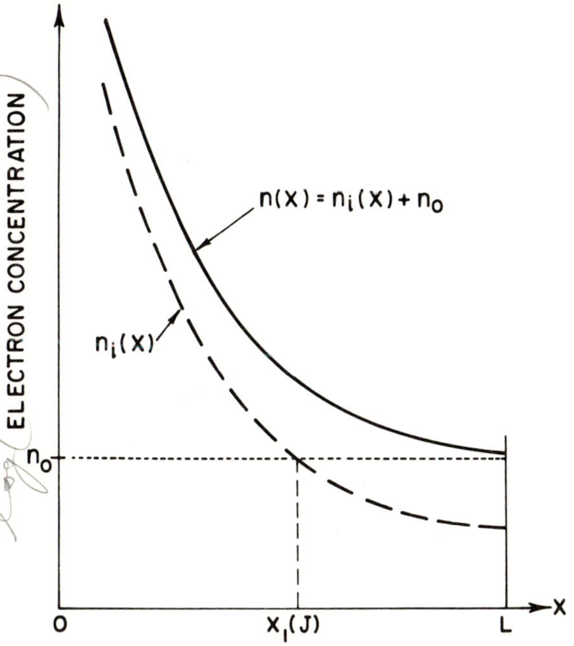

FIG. 4.3. Schematic spatial variation, at not-too-high injection level, of n_i and $n = n_i + n_0$ for SCL current injection into an insulator.

the exact solution above (see Fig. 4.2), $n_i(x)$ is a monotonic decreasing function of x. At not too high currents there will be a plane x_1 inside the insulator where $n_i(x_1) = n_0$; the plane x_1 is itself a function of the current as indicated in the figure. To the left of plane x_1, $n_i > n_0$ and to the right of plane x_1, $n_0 > n_i$. It would therefore appear to be a reasonable approximation to neglect n_0 to the left of plane x_1, and n_i to the right of plane x_1. This is the Regional Approximation. The insulator has been divided into two regions, Region I defined by $0 \leqslant x \leqslant x_1$ and Region II by $x_1 \leqslant x \leqslant L$, where $n_i(x_1) = n_0$. In each region an appropriate simplifying approximation is made. This approximation, in one version or another, has no doubt been employed in every branch of mathematical physics. It was first used to discuss injection currents in insulators by Patrick (1957). It is extensively reviewed by Lampert and Schilling (1970).

The defining equations in the different regions of the insulator are:

Region I ($0 \leqslant x \leqslant x_1$):

$$J = e\mu n\mathscr{E}, \tag{4.21}$$

$$(\epsilon/e)\,(d\mathscr{E}/dx) = n; \tag{4.22}$$

Region II ($x_1 \leqslant x \leqslant L$):

$$J = e\mu n_0 \mathscr{E}, \tag{4.23}$$

$$(\epsilon/e)(d\mathscr{E}/dx) = 0. \tag{4.24}$$

Equation (4.24) is, of course, not independent of (4.23) in this case. Further, plane x_1 is characterized by

$$n(x_1) = n_i(x_1) = n_0 . \tag{4.25}$$

Finally, in addition to the boundary condition (4.4), we need a specification on how to join the solutions in Regions I and II. Here we require that $\mathscr{E}$ be continuous crossing x_1 :

$$\mathscr{E}(x_1^-) = \mathscr{E}(x_1^+), \tag{4.26}$$

where $\mathscr{E}(x_1^-)$ denotes the value of $\mathscr{E}$ at x_1 coming from the left, that is, Region I, and $\mathscr{E}(x_1^+)$ the value of $\mathscr{E}$ at x_1 coming from the right, that is, Region II.

For further discussion it is convenient to use the dimensionless variables (4.15). The Poisson equations, Eq. (4.22) in Region I and Eq. (4.24) in Region II become, respectively,

Region I:

$$du/dw = (1/u) \rightarrow u\,du = dw, \tag{4.27}$$

Region II:

$$du/dw = 0. \tag{4.28}$$

From (4.15) and (4.25), the transition plane $x_1(w_1)$ corresponds to

$$u_1 = u(w_1) = 1. \tag{4.29}$$

The integrals of Eqs. (4.27) and (4.28) are, respectively,

Region I:

$$w = \tfrac{1}{2}u^2, \tag{4.30}$$

satisfying the boundary condition $u = 0$ at $w = 0$,

Region II:

$$u = u_1 = 1 \qquad \text{(independent of } w\text{)}, \tag{4.31}$$

satisfying the boundary condition (4.26). From (4.30) and (4.29) it follows that

$$w_1 = w(x_1) = \tfrac{1}{2}. \tag{4.32}$$

From (4.32) and (4.15),

$$x_1 = \epsilon J/2e^2n_0{}^2\mu, \tag{4.33}$$

which states that x_1 increases linearly with J. As J increases, a critical value J_{cr} is eventually reached at which $x_1 = L$ ($w_a = \frac{1}{2}$). From (4.33),

$$J_{cr} = 2e^2n_0{}^2\mu L/\epsilon. \tag{4.34}$$

For $J \geqslant J_{cr}$, the Region I fills the entire insulator. The potential in Region I is given, using (4.27), by

Region I:

$$v = \int_0^w u\,dw = \int_0^u u^2\,du = \tfrac{1}{3}u^3, \tag{4.35}$$

so that

$$v_1 = v(w_1) = \tfrac{1}{3}u_1{}^3 = \tfrac{1}{3}. \tag{4.36}$$

The critical voltage corresponding to $x_1 = L$ and $J = J_{cr}$, namely $V = V_{cr}$, is obtained using (4.36) and (4.15):

$$V_{cr} = \epsilon J_{cr}^2/3e^3n_0{}^3\mu^2 = 4en_0L^2/3\epsilon. \tag{4.37}$$

Note that V_{cr}, given by (4.37), is just $\frac{4}{3}$ times V_X, given by (2.12). Since V_{cr} and V_X have the same physical significance they should, of course, be approximately equal.

With Region I filling the crystal we have, at the anode, from (4.30) and (4.35),

$J \geqslant J_{cr}$:

$$w_a = \tfrac{1}{2}u_a{}^2, \qquad v_a = \tfrac{1}{3}u_a{}^3. \tag{4.38}$$

Here u_a can be explicitly eliminated giving

$J \geqslant J_{cr}$:

$$\frac{1}{w_a} = \frac{9}{8}\left(\frac{v_a}{w_a{}^2}\right)^2, \tag{4.39}$$

which, using (4.19), becomes the trap-free square law (4.9).

For $J < J_{cr}$ it remains to find the potential in Region II:

Region II:

$$v = v_1 + \int_{w_1}^{w} u\,dw = v_1 + (w - w_1) = \tfrac{1}{3} + (w - \tfrac{1}{2}) = w - \tfrac{1}{6}. \tag{4.40}$$

For the full applied voltage take $w = w_a$:

$$v_a = w_a - \tfrac{1}{6}. \tag{4.41}$$

The dimensionless current-voltage relationship is

$J < J_{cr}$:

$$v_a/w_a^2 = (1/w_a) - \tfrac{1}{6}(1/w_a)^2. \tag{4.42}$$

This relationship is plotted as the dotted line on Fig. 4.1. We see that the Regional Approximation solution lies closer to the exact solution than does the simple phenomenological solution so long as $V < V_{cr}$ ($v_a/w_a^2 < \frac{4}{3}$). Above V_{cr}, the two approximations coincide, namely the trap-free square law applies.

There is an alternative and simpler two-region solution to this problem which is, however, not general. It is obtained by the satisfaction of current continuity at the boundary between the two regions. The current in Region I is space-charge-limited, that in Region II is ohmic:

$$J = (9/8)\,\epsilon\mu(V_{\mathrm{I}}^2/x_1^3) = en_0\mu[V_{\mathrm{II}}/(L - x_1)]. \tag{4.43a}$$

Further, the electric field is also continuous at x_1:

$$(3/2)\,(V_{\mathrm{I}}/x_1) = V_{\mathrm{II}}/(L - x_1), \tag{4.43b}$$

and $V = V_{\mathrm{I}} + V_{\mathrm{II}}$. Considerable algebraic manipulation of these relations ultimately produces (4.42).

4.3 The Perfect Insulator with Shallow Traps

Shallow traps, which have already been discussed in Chapters 2 and 3, are characterized by a fixed ratio of free to trapped carriers; $n(x)/n_t(x) = \theta$, where θ is a constant given by (3.32) for a single set of traps. In this case (4.2) reduces to

$$(\epsilon/e)(d\mathscr{E}/dx) = (n - n_0) + (n_t - n_{t,0}) = [(1/\theta) + 1](n - n_0) \simeq (1/\theta)(n - n_0). \tag{4.44}$$

Since ϵ appears only in the Poisson's equation, it is clear that all results derived for the preceding trap-free problem apply also to the shallow-trap problem if only ϵ is replaced by $\theta\epsilon$. (This is simply a formal substitution, convenient for this problem, and is without physical significance as regards other dielectric properties of the insulators.) The phenomenologically derived results of Section 2.4 for the shallow-trap problem are now justified mathematically. The exact shallow-trap square law is, from (4.9),

$$J = (9/8)\,\theta\epsilon\mu(V^2/L^3). \tag{4.45}$$

If there are several sets of shallow traps N_{tj}, well separated in energy, then it is the smallest corresponding θ_j which appears in Eqs. (4.44) and (4.45).

4.4 The Trap-Filled Limit (TFL) Problem

This problem is the apparently unphysical one in which one imagines that all of the traps, of concentration N_t, are filled prior to the application of voltage. It is a particularly useful problem to solve because, simple as it is, it nonetheless furnishes a good, limiting description of what happens when the quasi-Fermi level passes through a trap level. This is the physical situation, already examined phenomenologically in Section 2.5, which results in a nearly vertical rise in current. Mathematically this problem is very similar to the preceding problem, the only difference being the replacement of $(-n_0)$ in (4.14) by $(+N_t)$. However, the behavior of the solution, resulting from this mere change of sign, is quite different. The defining equations are (4.1) and

$$(\epsilon/e)\,(d\mathscr{E}/dx) = n + N_t\,. \tag{4.46}$$

It is convenient to employ dimensionless variables analogous to those defined by (4.15):

$$u_t = \frac{N_t}{n(x)} = \frac{eN_t\mu\mathscr{E}(x)}{J}, \qquad w_t = \frac{e^2N_t^2\mu x}{\epsilon J}, \qquad v_t = \frac{e^3N_t^3\mu^2V(x)}{\epsilon J^2}. \tag{4.47}$$

The differential equation in $\mathscr{E}$, obtained from (4.1) and (4.46), becomes, in the new dimensionless variables,

$$u_t\,du_t/(1 + u_t) = dw_t \tag{4.48}$$

with solution

$$w_t = u_t - \ln(1 + u_t), \tag{4.49}$$

where again $u_t = 0$ at $w_t = 0$, corresponding to the boundary condition (4.4). The potential integral is

$$v_t = \int_0^{w_t} u_t\,dw_t = \tfrac{1}{2}u_t^2 - u_t + \ln(1 - u_t). \tag{4.50}$$

The current J and voltage V are given by

$$J = \frac{e^2 N_t^2 \mu L}{\epsilon} \frac{1}{w_{ta}}, \qquad V = \frac{e N_t L^2}{\epsilon} \frac{v_{ta}}{w_{ta}^2}, \tag{4.51}$$

where, again, the subscript a corresponds to taking $x = L$.

A log–log plot of $1/w_{ta}$ vs v_{ta}/w_{ta}^2, obtained by numerical tabulation of w and v vs u is given in Fig. 4.4. There is a threshold voltage for

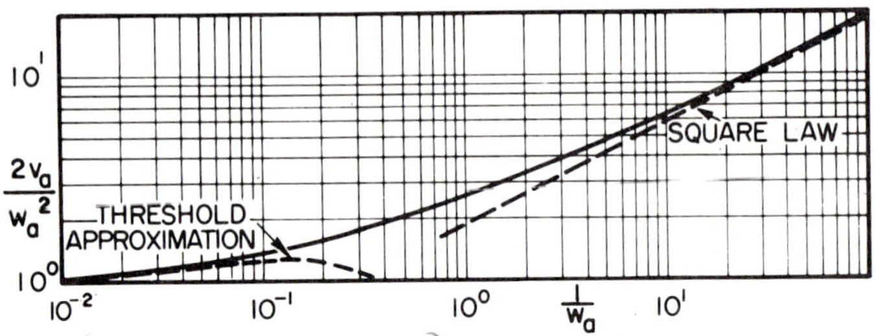

FIG. 4.4. Universal current-voltage characteristic for the trap-filled-limit, SCL current injection problem. $J \propto 1/w_a$ and $V \propto 2v_a/w_a^2$.

current flow, $V_{TH} = eN_tL^2/2\epsilon$, corresponding to $2v_{ta}/w_{ta}^2 = 1$, as discussed earlier in Section 2.5. There are two limiting regions in the characteristic. At low injection levels, where $u_{ta} \gg 1$, $u_{ta} \simeq w_{ta} + \ln w_{ta}$, $v_{ta} \simeq (w_{ta}^2)/2 + w_{ta}(\ln w_{ta} - 1)$ or equivalently,

$$2v_{ta}/w_{ta}^2 = 1 - (2/w_{ta})[1 + \ln(1/w_{ta})]$$

which is plotted in Fig. 4.4 as a dashed line labeled "Threshold Approximation." It is seen that this is a good approximation for $w_{ta} > 10$. The second regime is valid in the limit of high injection levels where $u_{ta} \ll 1$. This is identical to the square-law regime common to the preceding problems. The corresponding trap-free square law is plotted as a dashed line in Fig. 4.4.

Other quantities of interest are: $\mathscr{E}_{ta}/\mathscr{E}_\Omega = u_{ta}w_{ta}/v_{ta}$, $n_{ta}/\bar{n} = w_{ta}/u_{ta}(u_{ta} - w_{ta})$ and $e\bar{n}\mu\mathscr{E}_\Omega/J = v_{ta}(u_{ta} - w_{ta})/w_{ta}$. Representative values of these quantities are given in Table 4.3. As in the case of Table 4.2, the trap-free square law obtains in the limit $w_{ta} \to 0$. On the other hand, the limit $w_{ta} \to \infty$ here corresponds to the nearly vertical TFL current in the threshold regime. Of particular interest is the fact that, at very low currents near threshold, $n_{ta}/\bar{n}$ and consequently $J/e\bar{n}\mu\mathscr{E}_\Omega$

TABLE 4.3

NUMERICAL VALUES FOR SOME QUANTITIES OF INTEREST IN THE TRAP-FILLED-LIMIT PROBLEM

w_a	0	0.10	1.0	10	94	∞
$u_a w_a / v_a = \mathscr{E}_a / \mathscr{E}_\Omega$	3/2	1.55	1.65	1.82	1.94	2
$\dfrac{w_a}{u_a(u_a - w_a)} = \dfrac{n_a}{\bar{n}}$	1/2	0.465	0.405	0.302	0.207	0
$\dfrac{v_a(u_a - w_a)}{w_a^2} = \dfrac{e\bar{n}\mu\mathscr{E}_\Omega}{J}$	4/3	1.38	1.50	1.82	2.48	∞

also, approaches zero. Thus a "casual" use of averages, such as employed in the arguments of Chapter 2, would give incorrect results in this very restricted domain. In fact, however, this restricted domain is unphysical since, in real materials, Ohm's law currents dominate at sufficiently low currents.

The insulator TFL problem, first discussed by Lampert (1956) is mathematically identical to the one-carrier, SCL current punch-through problem for semiconductors discussed earlier by Shockley and Prim (1953). The choice of dimensionless variables and the analytical procedures are somewhat different in the two papers. More important, the physical interpretation of the theory is quite different for the semiconductor problem, where N_t refers to "built-in" ionic charge in a depletion layer. In the semiconductor treatment, trapping plays no role whatever.

4.5 THE OHM'S LAW—TFL AND TFL-TRAP-FREE SQUARE-LAW TRANSITIONS

Since the TFL law as derived in the preceding section corresponds, strictly speaking, to an artificial situation, it is desirable to have a picture of the real transition from a preceding regime of the current-voltage characteristic to the TFL regime. This can be obtained for two different, clearcut situations: (i) a single set of traps lying well below the thermal-equilibrium Fermi level, and (ii) a single set of traps lying well above this Fermi level. We treat first the former case where the crossover is directly from Ohm's law to the TFL regime and give the exact numerical solution as well as the solution obtained by the Regional Approximation.

4.5.1 *Exact Analytic Solution*

The defining equations are Eq. (4.1) and

$$(\epsilon/e)(d\mathscr{E}/dx) = (n - n_0) + (n_t - n_{t,0}) = (n - n_0) + (p_{t,0} - p_t)$$
$$= [n - (N_t N/gn)] + [(N_t N/gn_0) - n_0]. \tag{4.52}$$

In writing (4.52) we have made use of the deep-trap relation (3.33).

Now, using the dimensionless variables (4.15), the differential equation in $\mathscr{E}$ obtained from (4.1) and (4.52) can be written as

$$\frac{1}{1+A}\left[\frac{1}{1-u} - \frac{1}{1+Au}\right] du = dw, \tag{4.53}$$

where

$$A = N_t N/gn_0^2 = p_{t,0}/n_0\,, \tag{4.54}$$

with solution

$$w = [1/(1+A)][-\ln(1-u) - (1/A)\ln(1+Au)], \tag{4.55}$$

taken so that $u = 0$ at $w = 0$ (our usual boundary condition). The potential integral is

$$v = [1/(1+A)]\{-[1+(1/A)]\,u - \ln(1-u) + (1/A^2)\ln(1+Au)\}. \tag{4.56}$$

The current and the voltage are as given in (4.19). Log–log plots of $1/w_a$ vs v_a/w_a^2 are given by the solid curves in Fig. 4.5 for $A = 10^2$, 10^3, and 10^4. The dotted curves are the solutions according to the Regional Approximation, as derived in Section 4.6.2. The lower straight line is Ohm's law and the upper straight line is the trap-free square law. With $A = 10^4$, the current rises very steeply, by about three decades, during which the voltage only doubles, in the transition from Ohm's law to the trap-free square law. The latter is, of course, the same for all A. The Ohm's law region corresponds to $u_a \simeq 1$, the transition (TFL) region to $u_a \ll 1$, $Au_a \gg 1$, and the trap-free square-law region to $Au_a \ll 1$.

We show in Fig. 4.6 a semilogarithmic plot of $(n - n_0)/(n_a - n_0)$ vs x/L for various values of $u_a(w_a)$ and for $A = 10^4$. For $u_a \simeq 1$, corresponding to Ohm's law, we have $(n - n_0)/(n_a - n_0) \simeq \exp Aw_a[1 - (x/L)]$, which differs from the corresponding equation at the end of Section 4.2.1 by the appearance of the factor A in the exponential. The above equation corresponds to a scale distance for falloff of injected free-carrier concentration given by L/Aw_a. For the topmost curve in Fig. 4.6, $u_a = 0.99$

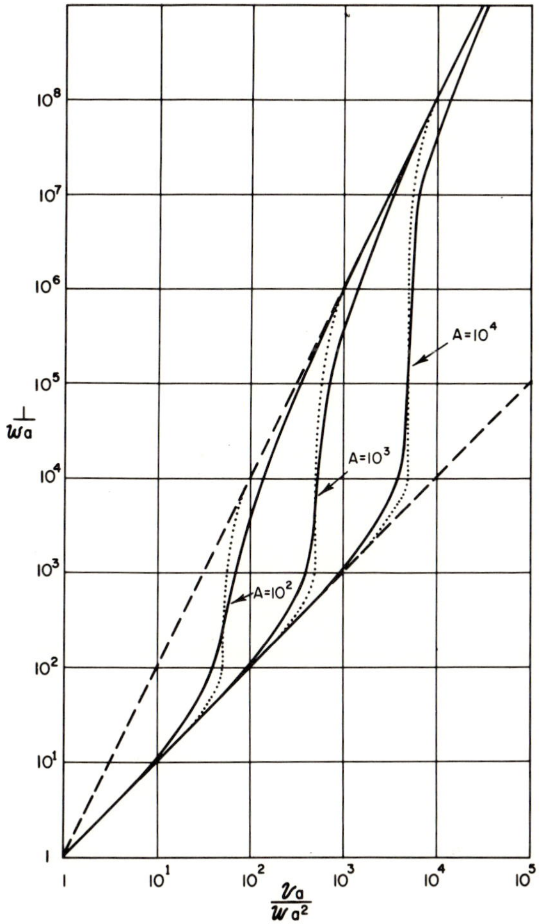

FIG. 4.5. Universal current-voltage characteristics for SCL current injection into an insulator with a single set of traps lying below the Fermi level. $J \propto 1/w_a$ and $V \propto v_a/w_a^2$. $A = p_{t,0}/n_0$.

and $w_a = 4.6 \times 10^{-4}$, the latter value being much smaller than the corresponding value on Fig. 4.2. This reflects the fact that the transition voltage to the trap-free square law is raised by the factor A through the presence, in thermal equilibrium, of the unfilled traps. As u_a becomes smaller with increasing injection level, the fraction of the injected space-charge that is free increases rapidly and the space-charge distribution undergoes a change similar to that illustrated by Fig. 4.2. Ultimately, when $Au_a \ll 1$, which is to say when the traps have been filled and

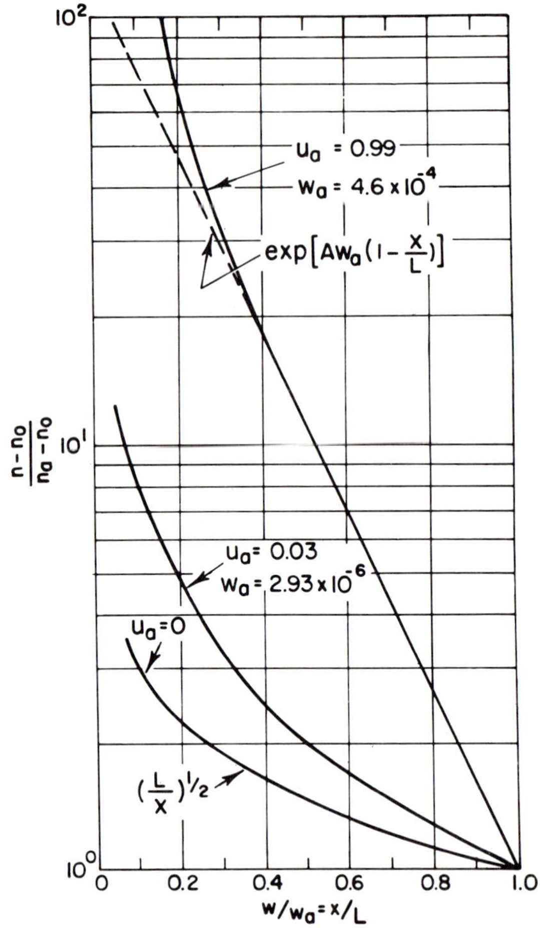

FIG. 4.6. Normalized spatial distribution of injected carriers for SCL current injection into an insulator with deep traps, with $A = p_{t,0}/n_0 = 10^4$. The quantities $u_a = 0.99$ and 0.03 are representative values for the Ohm's law regime and the transition (TFL) regime respectively; $u_a = 0$ corresponds to the pure trap-free, square-law regime.

practically all the injected space-charge contributes to the current, the injected charge distribution approaches the $(x/L)^{-1/2}$-dependence typical of trap-free SCL current flow.

4.5.2 *Regional Approximation*

We next obtain a solution to this problem through the Regional Approximation. We follow a path similar to that developed in Section 4.2.2, except that here three regions are required for an accurate approximation to the exact solution. As in the earlier problem, the injected free carrier concentration n_i decreases monotonically from ∞ at $x = 0$. At not too high currents there will be a plane $x_2(J)$ where $n_i(x_2) = n_0$. To the right of plane x_2, $n_0 > n_i$, and we neglect n_i; to the left, $n_i > n_0$ and we neglect n_0. However, because of the additional term $p_{t,0} - p_t$ in the Poisson equation (4.52), it is necessary to take a closer look at the region to the left of the plane x_2. First we note that $n_i = n_0$ or $n = n_0 + n_i = 2n_0$, corresponds to $p_t = p_{t,0}/2$ (upward motion of the Fermi level by $0.7kT$). Thus, to the left of plane x_2, $p_t < p_{t,0}/2$ and p_t can be neglected in (4.52). Finally, there will be a plane x_1 where $n_i(x_1) = n(x_1) = p_{t,0}$. To the left of plane x_1, $n > p_{t,0}$ and $p_{t,0}$ can be neglected in (4.52); to the right of plane x_1, up to plane x_2, $p_{t,0} \simeq p_{t,0} - p_t > n - n_0$ so that $n - n_0$, as well as p_t, can be neglected in (4.52). We see that three regions are required to characterize the problem. On an energy-band diagram, Fig. 4.7, the regions and their boundaries are delineated through the position of the quasi-Fermi level $F(x)$. (Note

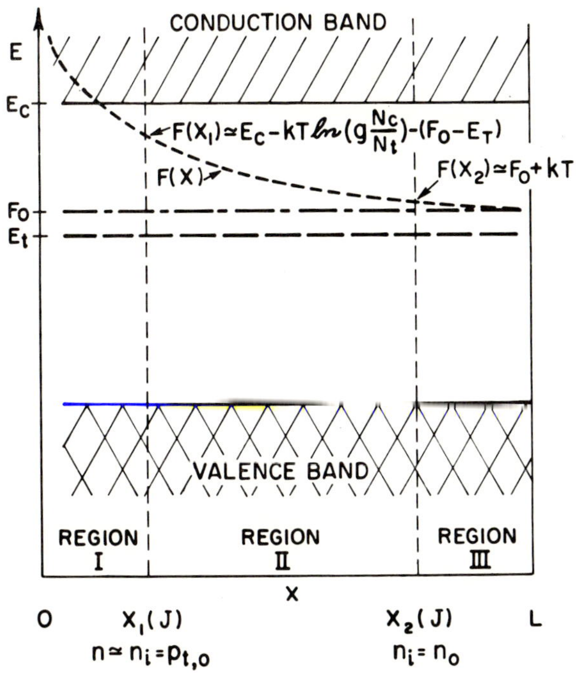

FIG. 4.7. Schematic energy-band, Regional Approximation diagram for the problem of SCL current injection into an insulator with deep traps.

that where $F(x) > E_c$, we ignore the required changeover to degenerate statistics. In practice this will almost always be a region extremely close to the cathode interface and this region is already being treated unphysically through our ignoring the real structure of the contact.)

The defining equations in the separate regions are:

Region I ($0 \leqslant x \leqslant x_1$):

$$J = e\mu n\mathscr{E}, \tag{4.57}$$

$$(\epsilon/e)\,(d\mathscr{E}/dx) = n, \tag{4.58}$$

$$n(x_1) = p_{t,0}\,; \tag{4.59}$$

Region II ($x_1 \leqslant x \leqslant x_2$):

$$J = e\mu n\mathscr{E}, \tag{4.60}$$

$$(\epsilon/e)\,(d\mathscr{E}/dx) = p_{t,0}\,, \tag{4.61}$$

$$n(x_2) = n_0\,; \tag{4.62}$$

Region III ($x_2 \leqslant x \leqslant L$):

$$J = e\mu n_0\mathscr{E}, \tag{4.63}$$

$$(\epsilon/e)\,(d\mathscr{E}/dx) = 0. \tag{4.64}$$

Finally, in addition to the boundary condition (4.4), we require continuity of $\mathscr{E}$ in crossing a boundary between two adjacent regions:

$$\mathscr{E}(x_1^-) = \mathscr{E}(x_1^+) \qquad \text{and} \qquad \mathscr{E}(x_2^-) = \mathscr{E}(x_2^+). \tag{4.65}$$

For further discussion, it is convenient to use the dimensionless variables (4.15). The Poisson equation, (4.58) in Region I, (4.61) in Region II, and (4.63) in Region III, becomes, respectively,

Region I:

$$du/dw = (1/u) \rightarrow u\,du = dw, \tag{4.66}$$

Region II:

$$du/dw = A, \tag{4.67}$$

Region III:

$$du/dw = 0, \tag{4.68}$$

From (4.15), (4.59), and (4.62), the transition planes x_1 and x_2 correspond to

$$x_1:\ u_1 = u(w_1) = 1/A; \qquad x_2:\ u_2 = u(w_2) = 1. \tag{4.69}$$

The integral of (4.66) is

Region I:

$$w = \tfrac{1}{2}u^2, \tag{4.70}$$

satisfying the boundary condition $u = 0$ at $w = 0$. From (4.69) and (4.70)

$$w_1 = 1/2A^2. \tag{4.71}$$

The integral of (4.67) is, using (4.69) and (4.71),

Region II:

$$u - u_1 = A(w - w_1) \rightarrow w = (1/A)\,u - (1/2A^2), \tag{4.72}$$

satisfying boundary condition (4.65). From (4.69) and (4.72),

$$w_2 = (1/A) - (1/2A^2). \tag{4.73}$$

The integral of (4.68), satisfying boundary condition (4.65), is

Region III:

$$u = u_2 = \text{constant} \rightarrow u = 1. \tag{4.74}$$

From (4.15), (4.71), and (4.74) we can write

$$x_1(J) = \frac{\epsilon J}{2A^2e^2n_0{}^2\mu} \qquad \text{and} \qquad x_2(J) = \frac{\epsilon J}{e^2n_0{}^2\mu}\left(\frac{1}{A} - \frac{1}{2A^2}\right) \simeq \frac{\epsilon J}{Ae^2n_0{}^2\mu}. \tag{4.75}$$

Corresponding to the three separate regions in the insulator, delineated by the bounding planes x_1 and x_2, there will be three separate regimes in the current-voltage characteristic delineated by the currents $J_{\text{cr},1}$ and $J_{\text{cr},2}$, corresponding respectively to $x_2(J_{\text{cr},1}) = L$ and $x_1(J_{\text{cr},2}) = L$. Thus, from (4.75):

$A \gg 1$:

$$J_{\text{cr},1} = Ae^2n_0{}^2\mu L/\epsilon \qquad \text{and} \qquad J_{\text{cr},2} = 2A^2e^2n_0{}^2\mu L/\epsilon. \tag{4.76}$$

For $J < J_{\text{cr},1}$, $x_2 < L$ and all three Regions I, II, and III are present in the insulator. This corresponds to the low current Ohm's law regime. In the range $J_{\text{cr},1} < J < J_{\text{cr},2}$, $x_1 < L < x_2$ and only Regions I and II are present in the insulator. This corresponds to the TFL regime. Finally, at still higher currents $J > J_{\text{cr},2}$, $x_1 < L$ and there is only Region I in the insulator. This is the perfect insulator, trap-free square-law regime.

The potentials in the three separate regions are:

Region I:

$$v = \tfrac{1}{3}u^3. \tag{4.77}$$

Using (4.69), this gives

$$v_1 = v(w_1) = 1/3A^3 \tag{4.78}$$

Region II:

$$v = v_1 + \int_{u_1}^{u} u \frac{dw}{du}\, du = \frac{1}{2A}\, u^2 - \frac{1}{6A^3}\,. \tag{4.79}$$

Using (4.69), this gives

$$v_2 = v(w_2) = \frac{1}{2A} - \frac{1}{6A^3}\,. \tag{4.80}$$

Region III:

$$v = v_2 + (w - w_2) = w - \frac{1}{2A} + \frac{1}{2A^2} - \frac{1}{6A^3}\,. \tag{4.81}$$

For the full applied voltage, take $w = w_a$:

$$v_a = w_a - \frac{1}{2A} + \frac{1}{2A^2} - \frac{1}{6A^3}\,, \tag{4.82}$$

and the dimensionless current-voltage relationship is

$J < J_{\mathrm{cr},1}$:

$$\frac{v_a}{w_a^{\,2}} = \frac{1}{w_a} - \frac{1}{2A}\left(1 - \frac{1}{A} + \frac{1}{3A^2}\right)\left(\frac{1}{w_a}\right)^2 \simeq \frac{1}{w_a} - \frac{1}{2A}\left(\frac{1}{w_a}\right)^2. \tag{4.83}$$

At this point, we make an observation which is quite useful for the further developments in this chapter, and that is that Region I makes a negligible contribution to the current-voltage characteristic (4.83). In fact, if we neglect Region I altogether, that is, take the left-hand boundary of Region II to be at $w = 0$ (still taking $u = 0$ at $w = 0$), we obtain for the current-voltage characteristic, in place of (4.83),

$J < J_{\mathrm{cr},1}$:

$$\frac{v_a}{w_a^{\,2}} = \frac{1}{w_a} - \frac{1}{2A}\left(\frac{1}{w_a}\right)^2, \tag{4.84}$$

which, for A large, is indistinguishable from (4.83). [Note that in deriving (4.84), we must replace (4.73) by $w_2 = 1/A$.] The reason why Region I is utterly negligible for $J < J_{cr,1}$ is that $x_1(J)/x_2(J) \simeq 1/2A \ll 1$, from (4.75).

For $J_{cr,1} < J < J_{cr,2}$, only Regions I and II are present in the insulator. Now, at the anode, from (4.79),

$$v_a = \frac{1}{2A} u_a^2 - \frac{1}{6A^3}. \tag{4.85}$$

From (4.72), $u_a = Aw_a + (1/2A)$, which substituted into (4.85) gives

$J_{cr,1} < J < J_{cr,2}$:

$$\frac{v_a}{w_a^2} = \frac{A}{2} + \frac{1}{2A}\left(\frac{1}{w_a}\right) - \frac{1}{24A^3}\left(\frac{1}{w_a}\right)^2 \simeq \frac{A}{2} + \frac{1}{2A}\left(\frac{1}{w_a}\right), \tag{4.86}$$

which is the TFL law.

For $J > J_{cr,2}$, only Region I is present in the insulator, described by (4.70), (4.71), (4.77), and (4.78). The current-voltage characteristic in this regime is (4.39), which corresponds to the trap-free square law.

Corresponding to $J = J_{cr,1}$, from (4.73) and (4.82), $v_{a,cr,1} \simeq 1/2A$; corresponding to $J = J_{cr,2}$, from (4.78), $v_{a,cr,2} = 1/3A^3$. Using these results in (4.15), we obtain for the critical voltage corresponding to the critical currents in (4.76):

$$V_{cr,1} = \frac{Aen_0L^2}{2\epsilon} = \frac{ep_{t,0}L^2}{2\epsilon}; \qquad V_{cr,2} = \frac{4Aen_0L^2}{3\epsilon} = \frac{4ep_{t,0}L^2}{3\epsilon}. \tag{4.87}$$

In terms of ratios, (4.76) and (4.87) are expressed as follows:

$$\frac{J_{cr,2}}{J_{cr,1}} = 2A = \frac{2p_{t,0}}{n_0}, \qquad \frac{V_{cr,2}}{V_{cr,1}} = \frac{8}{3}. \tag{4.88}$$

Equation (4.88) is the hallmark of the TFL law: a very large change in current occurring over a very restricted change in voltage.

The solutions obtained by the Regional Approximation, for the cases $A = 10^2$, 10^3, and 10^4, are plotted in Fig. 4.5 as the dotted curves. They differ by less than a factor of 2 from the corresponding exact solutions (solid curves).

4.6 Full Single-Level Theory

We are now ready to treat the full theory for a single trap level, corresponding to the case where the traps, of concentration N_t, lie well above the Fermi level: $E_t - F_0 \gg kT$. This is the second case mentioned in the introduction to Section 4.5. A rigorous analytic solution is available for this problem but it is excessively unwieldy and so its presentation is deferred until the end of this section. Rather we proceed directly to the approximate solution.

4.6.1 *Regional Approximation*

There are four distinct regions in this problem, as illustrated in the energy-band diagrams of Fig. 4.8. These four regions are characterized as follows:

Region I ($0 \leqslant x \leqslant x_1$):

$$J = e\mu n\mathscr{E}, \tag{4.89}$$

$$(\epsilon/e)(d\mathscr{E}/dx) - n, \tag{4.90}$$

$$n(x_1) = N_t\,, \tag{4.91}$$

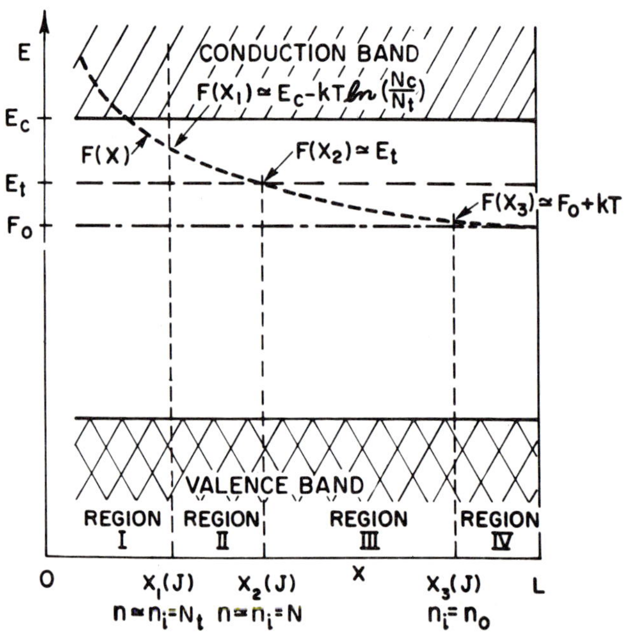

FIG. 4.8. Schematic energy-band, Regional Approximation diagram for the problem of SCL current injection into an insulator with a single set of traps, initially shallow ("full" single-level theory).

Region II ($x_1 < x < x_2$):

$$J = e\mu n\mathscr{E}, \tag{4.92}$$

$$(\epsilon/e)(d\mathscr{E}/dx) = N_t, \tag{4.93}$$

$$n(x_2) = N/g = (N_c/g)\exp[(E_t - E_c)/kT]. \tag{4.94}$$

Region III ($x_2 < x < x_3$):

$$J = e\mu n\mathscr{E}, \tag{4.95}$$

$$(\epsilon/e)(d\mathscr{E}/dx) = n/\theta, \qquad (1/\theta) = 1 + (gN_t/N), \qquad \theta \ll 1, \tag{4.96}$$

$$n(x_3) = n_0. \tag{4.97}$$

Region IV ($x_3 < x < L$):

$$J = e\mu n_0\mathscr{E}, \tag{4.98}$$

$$(\epsilon/e)(d\mathscr{E}/dx) = 0. \tag{4.99}$$

The boundary conditions, in addition to (4.4), are the usual continuity equations

$$\mathscr{E}(x_1^-) = \mathscr{E}(x_1^+), \qquad \mathscr{E}(x_2^-) = \mathscr{E}(x_2^+), \qquad \mathscr{E}(x_3^-) = \mathscr{E}(x_3^+). \tag{4.100}$$

In writing the above we have simplified the exact Fermi–Dirac occupation function (4.3) relating free to trapped charge concentrations. Namely, we have assumed the shallow-trap relation, $n \simeq Nn_t/gN_t$, to hold right up to the point that $n_t = N_t$, that is, for $n \leqslant N/g$. For $n > N/g$ we simply take $n_t = N_t$. This approximation constitutes a "sharpening up" of the occupation function where the Fermi level crosses the trap level.

Again it is convenient to use the dimensionless variables (4.15). Equations (4.90), (4.93), (4.96), and (4.99) become, respectively,

Region I:

$$du/dw = (1/u) \rightarrow u\,du = dw, \tag{4.101}$$

Region II:

$$du/dw = B, \qquad B = N_t/n_0, \tag{4.102}$$

Region III:

$$du/dw = (1/\theta u) \rightarrow \theta u\,du = dw, \tag{4.103}$$

Region IV:

$$du/dw = 0. \tag{4.104}$$

Corresponding to each of the transition planes x_3, x_2, x_1, there is a critical current $J_{\text{cr},1}$, $J_{\text{cr},2}$, $J_{\text{cr},3}$ defined, respectively, by

$$x_3(J_{\text{cr},1}) = L, \qquad x_2(J_{\text{cr},2}) = L, \qquad x_1(J_{\text{cr},3}) = L. \tag{4.105}$$

Concomitantly there are four separate regimes in the current-voltage characteristic:

$J < J_{\text{cr},1}$:	Ohm's law regime,
$J_{\text{cr},1} < J < J_{\text{cr},2}$:	shallow-trap square-law regime,
$J_{\text{cr},2} < J < J_{\text{cr},3}$:	TFL regime,
$J_{\text{cr},3} < J$:	trap-free square-law regime.

We discuss these regimes in order.

$J < J_{\text{cr},1}$ (Ohm's Law Regime)

Since $x_3(J) < L$, all four regions, I–IV, are in the insulator. Although we could proceed to solve the problem taking into account all four of the regions, we expect that only Regions III and IV are important in this regime. Indeed it comes out that $x_1/x_2 = N/2gN_\text{t} \simeq \theta/2 \ll 1$, and $x_2/x_3 = 2(gn_0/N)^2 \ll 1$ since $E_\text{t} - F_0 \gg kT$, so that Regions I and II contribute negligibly in this regime. Accordingly, we ignore Regions I and II and deal with the simpler two-region problem in which only Regions III and IV occupy the insulator, Region III extending right to the cathode. With a minimum of algebra, we then obtain for the current-voltage characteristic:

$$\frac{v_a}{w_a{}^2} = \frac{1}{w_a} - \frac{\theta}{6}\left(\frac{1}{w_a}\right)^2, \tag{4.106}$$

which is Ohm's law with a small correction.

This regime terminates at the critical current and critical voltage given, respectively, by

$x_3(J_{\text{cr},1}) = L$:

$$J_{\text{cr},1} = (2e^2 n_0{}^2 \mu L/\theta\epsilon), \qquad V_{\text{cr},1} = (4en_0 L^2/3\epsilon\theta). \tag{4.107}$$

Comparing (4.107) with (4.34) and (4.37), we see that the presence of shallow trapping delays the departure from Ohm's law to a critical current and voltage that are each $1/\theta \times$ the corresponding trap-free critical current and voltage.

$J_{cr,1} < J < J_{cr,2}$ (Shallow-Trap Square-Law Regime)

Since $x_3(J) > L$, there are the three regions, I, II, and III in the insulator. However, the result cited above, $x_1/x_2 \simeq \theta/2 \ll 1$ is still valid so that Region I contributes negligibly to this regime. Accordingly we again ignore Region I and deal with a simple two region problem, Region II extending right to the cathode. Straightforward algebra now gives for the current-voltage characteristic:

$$\left(\frac{v_a}{w_a{}^2} - \frac{1}{6BC^2}\frac{1}{w_a{}^2}\right)^2 = \frac{8}{9\theta}\frac{1}{w_a}\left(1 - \frac{1}{2BC}\frac{1}{w_a}\right); \qquad C = \frac{N}{gn_0}, \tag{4.108}$$

with B given in (4.102). (Note that $\theta = C/B \ll 1$.) This is the shallow-trap square law with a small correction.

This regime terminates at the critical current and critical voltage given, respectively, by

$x_2(J_{cr,2}) = L$:

$$J_{cr,2} = \frac{e^2 n_0{}^2 \mu LBC}{\epsilon}, \qquad V_{cr,2} = \frac{en_0L^2}{\epsilon}\frac{B}{2} = \frac{eN_tL^2}{2\epsilon}. \tag{4.109}$$

Observe that $V_{cr,2}$ is essentially V_{TFL} of the phenomenological theory. (The latter voltage contains the factor 2 in the denominator if one takes for the capacitance C in the $Q = CV$ relation the true capacitance $C = 2C_0 = 2\epsilon/L$.)

$J_{cr,2} < J < J_{cr,3}$ (TFL Regime)

Since $x_2(J) > L$, there are now only the two regions, I and II, in the insulator. This problem is now formally identical to the TFL regime for the deep-trap problem of Section 4.5 and consequently the current-voltage characteristic is given precisely by (4.86) with B replacing A:

$$\frac{v_a}{w_a{}^2} = \frac{B}{2} + \frac{1}{2B}\frac{1}{w_a} - \frac{1}{24B^3}\left(\frac{1}{w_a}\right)^2. \tag{4.110}$$

Corresponding to (4.76) and (4.87) we have here, replacing A by B, the terminal critical current and critical voltage for this regime:

$$J_{cr,3} = \frac{2B^2e^2n_0{}^2\mu L}{\epsilon} = \frac{2e^2N_t{}^2\mu L}{\epsilon}, \qquad V_{cr,3} = \frac{4eN_tL^2}{3\epsilon}. \tag{4.111}$$

$J_{cr,3} < J$ (Trap-Free Square Law Regime)

Now only Region I is in the insulator, and this is a perfect insulator region. The corresponding current-voltage characteristic is the trap-free square law (4.39).

The various regimes in the current-voltage characteristic for the full single-level problem join up to form a continuous curve which is plotted as the solid curve of Fig. 4.9 for the following choices of

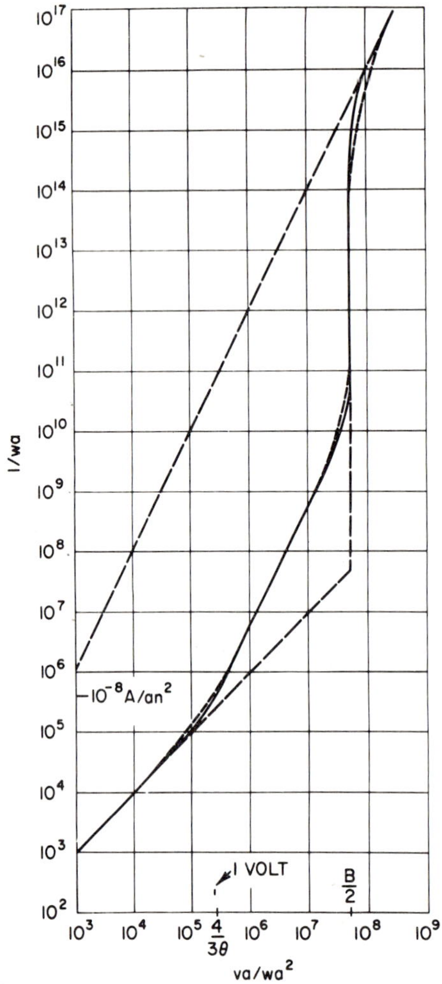

FIG. 4.9. Current-voltage characteristic for a particular case of SCL current injection into an insulator with a single set of traps lying above the Fermi level. $\theta = \{N_c \exp(E_t - E_c)/kT\}/gN_t = 5 \times 10^{-6}$ and $B = N_t/n_0 = 10^8$.

parameters: $n_0 = 10^6\ \text{cm}^{-3}$, $N_t = 10^{14}\ \text{cm}^{-3}$, $E_c - E_t = 0.6$ eV, $\theta = 5 \times 10^{-6}$ ($g = 2$), $N_c = 10^{19}\ \text{cm}^{-3}$ [$T = 300$°K, $(m^*/m) = 0.60$], $(\epsilon/\epsilon_0) = 11$, $\mu = 200\ \text{cm}^2/\text{V sec}$ ($\rho = 3 \times 10^{10}\ \Omega$ cm). These parameters correspond to $B = 10^8$, $C = 500$. They are the same parameters used by Lampert (1956) in his initial study of the full single-level problem. The various critical values of $1/w_a$ and v_a/w_a^2 are also indicated. The curve shows all the features of the phenomenological solution (Curve II in Fig. 2.2a). It is worth commenting that the Regional Approximation frequently involves the same mathematical formalism for the treatment of different physical situations. All that is necessary in such cases is a suitable change in the notation and the solutions can then be used intact in the new situation.

The analytic solution for this problem employing the exact Fermi–Dirac function to describe the occupation of the trap level is also plotted, for purposes of comparison, using the same values for the parameters. It is shown as the dotted curve. We shall formulate this problem in the next section. At this point, we merely wish to comment that the solution according to the Regional Approximation differs from the exact solution by at most a factor of 2 and that the former is "sharpened up" at the transition points with respect to the latter. This is similar to our experience with the TFL problem.

4.6.2 *Formal Analytic Solution*

For the sake of completeness, we present the formal analytic solution to the full single-level theory (Lampert, 1956). The defining equations are the set of equations (4.1)–(4.3) for a single set of traps (with only a single value for the j subscript it can be omitted altogether for convenience). In the dimensionless variables (4.15), the Poisson equation (4.2) is

$$\frac{du}{dw} = \frac{(1-u)(1+Gu)}{u(1+Cu)}, \tag{4.112}$$

where $G = C + D$, $D = BC/(1 + C)$ and, as before $B = N_t/n_0$ and $C = N/gn_0$. Equation (4.112) is integrated using the well-known method of expansion in partial fractions; so, likewise, is the potential integration, $v = \int u\, dw = \int u(dw/du)\, du$ carried out. The resulting parametric solutions for v_a and w_a are

$$v_a = -\frac{C}{2G} u_a^2 - \left(R + \frac{S}{G}\right) u_a - R \ln(1 - u_a) + \frac{S}{G^2} \ln(1 + Gu_a), \tag{4.113}$$

$$w_a = -\frac{C}{G} u_a - R \ln(1 - u_a) - \frac{S}{G} \ln(1 + Gu_a) \tag{4.114}$$

where $R = (C + 1)/(G + 1)$ and $S = D/G(G + 1)$. The dashed curve in Fig. 4.9 is the current-voltage characteristic computed from this pair of equations using (4.19).

Note that the parametric solution (4.113) and (4.114) is completely general for a single set of traps, independent of their concentration and position with respect to the Fermi level. It is this very generality of the solution which makes it unwieldy and of correspondingly limited practical significance.

4.7 Traps Distributed in Energy

We treat next the influence on the space charge-limited current, of traps that can be considered distributed in energy rather than located at a discrete energy E_t. The plausibility of such trap distributions has already been discussed in Chapter 2. We shall consider two different distributions: (i) the exponential distribution in which the concentration per unit energy decreases exponentially with energy from the band edge, and (ii) the uniform distribution which has the traps uniformly distributed in energy. We shall see that the Regional Approximation is a very useful one for obtaining analytic solutions of these problems.

4.7.1 *The Exponential Trap Distribution*

Mathematically, a problem involving a continuous distribution of traps differs from one involving discrete traps in that the sum over the j trap levels on the right side of (4.2) is replaced by an integral over a continuous function in energy, namely the trap distribution function. For an exponential distribution of traps, the distribution function is conveniently written as

$$\mathscr{N}_t(E) = (N_t/kT_t) \exp[(E - E_c)/kT_t], \tag{4.115}$$

where $\mathscr{N}_t(E)$ is the concentration of traps per unit energy range, E is the energy, and T_t is a characteristic constant of the distribution. Only values of $E < E_c$ are considered. Note that the quantity N_t/kT_t is the same as the quantity called $\mathscr{N}_0$ in the phenomenological discussion of Chapter 2, Eq. (2.26). The total concentration of traps $N_t[F(x), F_0]$ lying between the thermal position F_0 of the Fermi level and the steady-state position-dependent quasi-Fermi level $F(x)$ is given by the integral of (4.115):

$$N_t[F(x), F_0] = \int_{F_0}^{F(x)} \mathscr{N}_t(E)\, dE = N_t \left\{ \exp\left[\frac{F(x) - E_c}{kT_t}\right] - \left(\frac{n_0}{N_c}\right)^{(T/T_t)} \right\}, \tag{4.116}$$

where we have replaced $\exp[(F_0 - F_c)/kT_t]$ by $(n_0/N_c)^{T/T_t}$.

We now make the important assumption that the concentration of injected trapped electrons $n_{ti}(x)$ is adequately approximated by the trap concentration $N_t[F(x), F_0]$. This assumption is equivalent to sharpening up the Fermi–Dirac occupation function into a step function of value unity for E below $F(x)$ and of value 0 for E above $F(x)$ (as if we took $T = 0$). This is a good approximation if $T_c > T$. Further, we assume that $n_{ti}(x) \gg n_i(x)$, the injected free electron concentration. Thus the Poisson equation (4.2) becomes

$$(\epsilon/e)(d\mathscr{E}/dx) = n_{ti}(x), \qquad n_{ti}(x) = N_t[F(x), F_0]. \tag{4.117}$$

The relation between the free electron concentration $n(x)$ and the total injected electron concentration $n_{ti}(x)$ is obtained by eliminating $F(x)$ from (4.3) for $n(x)$ and (4.116). One obtains

$$n_{ti}(x) = N_t[(n(x)/N_c)^{(1/l)} - (n_0/N_c)^{(1/l)}], \qquad \text{with} \quad l = T_t/T. \tag{4.118}$$

We now formulate the problem in the dimensionless variables (4.15). Poisson's equation (4.117) becomes

$$u^{(1/l)}\, du/(1 - u^{1/l}) = \Gamma\, dw, \tag{4.119}$$

where

$$\Gamma = (N_t/n_0)(n_0/N_c)^{(1/l)}. \tag{4.120}$$

Equation (4.119) cannot be integrated and so we are led to use the Regional Approximation Method. At low currents there will be three regions in the insulator: Region I, the usual perfect-insulator region adjacent to the cathode, Region II which is a region dominated by trapped charge, and Region III, the usual ohmic region. These regions are characterized as follows.

Region I ($0 \leqslant x \leqslant x_1$)

$$du/dw = 1/u, \tag{4.121}$$

$$n(x_1) = n_{ti}(x_1) \rightarrow \frac{1}{u_1} = \frac{\Gamma}{u_1^{(1/l)}}, \qquad u_1 = u(w_1), \tag{4.122}$$

with $n_{ti}(x_1)$ given by (4.118).

Region II ($x_1 \leqslant x \leqslant x_2$)

$$du/dw = \Gamma u^{-(1/l)}, \tag{4.123}$$

$$n(x_2) = n_0 \rightarrow u_2 = 1, \qquad u_2 = u(w_2). \tag{4.124}$$

In this region, which is dominated by injected, trapped charge, we have made the usual approximation of neglecting the thermal free carriers n_0, that is, taking $n_0 = 0$ in (4.118) and hence replacing (4.119) by (4.123). More properly, this replacement is equivalent to substituting $-\infty$ for the lower limit F_0 in (4.116), i.e., neglecting the thermal trapped charge compared to the injected trapped charge. This is a valid approximation if $n_i \geqslant n_0$ and $T_t > T$.

Region III ($x_2 \leqslant x \leqslant L$)

$$du/dw = 0. \tag{4.125}$$

In addition, the usual boundary conditions, Eq. (4.4) and continuity of the electric field at x_1 and x_2, obtain.

From previous experience we know that we need deal with only two regions at a time in constructing the current-voltage characteristic. At low current, before Region III has left the insulator, we neglect Region I and extend Region II up to the cathode. Region II is then characterized by

$$u = \left(\frac{\Gamma(l+1)}{l}\right)^{[l/(l+1)]} w^{[l/(l+1)]}, \tag{4.126}$$

which is the integral of (4.123) satisfying $u = 0$ at $w = 0$, and

$$v = \left[\frac{\Gamma(l+1)}{l}\right]^{[l/(l+1)]} \left(\frac{l+1}{2l+1}\right) w^{[(2l+1)/(l+1)]}, \tag{4.127}$$

obtained from $v = \int_0^w u\, dw$ and (4.126). Also

$$u_L = u(w_2) = 1, \qquad w_2 = \frac{l}{\Gamma(l+1)}, \qquad v_2 = v(w_2) = \frac{l}{\Gamma(2l+1)}. \tag{4.128}$$

In Region III, $u = \text{constant} = u_2 = 1$, so that the potential is $v = v_2 + (w - w_2) = w - l^2/\Gamma(l+1)(2l+1)$, using (4.127) and (4.128). This gives for the current-voltage characteristic,

$$\frac{v_a}{w_a^2} = \frac{1}{w_a} - \frac{1}{w_a^2} \frac{l^2}{\Gamma(l+1)(2l+1)}, \tag{4.129}$$

which is Ohm's law with a correction term, as expected.

Setting $w_2 = w_a$ (i.e., $x_2 = L$), we obtain the critical current and voltage at which the ohmic Region III leaves the insulator:

$$\left(\frac{1}{w_a}\right)_{\text{cr},1} = \frac{\Gamma(l+1)}{l}, \qquad \left(\frac{v_a}{w_a^2}\right)_{\text{cr},1} = \frac{\Gamma(l+1)^2}{l(2l+1)}. \tag{4.130}$$

For injection levels exceeding (4.130), there are only Regions I and II in the insulator. As usual, Region I is characterized by

$$w = \tfrac{1}{2}u^2 \qquad \text{and} \qquad v = \tfrac{1}{3}u^3. \tag{4.131}$$

Equation (4.122) gives

$$u_1 = \left[\frac{2l}{\Gamma(l+1)}\right]^{[l/(l-1)]}, \qquad w_1 = \frac{1}{2}\left[\frac{2l}{\Gamma(l+1)}\right]^{[2l/(l-1)]},$$
$$v_1 = \frac{1}{3}\left[\frac{2l}{\Gamma(l+1)}\right]^{[3l/(l-1)]}. \tag{4.132}$$

In Region II, the integral of (4.123) gives

$$w = w_1 + \frac{l}{\Gamma(l+1)}\,[u^{[(l+1)/l]} - u_1^{[(l+1)/l]}], \tag{4.133}$$

and the potential integral $v = v_1 + \int_{u_1}^{u} u(dw/du)\,du$ becomes

$$v = v_1 + \frac{l}{\Gamma(2l+1)}\,[u^{[(2l+1)/l]} - u_1^{[(2l+1)/l]}]. \tag{4.134}$$

At the anode, $u = u_a$, $w = w_a$, $v = v_a$, and (4.133) and (4.134) give

$$w_a = \frac{1}{2}\left[\frac{2l}{\Gamma(l+1)}\right]^{[2l/(l-1)]} + \frac{l}{\Gamma(l+1)}\left\{u_a^{[(2l+1)/l]} - \left[\frac{2l}{\Gamma(l+1)}\right]^{[(l+1)/(l-1)]}\right\}, \tag{4.135}$$

$$v_a = \frac{1}{3}\left[\frac{2l}{\Gamma(l+1)}\right]^{[3l/(l-1)]} + \frac{l}{\Gamma(2l+1)}\left\{u_a^{[(2l+1)/l]} - \left[\frac{2l}{\Gamma(l+1)}\right]^{[(2l+1)/(l-1)]}\right\}. \tag{4.136}$$

Equations (4.135) and (4.136) are parametric equations in u_a determining the dimensionless current-voltage characteristic in this regime. We obtain the critical current and voltage at which Region II leaves the insulator by taking $w_a = w_1$, $v_a = v_1$ as given by (4.132):

$$\left(\frac{1}{w_a}\right)_{\mathrm{cr},2} = 2\left[\frac{\Gamma(l+1)}{2l}\right]^{[2l/(l-1)]} \quad \text{and} \quad \left(\frac{v_a}{w_a^2}\right)_{\mathrm{cr},2} = \frac{4}{3}\left[\frac{\Gamma(l+1)}{2l}\right]^{[l/(l-1)]}. \tag{4.137}$$

Above this current and voltage, the trap-free square law prevails.

It turns out that Region II above dominates over most of the interval between the two critical currents (or voltages), (4.130) and (4.137). In this case, an analytic expression is easily obtained for the current-voltage characteristic by simply extending (4.127) to the anode. This gives

$$\frac{v_a}{w_a^2} = \left[\frac{\Gamma(l+1)}{l}\right]^{[l/(l+1)]}\left(\frac{l+1}{2l+1}\right)\left(\frac{1}{w_a}\right)^{[l/(l+1)]}. \tag{4.138}$$

In terms of the direct, physical variables, this equation becomes

$$J = N_{\mathrm{c}}\mu e^{(1-l)} \left[\frac{\epsilon l}{N_{\mathrm{t}}(l+1)}\right]^{l} \left(\frac{2l+1}{l+1}\right)^{(l+1)} \frac{V^{(l+1)}}{L^{(2l+1)}}. \tag{4.139}$$

The $(l+1)$-power-law dependence of J on V was first derived by Rose (1951, 1955). The equation (4.139) was first derived by Mark and Helfrich (1962). A simplified version of this relation was derived on physical grounds in Chapter 2, namely Eq. (2.29). To this same approximation, one can also obtain an expression for the position dependence of the quasi-Fermi level. It is

$$F(x) = kT_{\mathrm{t}} \ln \left\{\frac{(l+1)^2}{l(2l+1)} \frac{eN_{\mathrm{t}}L^2}{\epsilon V} \left(\frac{x}{L}\right)^{[1/(l+1)]}\right\}. \tag{4.140}$$

The complete current-voltage characteristic, computed using the Region Approximation is plotted as the solid curve of Fig. 4.10. Ohm's law and the trap-free square law are also shown by dashed lines. The curve was computed using the following values for the parameters: $N_{\mathrm{t}} = 10^{14}\ \mathrm{cm}^{-3}$, $n_0 = 10^6\ \mathrm{cm}^{-3}$, $N_{\mathrm{c}} = 10^{19}\ \mathrm{cm}^{-3}$ for $l = 2$ ($\Gamma = 3.16 \times 10^1$) and $l = 3$ ($\Gamma = 4.64 \times 10^3$). Note how sensitive the location of the trap dominated current is to the value of l alone, without changing any of the concentrations. It is apparent from the figure that the $(l+1)$-power law of Eq. (4.139) describes the trap-limited SCL current over most of the interval between the two critical values.

Very frequently in experimental work, the range of measurements is insufficient to characterize accurately the transition regions. Thus, the available current range which is amenable to measurement usually contains only one transition region and sometimes contains neither. However, either the trap-free square law or Ohm's law (or both) are known from independent measurements. In this case, the parameter Γ can be evaluated by a linear extrapolation of the $(l+1)$-power law to the intersection with these curves. We include these crossover voltages here for convenience. The crossover with Ohm's law is

$$V_{\Omega\text{-T}} = \frac{en_0L^2}{\epsilon} \left(\frac{l+1}{2l+1}\right)^{[(l+1)/l]} \frac{\Gamma(l+1)}{l}, \tag{4.141}$$

and the crossover with the trap-free square law is

$$V_{\text{T-C}} = \frac{en_0L^2}{\epsilon} \Gamma^l \left[\frac{9}{8}\left(\frac{l+1}{l}\right)^{l}\left(\frac{l+1}{2l+1}\right)^{l+1}\right]^{[1/(l-1)]}. \tag{4.142}$$

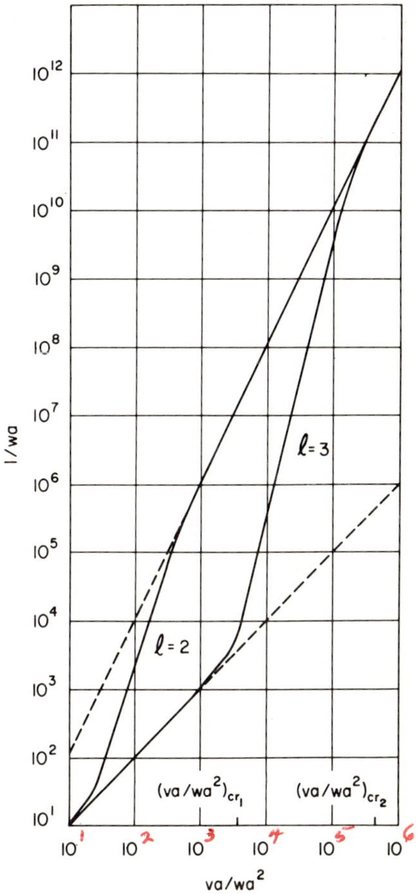

FIG. 4.10. Dimensionless current-voltage characteristics for the problem of SCL current injection into an insulator with an exponential distribution of traps, with $N_t = 10^{14}$ cm^{-3}, $n_0 = 10^6$ cm^{-3}, and $N_c = 10^{19}$ cm^{-3} for the two cases $l = T_t/T = 2$ and 3. The critical voltages, $(v_a/w_a^2)_{cr,1}$ and $(v_a/w_a^2)_{cr,2}$ pertain to the case, $l = 3$.

4.7.2 *The Uniform Trap Distribution*

In Chapter 2, we considered the uniform trap distribution as the limiting form of the exponential trap distribution for large T_c. We noted there that the exponential trap distribution may be considered uniform if $\bar{F} - F_0 < kT_t$ where $\bar{F}$ is the mean (spatial average) excursion of the quasi-Fermi level. In the present mathematical treatment of the problem, we shall treat the distribution as strictly constant:

$$N_t(E) = \mathscr{N}_n = \text{constant}. \tag{4.143}$$

Note that the quantity $\mathscr{N}_n$ is the same as that which appears in the phenomenological discussion of Chapter 2, Eqs. (2.30) and (2.31).

The total density of traps $N_T[F(x), F_0]$ lying between F_0 and $F(x)$ is given by

$$N_t[F(x), F_0] = \mathcal{N}_n[F(x) - F_0] = n_{ti}(x). \tag{4.144}$$

Here, as in the preceding problem of the experimental distribution, we have made the assumption that the concentration of injected trapped electrons $n_{ti}(x)$ is adequately represented by the trap concentration $N_t[F(x), F_0]$. We now treat this problem in terms of the same three-region approximation we used in dealing with the exponential trap distribution. The Poisson equation in Region II is given by (4.117). Elimination of $F(x)$ from (4.144), using (4.3), gives

$$n_{ti}(x) = \mathcal{N}_n kT \ln[n(x)/n_0]. \tag{4.145}$$

In the dimensionless variables, Eqs. (4.15), (4.117), and (4.145) reduce to

$$du/dw = D \ln(1/u), \qquad D = \mathcal{N}_n kT/n_0 . \tag{4.146}$$

The integral of this equation satisfying $n = n'$ at $w = w'$ is

$$w - w' = \frac{1}{D} \int_{u'}^{u} \frac{du}{\ln(1/u)} = \frac{1}{D} \left[E_1 \left(\ln \frac{1}{u}\right)\right] - \frac{1}{D} \left[E_1 \left(\ln \frac{1}{u'}\right)\right], \tag{4.147}$$

where the function E_1 is the tabulated exponential integral:

$$E_1(y) = \int_y^\infty (e^{-x}/x)\, dx.$$

The corresponding potential integral is

$$v - v' = \int_{u'}^{u} u(dw/du)\, du = (1/D) \int_{u'}^{u} u\, du/\ln(1/u),$$

which can also be written directly in terms of the function E_1 as

$$v - v' = (1/D)[E_1(2 \ln(1/u))] - (1/D)[E_1(2 \ln(1/u'))]. \tag{4.148}$$

We may now follow the familiar prescription of considering only two regions at a time. In the transition from ohmic current flow to trap-limited SCL current flow, we neglect Region I and extend Region II to the cathode, that is, take $u' = 0$, $w' = 0$, and $v' = 0$ in (4.147) and (4.148). Note that in the present problem (4.146) is the direct analogue of (4.119), not of (4.123); that is, it has not been necessary to neglect the thermal free carriers in Region II, as was done in the preceding problem. Using the parametric equations (4.147) and (4.148), with $u' = w' = v' = 0$, and the trivially simple properties of the ohmic Region III, the transition from ohmic current flow to SCL current flow has been computed with results shown in Fig. 4.11.

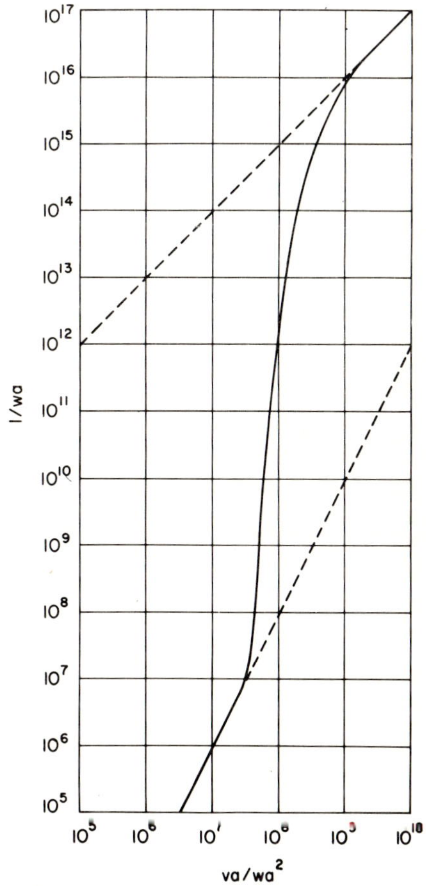

FIG. 4.11. Dimensionless current-voltage characteristic for the problem of SCL current injection into an insulator with a uniform distribution of traps.

The transition to the trap-free square law is obtained using only Regions I and II, since Region III has already passed out of the insulator. The bounding plane x_1 between Regions I and II is specified by the condition

$$n_1 = n(x_1) = n_{ti}(x_1) \rightarrow (1/u_1) = D \ln(1/u_1), \tag{4.149}$$

where we have used (4.145) and (4.146). Region I is the familiar perfect-insulator region characterized by $w_1 = \frac{1}{2}u_1^2$ and $v_1 = v(w_1) = \frac{1}{3}u_1^3$.

Region II is now described by (4.147) and (4.148) taking $u' = u_1$, $w' = w_1$, $v' = v_1$. Finally, at the anode we have

$$w_a = w_1 + \frac{1}{D}\int_{u_1}^{u_a} \frac{du}{\ln(1/u)} = \tfrac{1}{2}u_1^2 + \frac{1}{D}\left\{E_1\left(\ln\frac{1}{u_a}\right) - E_1\left(\ln\frac{1}{u_1}\right)\right\} \tag{4.150}$$

and

$$v_a = v_1 + \frac{1}{D}\int_{u_1}^{u_a} \frac{u\,du}{\ln(1/u)} = \tfrac{1}{3}u_1^3 + \frac{1}{D}\left\{E_1\left(2\ln\frac{1}{u_a}\right) - E_1\left(2\ln\frac{1}{u_1}\right)\right\}. \tag{4.151}$$

The current-voltage characteristic, $1/w_a$ vs v_a/w_a^2, for this transition regime can now be computed from Eqs. (4.149)–(4.151). It is also shown on Fig. 4.11.

Finally, in analogy with our treatment of the exponential trap distribution, we shall compute the trap-limited current-voltage characteristic on the assumption that Region II fills the entire interelectrode space. In this case, taking $u' = w' = 0$, we evaluate (4.147) and (4.148), at the anode with the assumption that we are already well into SCL current flow; i.e., that $u_a \ll 1$. In this case, $E_1(x) \simeq e^{-x}/x$ so that $1/w_a \simeq (-D \ln u_a)/u_a$ and $v_a/w_a^2 = (D \ln u_a)/2$. We may eliminate u_a from these two equations and obtain the dimensionless current-voltage characteristic

$$\frac{1}{w_a} = 2\left(\frac{v_a}{w_a^2}\right) \exp\left[\frac{2}{D}\left(\frac{v_a}{w_a^2}\right)\right], \tag{4.152}$$

which, in terms of the physical variables, is

$$J = 2en_0\mu(V/L)\exp(2\epsilon V/\mathcal{N}_n kTeL^2). \tag{4.153}$$

This equation has the same form as (2.31), the current-voltage characteristic computed from the phenomenological theory in the limit of an exponential distribution with large l. However, the above equation has the factor 2 in the exponent which is missing in (2.31). Thus we see that although the phenomenological theory gives the proper form of the current-voltage characteristic, it is greatly in error as regards the magnitude. The source of this discrepancy is pointed out following (2.31).

4.8 The Scaling Law

The simplified theory, for the general case, is specified by Eqs. (4.1)–(4.3) and boundary condition (4.4). These equations imply definite scaling relations as regards the role of the cathode–anode spacing L in the current-voltage characteristic. The scaling relations are obtained directly by going over to the dimensionless variables (4.15). Equations (4.1)–(4.3) then yield

$$\frac{du}{dw} = \frac{(1-u)}{u}\left\{1 + \sum_j \frac{n_{tj,0}}{n_0}\,\frac{u}{u + (g_j n_0/N_j)}\right\}. \tag{4.154}$$

Thus, for w taken at the anode, $w_a = w(L)$, we have

$$w_a = \int_0^{u_a} u\left[(1-u)\left\{1 + \sum_j \frac{n_{tj,0}}{n_0}\,\frac{u}{u + (g_j n_0/N_j)}\right\}\right]^{-1} du. \tag{4.155}$$

The applied voltage $V = \int_0^L \mathscr{E}\, dx$, or dimensionlessly,

$$\begin{aligned} v_a &= \int_0^{w_a} u\, dw = \int_0^{w_a} u \frac{dw}{du}\, du \\ &= \int_0^{w_a} u^2 \left[(1-u)\left\{1 + \sum_j \frac{n_{tj,0}}{n_0}\,\frac{u}{u + (g_j n_0/N_j)}\right\}\right]^{-1} du. \end{aligned} \tag{4.156}$$

where (4.154) has been used.

Equations (4.155) and (4.156) give a parametric solution of the problem in terms of the parameter u_a (the field at the anode). Insofar as the scaling law is concerned, the key observation is that the only constants appearing in (4.155) and (4.156) are equilibrium constants of the material, namely, n_0, $n_{tj,0}$, g_j, and N_j. Neither the current density J nor the applied voltage V appears in these equations; that is, the functional relation between $1/w_a$ and v_a/w_a^2 does not have the parameters J and V appearing in it. Therefore, rewriting (4.19) as

$$\frac{1}{w_a} = \frac{\epsilon}{e^2 n_0^2 \mu}\,\frac{J}{L}, \qquad \frac{v_a}{w_a^2} = \frac{\epsilon}{e n_0}\,\frac{V}{L^2}, \tag{4.157}$$

we see that the dimensionless current-voltage characteristic, $1/w_a$ vs v_a/w_a^2, is essentially J/L vs V/L^2, for fixed values of the material constants. Symbolically

$$J/L = f(V/L^2). \tag{4.158}$$

Equation (4.158) is the universal scaling law of the simplified theory with field-independent mobility. It is likewise valid for the general simplified theory (diffusion neglected) of double injection if the mobilities of both carriers are field-independent, as shown in Section 14.7. The universality of the scaling law (4.158) for both single and double injection, under the stated conditions, was first noted by Warter (1967). It is physically obvious, for example, by visual comparison of Figs. 9.6a and 9.6b, that (4.158) cannot be a valid scaling law if diffusion is included in the analysis and with it, an exact treatment of the real contacts.

4.8.1 *The Scaling Law with Field-Dependent Mobility*

It is instructive to derive the scaling relations for the case that the mobility is field-dependent. Suppose that, instead of (4.1), the current equation is

$$J = e\mu_1 n\mathscr{E}_1^{(1-\alpha)}\mathscr{E}^{\alpha} = \text{constant}, \tag{4.159}$$

where μ_1 and $\mathscr{E}_1$ are fixed parameters with the dimensions of mobility and field strength, respectively, and α is a constant. Equation (4.2) is necessarily still valid. The quasi-thermal relation (4.3) between free and trapped electrons is likely to be somewhat altered by heating of the free carriers but, for purpose of the present discussion, we assume that the required alteration is a small one and neglect it altogether. Thus we assume that (4.3) holds unmodified.

As in the previous problem, we wish to go over to dimensionless variables so chosen that J no longer explicitly appears in the equations. Because of the right-hand side of the Poisson equation (4.2), it is convenient to define $u = n_0/n$ as above. Then it follows from (4.159) that

$$d\mathscr{E} = \frac{1}{\alpha}\left(\frac{J}{e\mu_1\mathscr{E}_1^{(1-\alpha)}n_0}\right)^{(1/\alpha)} u^{((1/\alpha)-1)}\,du. \tag{4.160}$$

Now substituting $x = x_1 w$ and (4.160) into (4.2), it is seen that the choice of x_1 that eliminates J from the dimensionless Poisson equation is that given in (4.61):

$$\begin{aligned} n = n_0\frac{1}{u}, \qquad x = x_1 w, \qquad &\text{with} \quad x_1 = \frac{\epsilon}{\alpha e n_0}\left(\frac{J}{e\mu_1\mathscr{E}_1^{(1-\alpha)}n_0}\right)^{(1/\alpha)}, \\ V = V_1 v \qquad &\text{with} \quad V_1 = \frac{\epsilon}{\alpha e n_0}\left(\frac{J}{e\mu_1\mathscr{E}_1^{(1-\alpha)}n_0}\right)^{(2/\alpha)}. \end{aligned} \tag{4.161}$$

With this choice, (4.2) becomes

$$\frac{du}{dw} = \frac{(1-u)}{u^{(1/\alpha)}}\left\{1 + \sum_j \frac{n_{tj,0}}{n_0}\frac{u}{u + (g_j n_0/N_j)}\right\}. \tag{4.162}$$

Further, with $V(x) = V_1 v$, the choice for V_1 that eliminates J from the voltage relation, $V = \int_0^L \mathscr{E}\,dx$, is that also given in (4.161).

The voltage relation then becomes

$$V(L) = V_1 v_a, \qquad v_a = \int_0^{w_a} u^{(1/\alpha)}\, dw = \int_0^{u_a} u^{(1/\alpha)} \frac{dw}{du}\, du. \tag{4.163}$$

The basic equations are now in suitable dimensionless form. The analog to (4.157), for the present problem, is

$$\left(\frac{1}{w_a}\right)^\alpha = \left(\frac{\epsilon}{\alpha e n_0}\right)^\alpha \frac{1}{e\mu_1 \mathscr{E}_1^{(1-\alpha)} n_0} \frac{J}{L^\alpha}, \qquad \frac{v_a}{w_a^2} = \frac{\epsilon}{\alpha e n_0} \frac{V}{L^2}, \tag{4.164}$$

and the corresponding analog to (4.158) is

$$J/L^\alpha = f(V/L^2). \tag{4.165}$$

This is the desired scaling relation. For $\alpha = 1$ it reduces to (4.158) as indeed it must. For $\alpha = \frac{1}{2}$ it is consistent with (2.37), namely $J/L^{1/2} \propto (V/L^2)^{3/2}$; and for $\alpha = 0$, it is consistent with (2.3a), namely $J \propto V/L^2$.

5

RESEARCH APPLICATIONS AND EXPERIMENTAL RESULTS

This chapter is devoted to a survey of the current experimental status of the steady-state single-injection problem in solids. Transient effects are reserved for the following chapter. We shall begin the survey with a brief history. This will be followed by a discussion of the information that can be extracted from single injection measurements with solids. Here we shall stress the necessity of self-consistent experimentation to establish the existence of SCL current flow. Then we proceed to the survey of experimental results according to materials. This survey is meant to illustrate the theoretical considerations of the earlier chapters as well as to bring out the importance of experimental confirmation in this branch of research. Because the survey is intended to serve only these rather limited purposes, it is not a definitive review of the body of experimental literature, which has already grown to formidable proportions. Such matters as circuitry, sample preparation, and contact fabrication techniques are not discussed. Finally we conclude with a very brief survey of recent research on technological applications of current injection in insulators.

5.1 A Brief History

The conceptual basis for the science of current injection into insulators, as already pointed out in the first chapter, was clearly presented in the classic book of Mott and Gurney, "Electronic Processes in Ionic Crystals," published in 1940. However, this book did not spark the initiation of any experimental program in this field. The explanation for this lies, no doubt, primarily in the circumstance of its appearance in wartime and perhaps also, in part, in its focus on a class of materials not particularly well-suited for such an investigation.

The interest in this field was born out of technological imperatives. In the course of study of photoconductive films that appeared attractive for use as target materials in the Vidicon (a photoconductive TV camera tube), current-voltage characteristics were observed in the "dark"; i.e., in the absence of illumination, in amorphous selenium by Weimer and Cope, in 1951, and in antimony trisulfide by Forgue *et al.*, also in 1951, that were strongly suggestive of SCL currents. Since these dark currents contribute a background noise level, and loss of signal by leakage, in Vidicon operation it was important to understand them and set theoretical limits on their control. This was the background for the theoretical and experimental study of SCL currents in insulators at the RCA Laboratories, which largely pioneered this field. This earlier work, which took place over the period of years from 1947 to 1951 is reviewed in a series of papers published in the September 1951 issue of the *RCA Review*. It was in this issue that the first quantitative picture of the effect of traps on the SCL current was developed by Rose.

In order to carry out a more systematic study of SCL currents, an experimental program using single crystals of CdS was initiated by Smith and Rose in 1952, culminating in their classic papers of 1955. Early results of this work were presented at the June 1953 meeting of the American Physical Society at Rochester, New York. Over this same period, Rose developed a comprehensive theoretical picture of the effects of trapping, with particular emphasis on traps distributed in energy in the forbidden gap, culminating in his paper of 1955. Rose's work was extended to the case of the discrete trap level by Lampert in 1956. At this point the theoretical picture of steady-state SCL currents was complete in its major outlines.

At about the same time, Böer and Kümmel (1952, 1954a,b, 1957, 1958) in Germany initiated a program of similar studies on single crystal CdS and obtained similar results. However, Böer and Kümmel gave an entirely different interpretation to their steep current-voltage characteristics, namely they attributed them to electrical breakdown. Quite certainly, the SCL current interpretation is correct, as shown by the incisive experiments of Smith and Rose that rule out breakdown. These experiments are cited in the following section. Since then, numerous investigators have obtained results similar to those of Smith and Rose and have similarly interprcted them, as outlined in the following sections. Because of the extreme structure sensitivity of SCL current phenomena, it is very difficult to make a definitive judgement on Böer and Kümmel's interpretation of their own data. (Only an exchange of crystals between the disputing parties can settle differences of this nature.) However the authors believe that it is still an open question, experimentally, as to just what field is required in a crystal of CdS, at room temperature, to produce impact ionization of occupied electron traps several tenths of a volt, or more, removed from the conduction band. Many (1965) has applied, via a blocking contact, pulsed uniform fields as high as 1×10^5 V/cm without observing any deviation from Ohm's law. Impact ionization across the forbidden gap in CdS is in an even more questionable state. An effort extending over many years, at RCA Laboratories, to observe this phenomena has not been successful. Williams (1961, 1962) has applied (nonuniform) fields as high as 2×10^6 V/cm, via a blocking contact, without any sign of even preavalanching (impact ionization), although breakdown due to field emission from the electrodes was observed at fields between 1 and 2×10^6 V/cm.

An experimental study contemporaneous with the CdS work of Smith and Rose is the observation by Dacey in 1953, of SCL hole currents in single-crystal Ge at 77°K. Since Dacey's study involved a special semiconductor effect, namely punch-through of a narrow n-type (or intrinsic) region sandwiched between two p-type regions, it has not had a major influence on the study of SCL currents in insulators.

5.2 Information Obtainable by Steady-State SCL Current Measurements

We turn next to a consideration of the information that can be extracted from single injection measurements with solids. We shall see that this is largely in the realm of defect structure although information concerning transport properties and electrical contacts is also furnished in a less direct manner.

If, on a log–log plot of current vs voltage, there is a nearly vertical section in the characteristic, as in Figs. 4.5 and 4.9, then, following the discussion in Sections 2.5, 4.5, and 4.6, this can be attributed to the filling of a discrete set of traps, and (4.87) for $V_{\text{cr},1}$ is applicable. Rewritten in practical units

$$p_{t,0} = \frac{1.1 \times 10^6 \kappa V_{\text{TFL}}}{L^2} \quad \text{cm}^{-3}, \tag{5.1}$$

where κ is the relative static dielectric constant, V is in volts, and L in centimeters. Note that the trap-filled-limit voltage measures that fraction of the total concentration of traps that is empty in thermal equilibrium. If the nearly vertical section of the characteristic is directly preceded by a square-law section, as in Fig. 4.9, one can conclude that the traps lay well above the thermodynamic Fermi level F_0: $(E_t - F_0)/kT \gg 1$, and so were essentially all empty at the outset: $p_{t,0} = N_t$. This particular result, which derives from the powerful combination of elementary electrostatics and Fermi–Dirac statistics, is an outstanding success of the SCL current technique in obtaining useful information about solids. Not only can the experimental effect be a very large and obvious one, but minimum knowledge of the material constants is required, namely, just the relative static dielectric constant. (Note that V_{TFL} will be independent of mobility, even under high field, hot-carrier conditions, so long as impact ionization and negative differential mobility effects do not take place. Under these conditions, $Q \simeq C_0 V$ and the essential features of Fermi–Dirac statistics are preserved, and these are the critical ingredients of the TFL–SCL current phenomenon.)

On the other hand, limitations on the upper limit of trap concentrations which can be determined in this manner are implicit in Eq. (5.1). If N_t is too large, the field V_{TFL}/L across the sample may cause breakdown, or the potential V_{TFL} may produce arcing at a contact. An upper limit on N_t appears to be about 10^{18} cm^{-3} for $L \approx 0.1\ \mu$, and is proportionately smaller if L is larger (thus about 10^{16} cm^{-3} for $L = 10\ \mu$). An upper limit on N_t might also be given by the onset of double injection due, say, to field-induced emission of minority carriers at the exiting electrode for majority carriers. Practically, there is virtually no lower limit on detectable N_t, the real limit being imposed by overall crystal purity, since the most abundant set of traps in the general vicinity of the Fermi level F_0 will dominate the current-voltage characteristic. Thus the SCL current technique has the unusual merit that, the purer the material, the easier it is to apply it.

If the nearly vertical section of the characteristic is directly preceded by Ohm's law, as in Fig. 4.5, one knows only that the traps lay at or below F_0: $(F_0 - E_t)/kT \gtrsim 1$, and only a lower limit can be set on N_t, namely, $N_t \geqslant p_{t,0}$. If the trap level E_t were determined by other means, say impurity optical absorption or impurity photoconductivity, N_t could be calculated knowing F_0 and E_t. The location of F_0 is normally determined simply by dc conductivity measurements, which implies a knowledge of free-carrier drift mobility and band density of states. The latter in turn requires a knowledge of band parameters, namely, free-carrier effective mass and the number of band extrema. Fortunately, since F_0 depends only logarithmically on these quantities, the degree of ignorance concerning them is only weakly reflected into an uncertainty in the location of F_0. Thus, a factor of 100 uncertainty in these quantities gives an uncertainty of only 0.1 eV in F_0 at room temperature.

The use of (5.1) presumes a plane-parallel geometry with planar current flow. This is effectively achieved if the cathode–anode spacing L is substantially less than the smallest linear dimension of the electrode area. If this geometric condition is not realized, say, if L is larger than any cross-sectional dimension, then (4.62) must be used in the form: $Q_{\text{TFL}} = CV_{\text{TFL}}$, where C is the appropriate capacitance. In this case, the trapped charge concentration at V_{TFL} is obtained with an uncertainty corresponding to that of C, and there is the additional complication that it is not known what fraction of the injected charge is contained in surface traps as opposed to volume traps. This geometry and the cylindrical and point-contact electrode configurations are discussed at length in Chapter 8.

Additional information, besides N_t, can frequently be obtained from SCL current measurements. If the nearly vertical section of the characteristic corresponding to the TFL law is directly preceded by a square-law section, the latter must be the shallow-trap square law (2.20) corresponding to the same set of traps N_t. This branch of the log I–log V plot is displaced below the trap-free square law (2.7) and (4.7) by the amount $1/\theta$ given by (2.18). With N_t already determined by (5.1), E_t is then given by

$$E_c - E_t = kT \ln(N_c/\theta g N_t). \tag{5.2}$$

Note that information about κ and μ must be available to determine θ, and about N_c and g to determine E_t, although g is not likely to differ greatly from $g = 2$.

A more reliable technique for obtaining E_t is to measure the θ pertaining to the shallow-trap square-law section of the characteristic as a function of temperature. Then the resultant plot of ln θ vs $1/T$ will be a straight line whose slope is $(E_t - E_c)/k$, and whose ordinate-intercept at $1/T = 0$ is $\ln(N_c/gN_t)$. The temperature dependence of μ enters these determinations; fortunately uncertainties about μ reflect only logarithmically into uncertainties in E_t. If the nearly vertical TFL section of the characteristic is also observed, the temperature measurements can serve as an independent check of N_t, as determined from (5.1), and E_t, as determined from (5.2). If the nearly vertical TFL sections cannot be reached because of voltage or field difficulties, the temperature measurements alone serve to determine N_t and E_t. A plot of ln θ vs $1/T$ for electrons in CdS is given by Bube (1962) and by Harnik (1965), for holes in ZnTe by Hinotani and Sugigami (1965), for holes in amorphous selenium by Hartke (1962), and for holes in iodine by Many *et al.* (1962).

Another point to be noted is that, if the shallow-trap, square-law section of the characteristic is directly preceded by an Ohm's law section, θ is also determined by the crossover voltage V_X, on a log–log plot, of the two extrapolated straightline sections. From (2.21), rewritten in practical units,

$$\theta = 1.8 \times 10^{-6} n_0 L^2 / \kappa V_X . \tag{5.3}$$

However, we should note that a careful study on CdS and ZnS by Russ (1963) has shown that the Ohm's law branch of a SCL I–V characteristic can be due, in the absence of guard rings, to surface currents.

Still another useful technique involving SCL currents was applied by Marlor and Woods (1965) to measure trap concentrations and depths in CdS crystals. It consists of measuring the thermally stimulated current subsequent to the passage of a SCL current. The method shares practically all the features of thermally stimulated currents observed after optical excitation (Bube, 1960, pp. 292–299).

A final point of interest is the obvious, but useful one, that where the trap-free square law has been reached in an I–V characteristic, a significant upper limit has been established on trap concentrations in that region of the forbidden gap traversed by the quasi-Fermi level.

If the current-voltage characteristic contains more than one shallow-trap, square-law section and more than one TFL, nearly vertical section, as in Fig. 2.2b, correspondingly more information can be extracted. A characteristic as rich in structure as either of the curves in Fig. 2.2b has not yet been observed, to the writers' knowledge. If such a characteristic is ever observed, it will represent an extraordinary application of the SCL current technique.

All of the above discussion refers to current-voltage characteristics dominated by one or two sets of traps at discrete energy levels. In materials with a high degree of structural disorder, even though chemically very pure, e.g., evaporated films or vitreous solids, we may expect to find traps smoothly distributed in energy. [A continuum of localized trapping states will presumably be present also in "good," pure single crystals, due to varying degrees of association of impurities or structural defect states with each other or with other types of structural defects, such as dislocations. However, these will constitute a weak, diffuse background in the presence of the dominant sets of discrete trap levels and will, accordingly, make a negligible contribution to SCL current characteristics. Nevertheless this weak background can still be detected by sensitive photoconductive techniques, as discussed by Rose (1963, Section III-7) and demonstrated for CdS by Niekisch (1961) and by Bube (1961)]. To the extent that the distribution can be characterized by an exponential with a characteristic temperature T_t, as in Eq. (4.115), it will yield a SCL current characteristic with J varying as $V^{(l+1)}$, $l = T_t/T$, namely, Eq. (4.139). Thus where a power-law dependence of J on V is observed, with exponent exceeding two, as in Fig. 5.7, one can derive directly from the plot of log J vs log V the characteristic temperature T_t of the exponential distribution of traps. Further, by actually fitting the complete expression (4.139) to the experimental plot, the trap parameter N_t of the exponential trap distribution (4.115) is likewise determined to the accuracy that μ and N_c are known. It should be kept in mind that the trap distribution so derived is "verified" only for the region in the forbidden energy gap lying between F_0 and F_h, where F_h denotes the position of the quasi-Fermi level at the highest voltage of measurement. Where the transition from Ohm's law to a power law is somewhat broad (see Fig. 5.4), a somewhat more sophisticated procedure of curve-fitting is indicated, as, for example, the one carried out by Lanyon (1963). In general, a useful observation is that the change of voltage needed to double the current is a good measure of the trap density per kT interval at the position of the quasi-Fermi level. It is also possible, in organic molecular crystals, to observe a discrete trap embedded in a background of an exponential distribution of traps. Such a level, for holes, has been found in anthracene by Helfrich and Mark (1963) and by Adolph *et al.* (1963), located more than 0.9 eV above the hole band, in concentration of 10^{11} cm^{-3}, and having a hole capture cross section on the order of 10^{-19} cm^2.

A further application of the SCL current technique was the recent determination of the effective density of states in the hole conduction band of anthracene by Helfrich and Mark (1963). This determination rested on the existence of an exponential trap distribution in anthracene, as previously determined by Mark and Helfrich (1962).

If measurements above V_{TFL} in the trap-free square law region are possible, information concerning the carrier mobility can be obtained. Measurements by Gregory and Jordan (1964), by Aiken and Jordan (1965), and by Denda and Nicolet (1966) with silicon have furnished the temperature dependence of the low-field electron and hole mobilities as well as their dependences on the electric field strength when the injected carriers become "warm."

Less directly, SCL currents also yield some information about electrical contacts. In the first place, the mere existence of SCL currents requires a reservoir of the carrier type being injected. The current level above which this reservoir is exhausted provides information about the contact potential (Mott and Gurney, 1940, pp. 169–173). Thus, by noting that an indium contact to CdS was able to sustain a pulsed SCL current of 20 A/cm^2, Smith and Rose (1955) concluded that the indium Fermi level must be less than 0.3 eV below the CdS conduction band at the contact interface. An excellent review of the role played by metal contacts in the injection of space-charge has been given by Stöckmann (1961). Charge transfer between electrolytes and solids can also be investigated by SCL current techniques, in particular with organic solids. It was shown by Kallmann and Pope (1960a,b) that oxidizing electrolytes can inject holes into anthracene, and by Mark and Helfrich (1962) that such contacts are ohmic below a saturation current whose magnitude is representative of the electrode–solid system. Electron-injecting electrolytic contacts to anthracene have been investigated by Helfrich and Schneider (1965, 1966) and by Büchner and Mehl (1965). An excellent review of electrolytic contacts has been given by Mehl and Hale (1967).

Up to this point, our discussion has been very "positive" in nature. However, in particular cases there may be a legitimate question as to whether a collection of experimental data that seems to be adequately interpreted by a one-carrier SCL current model is actually being correctly interpreted, whether there may not be an equally reasonable interpretation of the same data using a different physical model. For example, a nearly vertical section in a current-voltage characteristic is highly suggestive of an electrical breakdown in the material, deriving from, say, field-induced emptying of filled traps or impact ionization of filled traps or even impact ionization across the bandgap.

There are several ways to distinguish a SCL current TFL law from an electrical breakdown. For example, Smith and Rose, in Fig. 4 of their classic paper on CdS (1955b), showed that the peak current, under pulse conditions, followed the trap-free square law up to the voltage of "vertical" rise of current in the steady-state, indicating that no trap ionization was taking place. The peak current was seen for a time at least one order of magnitude longer than needed to achieve breakdown. (It should be noted that these transient measurements differ from the time-of-flight measurements discussed in the following chapter, namely, they were made over a longer time scale and concomitantly with less time resolution.) Likewise, under steady illumination Ohm's law is observed in the steady state well into the region of voltage where the current is rising steeply—again an observation incompatible with ionization. Another possible test to distinguish the two interpretations is the thickness dependence of the critical voltage for the steep current rise. This voltage varies as L^2 for the SCL current TFL law, (5.1), and varies as L for the onset of ionization. This was one of the tests used by Gregory and Jordan (1964) in their work on silicon to confirm the SCL current TFL interpretation of their I–V curves.

A more difficult problem is to distinguish one-carrier SCL current phenomena from two-carrier (double-injection) phenomena, which are discussed at length in Part II. Thus, if one is studying injected electron currents, we may be dealing with an anode material which appears to be ohmic for electrons at low fields but which "breaks down," through field emission, at higher fields, furnishing a copious injection of holes. Unfortunately the full range of double-injection phenomena has hardly been explored yet, so that a definitive comparison of the two realms is not possible at this time. Nonetheless it has been established, both experimentally and analytically, that the two realms can have certain nontrivial features in common. For example, the purely theoretical, double-injection characteristic of Fig. 1.5 exhibits a vertical section, at voltage V_{TH}, directly preceded by a square-law section. Yet the double-injection interpretation of these features, given in Section 14.1, is unrelated to the one-carrier TFL law. Likewise, the experimental current-voltage characteristic of Fig. 14.15 for CdS exhibits a very steep section in the electroluminescent, hence double-injection, range which looks much like the very steep section in the one-carrier SCL current, ZnS characteristic, Fig. 5.1. Note however, that both Figs. 1.5 and 14.15 have one feature in common that appears to be a usual, though certainly not universal, feature of double injection in insulators with trapping (as discussed in Chapter 14); this is a current-controlled negative resistance regime in the current-voltage characteristic. Assuming a good injecting

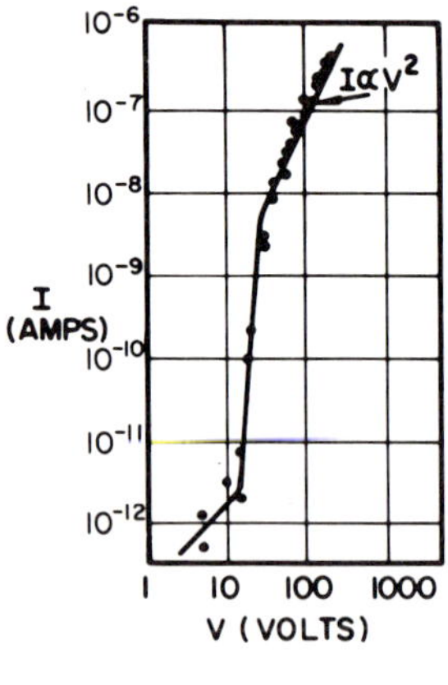

FIG. 5.1. Current-voltage characteristic for a ZnS crystal. $L = 5 \times 10^{-3}$ cm. The solid circles are experimental points. Some were taken with voltage increasing, the others with voltage decreasing. (From Ruppel, 1958, Fig. 1.)

contact, a current-controlled negative resistance is a highly unlikely phenomenon in a one-carrier current problem. In fact the only non-thermal mechanism known to the authors for achieving this involves a scattering or excitation of carriers from a low-mobility band or "pocket" in k-space to a higher-mobility band or pocket, e.g., as discussed by McWhorter and Rediker (1959, 1961) and Bok *et al.* (1961). Of course, one certain indication of double injection is the observation of distributed electroluminescence in the steady state at relatively low fields. This was the situation in Fig. 14.15 (see the photograph shown in Fig. 3 of Smith's paper (1957) corresponding to a field of about 1000 V/cm). However, note that electroluminescent emission has been observed by Hinotani and Sugigami (1965) in their investigation of SCL hole currents in ZnTe without encountering a negative resistance. (See Fig. 5.2.)

We conclude from this discussion that it is indeed prudent to pin down the interpretation of experimental current-voltage characteristics as one-carrier SCL current phenomena with the help of redundancy or self-consistency, and also with independent checks using other experimental techniques. We have already seen in the earlier discussion in this section that if there is sufficient structure in the I–V characteristic, redundancy is possible in the determination of the quantity θ. We also saw that measurement of θ over a range of temperatures yields redundant determinations of N_t and E_t. The usefulness of this additional set of measurements was demonstrated strikingly by Bube (1962) who showed that, in studies on a single crystal of CdS, the plot of $\ln \theta$ vs $1/T$ yielded completely different values for N_t and E_t than did a fixed-temperature, I vs V plot. The former values checked with independent measurements of thermally stimulated current peaks and photocurrent response times. The observed steep rise of current with voltage was presumably due to some cause other than trap-filling.

Thickness dependence of the I–V characteristic, where feasible, affords an excellent check of a one-carrier SCL current interpretation. An $I \propto V^2$ characteristic should exhibit an L^{-3} dependence. This was shown for SCL hole currents in evaporated CdS layers by Zuleeg and Muller (1964) and for SCL electron currents in KCl by Tödheide-Haupt *et al.* (1966). Where $I \propto V^{(l+1)}$, $l > 1$, the characteristic should exhibit an $L^{-(2l+1)}$ dependence, where the power law is attributed to an exponential distribution of traps, Eqs. (4.115) and (4.139). This was verified by Lanyon in work on vitreous selenium (1963), with $l = 2.8$, by Sussman (1967) in work on evaporated layers of copper-phthalocyanine with $l = 1.5$, 2, 2.5, 3, and 4, and by Helfrich (1964) with SCL electron currents in anthracene. A further check is the temperature variation of

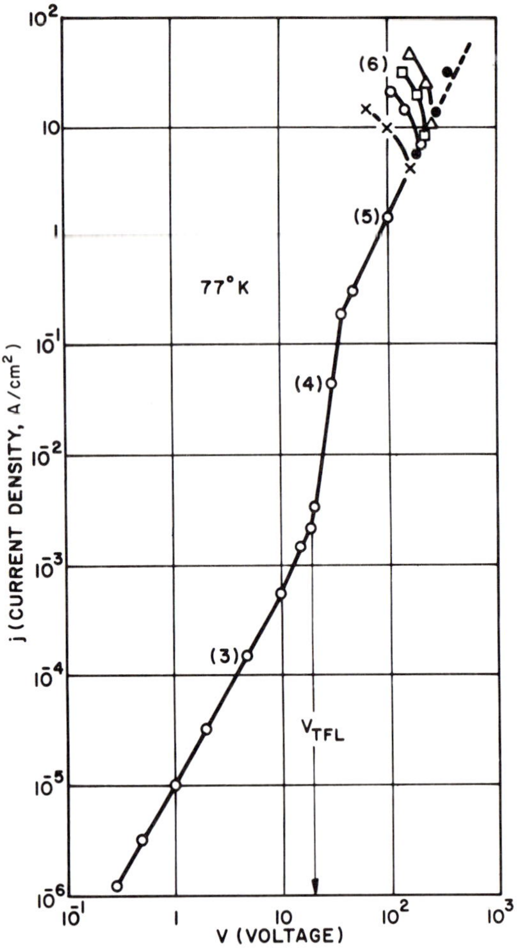

FIG. 5.2. Current-voltage characteristic for p-type ZnTe at 77°K. The regimes shown are (3) square-law, (4) transition, (5) square-law, and (6) breakdown. Pulse techniques were used for measurements above approximately 100 V. Pulse characteristics were: ●, $f = 200$ Hz, $\tau = 3$ μsec; △, $f = 1000$ Hz, $\tau = 100$ μsec; □, $f = 1000$ Hz, $\tau = 160$ μsec; ○, $f = 1000$ Hz, $\tau = 230$ μsec; ×, $f = 1000$ Hz, $\tau = 460$ μsec, with f the pulse repetition rate and τ the pulse width. The small open circles ($V \leqslant 100$ V) correspond to dc measurements. (From Hinotani and Sugugami, 1965, Fig. 1.)

$l = T_t/T$ in the power law, I–V characteristic. This was carried out by Sussman (1967) for one of his evaporated layers with $l = 1.5$ at 300°K. This same layer gave $l = 2.25$ at 200°K, as predicted. As noted previously, where a trap-filled-limit, nearly vertical section of the I–V characteristic has been observed, $V_{TFL} \propto L^2$.

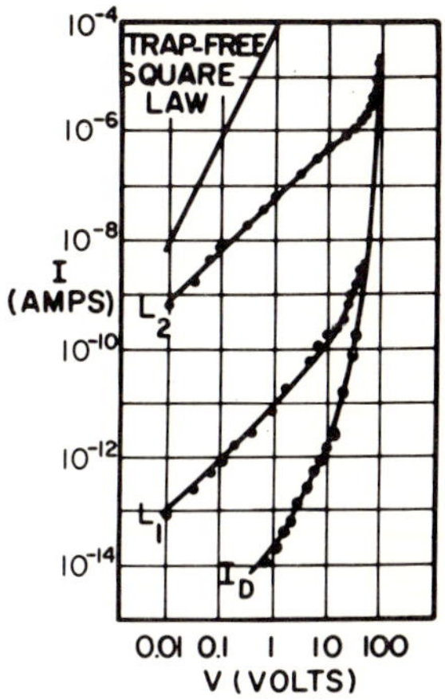

FIG. 5.3. Current-voltage characteristics for a CdS crystal. $L = 1.7 \times 10^{-3}$ cm. The curve labeled I_D is the dark characteristic; those labeled L_1, L_2 were obtained under photoillumination ($\lambda = 5200$ Å), L_1 corresponding to an incident flux of 1.6×10^4 photons/sec and L_2 to 3.4×10^7 photons/sec. The solid circles are experimental points. The top curve is theoretical. (From Smith, 1959, Fig. 1.)

The observation of current transients, particularly the transient, trap-free square law, as an aid in checking a SCL current interpretation has already been alluded to. This was used by Smith and Rose (1955), by Müller (1962), and by Zuleeg and Muller (1964) in their studies of CdS, by Helfrich and Mark (1962) with anthracene, by Mark and Helfrich (1962) with *p*-terphenyl, and by Gregory and Jordan (1964) and Aiken and Jordan (1965) with Si. Other aspects of transient behavior can also be useful for this purpose, e.g., measurements of current undershoot and overshoot by Smith and Rose (1955), measurement of the speed of transient response by Ruppel (1961a), and the observation of current flow with sudden reversal of bias by Helfrich and Mark (1962) and by Bässler *et al.* (1966). A discussion of transient SCL current measurements as a major technique in its own right is reserved for discussion in the following chapter.

Another electrical measurement useful in checking self-consistency, in addition to the obvious ohmic conductivity and Hall effect measurements, is that of the previously mentioned, thermally stimulated current, or conduction-glow peaks, discussed in Bube's book (1960). This has been used by Dresner and Shallcross (1962), by Bube (1962), by Franks and Keating (1961), and by Marlor and Woods (1965), all in studies of CdS.

Although we shall not review work on noise in SCL currents, mention must be made of the important paper by Shulman (1955) which confirmed the ohmic (reservoir) nature of the contacts of Ga and In to CdS by establishing that the photoconductive, Ohm's-law current noise was equal to the pure shot noise in the incident photon stream. Theoretical work on noise in SCL currents has been done by Fassett and

Van Vliet (1962). Further theoretical work has been published by Sergiescu (1965) and by van der Ziel (1966). Experimental work has been reported by Anderson and Meltzer (1961) and by Zijlstra and van der Ziel (1963).

Optical measurements are an obvious and powerful adjunct in SCL current studies. Thus, impurity optical absorption was used by Lanyon (1963) to check the trap distribution in vitreous Se obtained from his SCL current curves. Mark and Helfrich (1962) and Jansen *et al.* (1964) have made similar measurements with anthracene. Helfrich (1964) has developed a theory, based on the exponential trap distribution for holes found in anthracene, that accounts for the observed $I \propto V^2 f^{(l-1)/l}/L^3$ dependence in anthracene, where f is the illumination level. Likewise, photoconductivity measurements can be very useful. Where an exponential trap distribution has been obtained, with the SCL current given by $I \propto V^{(l+1)}$, the photoconductive current I_L with bandgap illumination (assuming the traps are not effective as recombination centers) should have a dependence on light intensity as $I_L \propto f^{l/(l+1)}$, as shown by Rose (1951, 1955). This was checked by Lanyon (1963), by Smith and Rose (1955), and by Sussman (1967) in their respective works mentioned above. Further, as a general relation, at any given light level, the ratio of photoconductive response time τ_0 to lifetime τ equals the ratio of trapped carrier concentration n_t to free carrier concentration n, for the "majority" carrier (Rose, 1951, 1955). The n_t so measured is a lower limit to the n_t in a SCL current determination, for the same position of the quasi-Fermi level. Such measurements were made by Smith and Rose (1955), by Bube (1962), and by Mark and Helfrich (1962).

In all of the preceding discussion, we have focused on the use of supplementary techniques as a means of obtaining verification and self-consistency of the SCL current interpretation of a steady-state current-voltage characteristic. It is obvious that many of these supplementary techniques have a usefulness beyond this limited function, namely in obtaining additional information about the defect states in the insulator. Steady-state, SCL I–V characteristics, since they involve a quasi-thermal equilibrium between free and trapped carriers, cannot, in themselves, provide information about dynamical interactions between free carriers and defect states. On the other hand, the characteristic times associated with current transients can provide such information as, for example, the free-carrier capture cross section for a trap and the thermal-release (attempt-to-escape) frequency for a trapped carrier. Such information was obtained for the electrons in CdS by Smith and Rose (1955), for holes in Si at 4°K by Gregory and Jordan (1964), for holes in p-terphenyl by Mark and Helfrich (1962), for holes in anthracene by Helfrich and Mark (1962), and Mark and Helfrich (1962), by Hoestery and Letson (1963), by Weisz *et al.* (1964), by Adolph *et al.* (1963), and by Helfrich and Schneider (1966), for electrons in anthracene by Helfrich and Schneider (1966), by Gränacher (1965), and by Büchner and Mehl (1965), for holes in iodine by Many *et al.* (1962) and by Simhony and Gorelik (1965). The combination of impurity optical absorption with steady-state SCL current measurements could, in principle, provide measurement of the absolute optical absorption cross section, though the authors are unaware of such work.

A combination of the SCL current characteristic with other steady-state electrical measurements (resistivity and Hall effect as a function of temperature) enabled Allen and Cherry (1961) to measure an almost exact compensation (deviation factor of 5×10^{-6}) in GaAs. Also, a combination of SCL currents and optical excitation enabled Mark (1965) to measure ambipolar free carrier diffusion in insulating CdS crystals.

5.3 Materials Studied with the SCL Current Technique

Steady-state, one-carrier injection currents have been observed in all the various types of nonmetallic solids: valence crystals (Ge, Si), ionic crystals (KCl), crystals of mixed covalent and ionic binding (the "classic" II–VI insulators), organic and inorganic, molecular (van der Waals binding) crystals (anthracene and iodine), and vitreous solids or glasses (Se). Both electron and hole injection currents have been observed, in some cases with the same material, the decisive factor in determining the carrier type being the availability of suitable injecting contacts.

5.3.1 *Cadmium Sulfide*

Cadmium sulfide has been the work horse of the field from the beginning up to the present. Originally the choice of CdS stemmed from its importance as a photoconductive material. The persistence of interest in CdS can be presumed to derive, in large measure, from its potentialities for the fabrication of insulator devices, as indicated in Section 5.4.

In general, it has been found that indium, gallium, and aluminum make good electron-injecting contacts, and with a minimum of care, at that. The definitive papers are those of Smith (1955) and Smith and Rose (1955), the former dealing largely with the behavior of contacts and the latter with experiments (steady-state, transient, and optical) establishing the SCL current interpretation. In one experiment, they directly measured the injected charge by dropping the sample, after application of voltage, onto an electrometer dish. Although the nearly vertical, TFL section of the I–V characteristic was observed in this early work, the shallow-trap square law preceding it was not seen. Thus, from Fig. 4 of the latter paper, it can only be concluded that there was a concentration of traps of at least 3×10^{14} cm^{-3}, lying at least 0.8 eV below the conduction band. Estimating the total concentration of effective traps to be 10^{15} cm^{-3}, the authors were able to deduce from their transient measurements an electron capture cross section for the traps on the order of 10^{-19} cm^2. In later studies of CdS, made photosensitive through incorporation of Cu and Cl, Smith (1959) and Lampert *et al.* (1959) reported the nearly vertical TFL current rise preceded by a square law, the curve labeled I_D in Fig. 5.3, taken in the dark. From these data, a trap concentration $N_t \simeq 3 \times 10^{14}$ cm^{-3}, lying 0.8 eV below the conduction band, was determined. The curves labeled L_1 and L_2 show the influence of light ($h\nu >$ bandgap) on the I–V characteristic. The trap-free square law curve is purely theoretical. Measurement of the photoconductive gain-bandwidth product for this sample established that the centers operative as electron traps in the dark functioned as recombination centers under illumination.

An interesting SCL current characteristic has been reported by Smith (1959, Fig. 1) for a thin crystal ($L = 10\,\mu$) with one contact ohmic, indium, and the other, blocking, tellurium. In the forward direction, the I–V characteristic exhibits an $I \propto V^{2.8}$ dependence, then a nearly vertical section, and finally an $I \propto V^2$ dependence, well below the theoretical trap-free square law.

Of unusual interest has been the repeated observation of the trap-free square law in thin crystals (L in the range of 5 to 50 μ) by Wright (1959) and co-workers and by Page *et al.* (1962). These measurements imply that the concentration of effective traps does not exceed 10^{12} cm^{-3}, approximately. This unusually low concentration is not taken to represent the overall degree of purity of the crystals; rather it is believed to result from an unusually high degree of compensation of deep-lying traps by shallow donors. Similar results have been reported by Allen and Cherry (1961) for GaAs, namely, an exceptionally low concentration of effective electron traps resulting from an exceptionally high degree of compensation of deep traps by shallow donors.

Bube's important paper (1962), already mentioned in Section 5.2, has spotlighted the desirability of self-consistent and independent checking of data derived from SCL I–V characteristics. Thus from his

SCL I–V characteristics (Fig. 1 of Bube's paper), taken on a crystal 75 μ thick, Bube deduced $N_t \simeq 10^{13}$ cm^{-3} and $E_c - E_t \simeq 0.5$ eV, using (5.1) and (5.2). On the other hand, his ln θ vs $1/T$ plot (Fig. 2 of his paper) gave $N_t \simeq 5 \times 10^{16}$ cm^{-3} and $E_c - E_t = 0.28$ eV, values confirmed by thermally stimulated current and photocurrent response-time measurements. Possible sources of the discrepancy, suggested by Bube, were field or impact ionization of the traps or double injection. We consider the former suggestion unlikely in view of his low maximum field, 2×10^4 V/cm. Bube's point is reinforced by the strange results of Dresner and Shallcross (1962) obtained with evaporated CdS films with one tellurium contact and one gold contact. Thus, with the Te contact positive (Fig. 4 of their paper), they see a nearly vertical TFL law followed by the *trap-free* square law corresponding to $\mu_n = 2$ cm^2/V sec, as obtained by Hall measurements. This corresponds to a trap concentration of 10^{13} cm^{-3}. On the other hand, thermally stimulated current measurements yielded trap concentrations $N_t \approx 10^{21}$ cm^{-3} (at levels 0.17, 0.29, and 0.31 eV below the conduction band), as are reasonable for evaporated films. The maximum field reached in their work was approximately 10^4 V/cm.

Other studies of SCL currents in CdS are those of Steinberger (1960), Ruppel (1961a,b), Muller (1962), Zuleeg (1963), Webb and Wright (1964), Marlor and Woods (1965), and Harnik (1965). The series of papers by Böer and Kümmel (1952, 1954a,b, 1957, 1958), interpreted otherwise by them, have already been mentioned in the previous section.

One-carrier SCL hole currents have not been reported for CdS as of this writing. The chief difficulty is certainly that of fabricating a good hole injecting contact, a very formidable technological problem whose successful solution would have profound technological consequences. (We may note here the possibility, suggested by Rose [private communication], that the currents observed by Broser and Broser–Warminsky (1958), when the anode of their crystal was illuminated by strongly absorbed light, were SCL hole currents rather than double-injection currents, as interpreted by the authors.)

5.3.2 *Zinc Sulfide*

The earliest reported indication of SCL currents in ZnS is by Alfrey and Cooke (1957). Ruppel (1958) obtained the very impressive SCL I-V characteristic, shown in Fig. 5.1, with a crystal 50 μ thick. From the ohmic section, taking $\mu_n \approx 10^2$ cm^2/V sec, $E_c - F_0 \simeq 0.85$ eV. The nearly vertical, TFL section corresponds (5.1) to $p_{t,0} \simeq 10^{13}$ cm^{-3}. Since there is no shallow-trap square law preceding the TFL section, we can conclude only that $F_0 - E_t \gtrsim 0$ and $N_t \gtrsim p_{t,0}$. The square-law section following the TFL-law is depressed by $\theta \approx 10^{-6}$ below the estimated trap-free square law, indicating a large concentration of shallow traps in the crystal.

5.3.3 *Zinc Telluride*

In its natural state, ZnTe is an extrinsic p-type semiconductor with a bandgap of 2.3 eV. In their study of SCL hole currents in this material, Hinotani and Sugugami (1965) prepared a semi-insulating n-type surface layer about 1 μ thick by the in-diffusion of indium which occupied the Zn vacancies of the p-type material. The SCL hole currents through this semi-insulating layer were observed by biasing the p-type bulk positive. Indium served as the other contact. The resulting I–V characteristic taken at 77°K is shown in Fig. 5.2. From V_{TFL}, $N_t = 10^{16}$ cm^{-3} and from the known hole mobility, 800 cm^2/V sec at 77°K, $\theta = 10^{-9}$, and $E_t = 0.17$ eV. Measurements of the temperature-dependent low-voltage square-law current yielded $E_t = 0.13$ eV and $N_t = 10^{17}$ cm^{-3}, in good agreement with the values obtained from V_{TFL}. The interpretation of the square law following the TFL current rise is not unambiguous. If a one-carrier phenomenon, it would indicate substantial shallow trapping, as in Ruppel's ZnS measurements, Section 5.3.2. However, light emission, in the range 2.0–2.1 eV, was observed at the onset of this square-law regime following V_{TFL}. This was interpreted as recombination radiation, via the indium centers near the conduction band, as a result of double injection, thus suggesting a possible key role of the double injection in this regime of the characteristic. The negative resistance at the very high currents is caused by thermal breakdown; it was estimated that ohmic dissipation produced a temperature rise exceeding 200°K at the lowest pulse duty cycle.

5.3.4 *Potassium Chloride*

SCL electron currents have been observed by Tödheide-Haupt *et al.* (1966) in zone refined KCl crystals provided with gold electrodes. Sample thicknesses ranged from 6 to 1000 μ. SCL ionic currents were apparently also observed. In both cases, the authors verified the L^{-3} thickness dependence.

5.3.5 *Gallium Arsenide*

SCL electron currents were reported in GaAs single crystal of thickness 120 μ, with tin contacts, by Allen and Cherry (1961). Their published I–V characteristic exhibits an Ohm's law, followed by a steep current rise, followed by, essentially, the trap-free square-law (to within a factor of three). The effective concentration of traps, $N_A - N_D$, computed from V_{TFL}, was 8×10^{10} cm^{-3}. In another sample, the corresponding I–V characteristic gave $N_A - N_D \simeq 4 \times 10^{12}$ cm^{-3}, which number, combined with other electrical measurements, gave $N_D \simeq 8 \times 10^{17}$ cm^{-3}—a remarkably small deviation factor, namely 5×10^{-6}, from exact compensation. This is even more extreme than the degree of compensation hypothesized by Page *et al.* (1962) for their CdS crystals.

5.3.6 *Silicon Carbide*

This material has been investigated by Ozarow and Hysell (1962) and by English and Drews (1963). The latter authors made an extensive study of the hexagonal phase of this material. They have observed, in high-resistance p-type crystals at 77°K, I–V characteristics of two kinds: $I \propto V$ going into $I \propto V^n$, and $I \propto V$ going into $I \propto V^2$ going into $I \propto V^n$, where $4 < n < 9$. From SCL current analysis of these characteristics they have identified, "with caution," for several of their crystals, trap concentrations $N_t \approx 10^{13}$ cm^{-3} located about 0.2 eV above the valence band. This paper also contains a theoretical analysis of SCL I–V characteristics dominated by two sets of discrete levels.

Earlier evidence for SCL currents in SiC, associated with reverse-biased junctions and a point-like injection geometry, is given by Patrick (1957).

5.3.7 *Aluminum Nitride*

SCL hole currents in single crystals of AlN have been investigated by Edwards *et al.* (1965) using gold electrodes. Several current-voltage characteristics of the sequence Ohm's law–TFL–square law are reported but there is some unreproducibility of the data. The high current square law would require a hole mobility ranging from 10 to 100 cm^2/V sec for interpretation as a trap-free SCL current. The lower value is probably the more realistic one since it is in better agreement with Hall measurements (14 cm^2/V sec) made by the same authors on doped AlN crystals.

5.3.8 *Titanium Dioxide* (*Rutile*)

Greener and Whitmore (1961), Cardon (1961), and von Hippel *et al.* (1962) have investigated this material. The measurements are plagued with time-dependent drifts, associated with ionic conduction, that make an unambiguous interpretation in terms of SCL currents difficult. The reader is referred to the literature for details.

5.3.9 *Barium Titanate*

The evidence for SCL current injection in this material rests on observed I–V characteristics (Branwood and Tredgold, 1960; Branwood *et al.*, 1962) of the form $I = a(V/L) + b(V/L)^2$. These experiments were done on single crystals, of thicknesses 100–600 μ, with applied fields up to 8 kV/cm. The coefficient a is a characteristic of the bulk material whereas b appears to be associated with the cathode material. It is a general characteristic of ferroelectrics ($BaTiO_3$ is ferroelectric below 120°C) that they have surface-layer electrical characteristics quite different from those of the bulk (Känzig, 1955; Lefkowitz, 1963). Therefore, homogeneous bulk SCL current theory, such as presented in this book, is presumably inapplicable to such materials. In any case, Roberts and Tredgold have interpreted the unusual thickness dependence, $I \propto L^{-2}$, as arising from limitations of the current by the diffusion field at the cathode. An extensive review of the experimental work with $BaTiO_3$ can be found in Tredgold's book (1965).

5.3.10 *Silicon*

SCL hole currents in Si p^+pp^+ structures have been investigated by Lemke (1963, 1966) and Lemke and Müller (1967). Steady-state measurements (1963), performed with 10^3 to 10^4 Ω cm Si, compensated with deep-lying Fe, Au, and Cu donors, indicate a shallow-trap-dominated SCL current. Pulse measurements at 77 to 300°K were used to verify the low-field L^{-3} thickness dependence (1963), and at 4.2°K to obtain the capture cross section for holes, 7×10^{-16} cm^2, of an Fe-center. Transient SCL current measurements (Section 6.3.8) in combination with steady-state measurements (1967) on Pt electroded, undoped crystals indicated trap-free behavior for steady-state SCL currents.

SCL hole currents have also been observed in very pure, p-type single crystals (resistivity $\rho_0 \simeq 2000\ \Omega$ cm at 300°K) by Gregory and Jordan (1964) at 4°K, with the hole drift velocity saturated (at about 10^7 cm/sec). This latter is a direct consequence of the extremely high mobility of free carriers, exceeding 10^5 cm^2/V sec, in very pure Si at 4°K, coupled with the field required to fill the traps, $\mathscr{E}_{TFL} \simeq 3 \times 10^3$ V/cm in their samples. Since this field was not high enough to produce impact ionization of the hole traps, they observed extremely sharp TFL curves corresponding to $N_{tp} = N_D \simeq 6 \times 10^{12}$ cm^{-3}. Note that this experiment measures the compensating (minority) impurity density. They further checked that $V_{TFL} \propto L^2$. Because of the extremely high resistivity of their Si at 4°K, they were unable to measure the Ohm's law current preceding the TFL section of the characteristic. At voltages $V > 4V_{TFL}$, the steady-state TFL characteristic merges with the trap-free characteristic (which is almost linear because of the saturated hole drift velocity) as measured by pulse techniques. Combining the SCL I–V data with measurements of transient current decay, they were able to determine a hole capture cross section for the acceptors of approximately 2×10^{-13} cm^2, an entirely reasonable number for a Coulomb attractive center and hot carriers (hole temperature T_h of many hundreds of degrees Kelvin). This work was done with p^+pp^+ samples of thicknesses 60 to 100 μ, with the thin p^+ regions degenerate to assure reservoir contacts at the low temperature.

SCL electron currents in Si at 78 and 300°K have been examined by Denda and Nicolet (1966). Trap-free SCL current flow was achieved which made it possible to study the temperature and electric field dependence of the electron drift mobility.

5.3.11 *Germanium*

SCL hole currents were observed by Dacey (1953) in a p–n–p structure at 77°K under punch-through conditions, i.e., with the thermally generated free electrons swept out from the entire n-region by the applied voltage. As pointed out in Section 4.4, this problem is formally identical to the TFL insulator problem in that they are both concerned with fixed, excess charge in the crystal. Thus Dacey's experimental curve, Fig. 6, looks like a typical TFL curve although, in actuality, it has nothing to do with trapping of injected carriers. His fit of theory to experiment was complicated by the field-dependence of the hole mobility which was not known accurately at 77°K at that time. Therefore, the theoretical curve in Dacey's Fig. 6 needs modification, as acknowledged by him. The observed SCL current was first predicted by Shockley and Prim (1953).

5.3.12 *Selenium*

The earliest reported observations of SCL currents in Se are those of Weimer and Cope (1951) on thin, evaporated, amorphous, films. The injected carriers (holes) were photogenerated and, at high light intensities, $I \propto V^2$ was observed. A very complete study correlating optical and electrical SCL current properties of vitreous Se has been

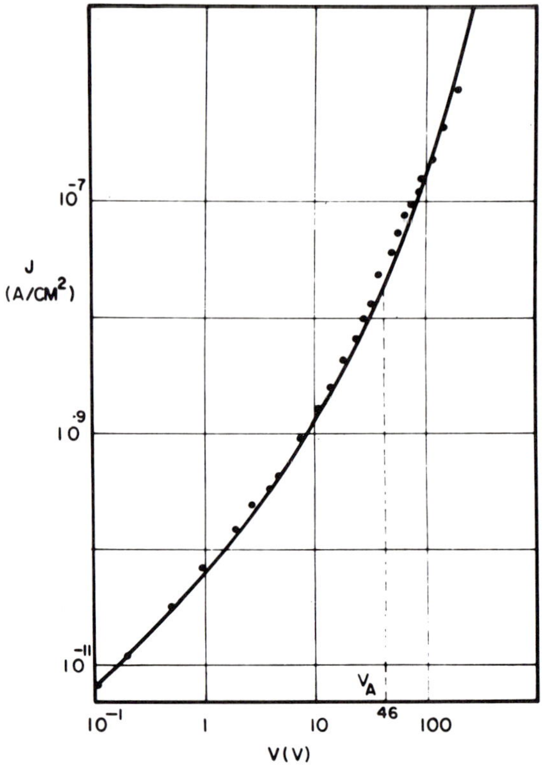

FIG. 5.4. Current-voltage characteristic for a vitreous Se sample. $L = 3.1 \times 10^{-3}$ cm. The solid circles are experimental points. The solid curve is theoretical, corresponding to $kT_t = 0.067$ eV. (From Lanyon, 1963, Fig. 6.)

published by Lanyon (1963). Figure 5.4, taken from this paper, shows an experimental run, on a sample 31 μ thick, fitted by a theoretical curve, solid line, of the form $J = \sigma_0(1 + V/V_T)^l\, V/L$ with $l = T_t/T = 2.8$. This corresponds to an exponential distribution of hole traps above the valence band of the form $\mathcal{N}_t(E) = \mathcal{N}_0 \exp\{-(E - E_V)/kT_t\}$ with $\mathcal{N}_0 = 10^{20}/\text{cm}^3$ eV and $kT_t = 0.067$ eV, holding at least over the range of hole quasi-Fermi level variation from 0.69 eV $< E - E_V <$ 1.0 eV. He also fitted Hartke's experimental data (1962) using the same T_t. Lanyon measured I–V characteristics on samples ranging in thickness from 2.4 to 60 μ, with platinum or tellurium electrodes. He obtained the current-thickness dependence of the I–V characteristic and also cross-checked the SCL-current-derived trap distribution by optical absorption measurements and by the photoconductivity-light intensity dependence, $I \propto L^{l/(l+1)}$. He also correlated the derived trap distribution with other experimental data on negative photoconductivity, electron-bombardment conductivity, and long wavelength quenching of photoconductivity. Lanyon points out that presumably an equally good fit to his data could be provided by a Gaussian distribution of traps over the limited region of the gap explored electrically. Earlier work on vitreous Se was reported by Spear and Lanyon (1960), by Lanyon and Spear (1961), by Hartke (1962), and by Lanyon (1963).

Space-charge-limited current measurements on single crystal Se have been reported by Polke *et al.* (1963) and by Polke (1964). These measurements were made with samples long compared to their cross-sectional dimensions, a subject that is discussed briefly in Chapter 8.

5.3.13 *Antimony Trisulfide*

The first reported observations of SCL hole currents are those of Forgue *et al.* (1951) who measured dark currents $I_D \propto V^2$ on thin, evaporated films. More recently, Bowlt (1962) has observed SCL I–V characteristics on thin films, thickness 3 to 8 μ, using gold electrodes. He measures Ohm's law going over to $I \propto V^n$ where $5.0 < n < 8.2$.

5.3.14 *Iodine*

SCL hole currents were observed in single crystals by Many *et al.* (1962). Holes available for injection were generated at the anode by intense steady illumination. Generally, in the steady state, a square-law I–V characteristic was observed with a θ corresponding to 0.02, the mobility $\mu_p = 0.7$ cm²/V sec (perpendicular to the *ac* crystallographic plane) being determined from transient SCL current data. The thickness dependence of the square law at fixed voltage is shown in Fig. 5.5. It is seen that $I \propto L^{-3}$ gives a good fit although there is much scatter because of variation of trap content among the samples used in the measurements. The scatter is much less for the transient, trap-free square law shown in the same figure because trapping cannot affect the *initial* transient current plotted in this figure. A plot of I vs $1/T$ gives $E_{tp} - E_v = 0.45$ to 0.6 eV for various samples, from which it is concluded that $N_{tp} \simeq 5 \times 10^{13}$ cm^{-3}. Finally, in combination with transient SCL current data, the above-derived steady-state data yield a hole capture cross section for the traps of $\approx 10^{-20}$ cm². The power of the combination of transient and steady-state SCL current measurements for obtaining information about insulators is beautifully illustrated by this work. The transient SCL current measurements, which have given much further information, are discussed more fully in the following chapter.

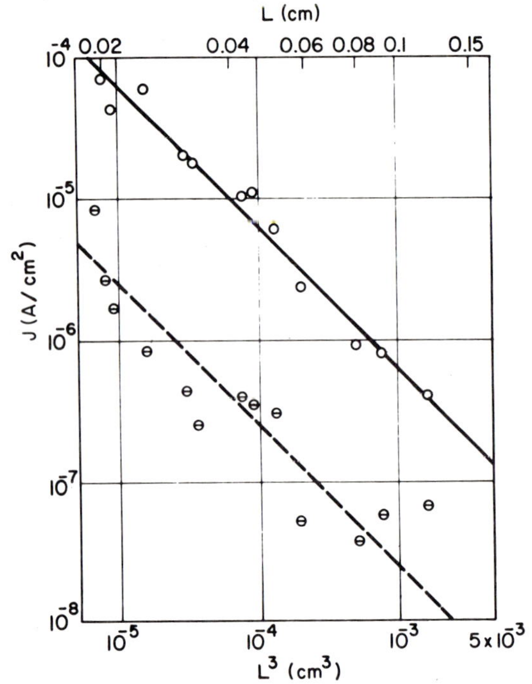

FIG. 5.5. Thickness dependence of transient and steady-state SCL currents in single crystal iodine. Open circles are transient measurements (initial amplitude), bisected circles are steady-state measurements. The solid line is the theoretical curve for the initial amplitude of the transient, taking $\mu = 2$, $\kappa = 2$. The dashed line is a best fit (with $J \propto L^{-3}$) to the steady-state data. (From Many *et al.* 1962b, Fig. 7.)

5.3.15 *Anthracene*

Injection current techniques have been very successfully applied to the study of the electronic properties of organic molecular crystals. Indeed, in prior times, electrical measurements with these materials were exceedingly unrewarding and unsatisfactory. It is perhaps not exaggerating to state that, in the current era of solid-state technology, injection current studies are the dominant, if not the only, technique for the examination of the electronic properties of these and related materials. An extensive review of this subject has recently been published by Helfrich (1968). In this section, we shall confine ourselves to a brief review of the information concerning defect structure that has been obtained from steady-state single injection measurements. Transport studies using transient techniques are the subject of the next chapter and double injection results are discussed in Chapter 14.

SCL hole currents in anthracene and in the lower polyphenyls (see below) were first studied by Mark and Helfrich (1962) in the form of single crystals grown by condensation from the vapor phase (Mark, 1961) and using aqueous electrodes: iodide solutions for the negative, non-injecting contact and an iodine-iodide oxidizing solution for the positive hole injecting contact. The use of oxidizing electrolytes to inject holes was introduced by Kallman and Pope (1960a, b).

The most prominent hole traps in anthracene are distributed exponentially in energy, as deduced from the commonly observed $I \propto V^{(l+1)}/L^{(2l+1)}$ dependences. The parameters l and N_t vary from sample to sample and are related, presumably, to structural defects in very pure crystals and to impurities in less pure crystals, since l is much larger in the latter than in the former. Usually $N_t \simeq 10^{16}$ cm^{-3}, which is not an unusually large structural defect concentration, considering the weak van der Waals binding forces and the likelihood of many types of stacking faults owing to the relatively complex shape of the anthracene molecule (Helfrich, 1968). Helfrich and Lipsett (1965) have also suggested that the trapping properties of structural defects may be due to associated local lattice compression in which the bands would be wider (within the frame work of the tight binding approximation) and thus form trapping states. The capture cross sections of the traps, roughly equal to the molecular areas, have been obtained from the trapping times and have been confirmed by determination of the release time from the traps (Mark and Helfrich, 1962). The trap release measurements have also furnished a direct measure of the injected space charge and thus confirmation of SCL currents (Helfrich and Mark, 1962). A deep hole trapping level at a discrete energy ($>$0.9 eV) and at small concentration, about 10^{11} cm^3, has also been observed (Helfrich and Mark, 1963 and Adolph *et al.*, 1963). Its origin is not known.

Weisz *et al.* (1966) have applied SCL current measurements to the study of radiation damage in anthracene. The high power I–V dependence was usually supplanted by a smaller square-law terminating in a trap-filled-limit under sufficiently large dosage (10^5 ergs from a ^{60}Co source). The energy required to form a trap is about 45 keV.

SCL electron currents in anthracene were first observed by Helfrich and Schneider (1965, 1966) using a solution of negative anthracene ions (anthracene and sodium in tetrahydrofuran) as an ohmic contact for electrons. With sufficiently pure crystals, it is usually possible to observe a trap-filled-limit current preceded by a less steep I–V curve characteristic of an exponential distribution. Because of the extreme sensitivity of the SCL electron current to crystal purity, it is thought that the electron traps are associated with impurities. SCL electron currents have also been observed by Gränacher (1965), by Büchner and Mehl (1965) using liquid electrodes, and by Mehl and Funk (1967) using a Na–K alloy, solid, ohmic cathode.

5.3.16 *Polyphenyl Single Crystals*

SCL hole currents have also been observed by Mark and Helfrich (1962) in single crystals of the lower polyphenyls.

For *p*-quaterphenyl, the measured power-law *I–V* characteristic corresponds to an exponential, trap-distribution temperature $kT_t \simeq 0.05$ eV. For *p*-terphenyl, they found the same trap-distribution temperature, $kT_t = 0.05$ eV and $\mathcal{N}_t \simeq 7 \times 10^{17}/\text{cm}^3$ eV. From the transient, trap-free square law, they determined $\mu_p \simeq 2.5 \times 10^{-2}$ cm²/V sec. From measurements of the transient current decay, they determined the hole capture cross section for the traps to be approximately 4×10^{-17} cm² and the attempt-to-escape frequency to be 10^{10} sec^{-1}.

5.3.17 *Phthalocyanine*

Phthalocyanine, $C_{32}H_{18}N_8$, is a planar molecule consisting of four pyrrole ring and four outer benzene rings. At the center of the molecule is a pair of hydrogen atoms. Its most stable polymorph is a base-centered monoclinic lattice with unit cell dimensions $a = 19.85$ Å, $b = 4.72$ Å, and $C = 14.8$ Å. It is a particularly interesting example of a molecular, organic crystal having semiconducting properties. SCL electron currents have been observed in single crystals, using silver paste electrodes and guard rings, by Heilmeier and Warfield (1963). They consistently observed *I–V* characteristics with Ohm's law followed by a square law, a typical curve being shown in Fig. 5.6. Some crystals exhibited more structure in the *I–V* characteristic, e.g., their sample 21-2 as shown in Fig. 5.6. Here a TFL curve is seen preceded by a shallow-trap square-law regime. This curve yields the information: $N_t = 2 \times 10^{12}$ cm^{-3} and $E_c - E_{tn} = 0.81$ eV, taking $\mu_n = 0.1$ cm²/V sec and $N_c = 10^{21}$ cm^{-3}, the former being known from Hall measurements and the latter estimated from the molecular density in the crystal.

5.3.18 *Cu-Phthalocyanine*

This material derives from the metal-free phthalocyanine by replacement of the two inner hydrogen atoms by one copper atom. SCL hole currents have been observed in very thin evaporated layers

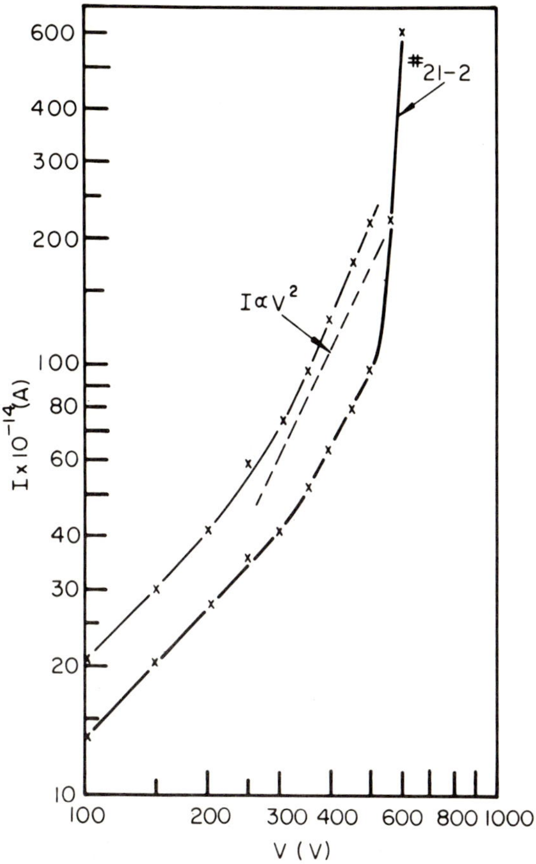

FIG. 5.6. Current-voltage characteristics for two metal-free phthalocyanine crystals. $L = 2 \times 10^{-2}$ cm. The crosses are experimental points. (From Heilmeier and Warfield, 1963, Figs. 3 and 8.)

($L \simeq 0.1$–$0.3\ \mu$) of this material by Sussman (1967) with $I \propto V^{(l+1)}$. An example of an I–V characteristic is shown in Fig. 5.7 (solid line). Also shown are the data (crosses) for another sample with a square-law characteristic going over into a cube-law characteristic. The temperature dependence of $l = T_t/T$ has been verified in one case, and so likewise has the photocurrent–light intensity dependence, $I \propto f^{l/(l+1)}$. From the highest voltage used, about 10 V, it can be concluded that these evaporated layers have hole trap concentrations N_{tp} at least of the order of 10^{17} to 10^{18} cm^{-3}. This material was also examined by Delacote *et al.* (1964) who measured SCL electron currents. A trap-filled-limit was reached, indicating a trap concentration of 10^{16} cm^{-3}. From the sub-

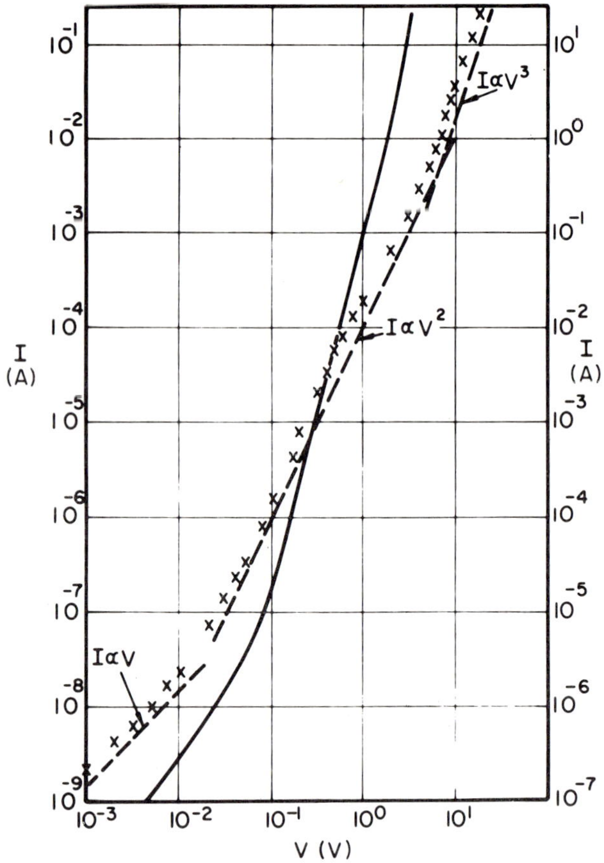

FIG. 5.7. Current-voltage characteristics for two evaporated-layer, Cu-phthalocyanine samples. $L \simeq 2 \times 10^{-5}$ cm. The solid curve is the experimental characteristic for one sample (right-hand scale), the crosses are the experimental points for another sample (left-hand scale). The dashed lines are merely reference lines for indications of slope. (From Sussman, unpublished.)

sequent square-law current, an electron mobility of 10^{-4} cm²/V sec was computed in good agreement with the transient measurements (see Section 6.4.8).

5.4 Applications

An early application to technology of the SCL current studies was the setting of an upper limit to the attainable gain-bandwidth product for photoconductors, due to the onset of space-charge currents, by

Rose (1957a) and by Rose and Lampert (1959a,b). Since the gain-bandwidth product is a figure of merit for device performance, this application has markedly affected device research. The gain-bandwidth product is discussed in some detail in Chapter 7.

There has been a continual interest in fabricating insulator devices in analogy with vacuum devices, for example, the diode rectifier and the triodc amplifier. Diode operation has been achieved with an asymmetric electrode configuration in which the insulator is sandwiched between one injecting and one noninjecting contact such that forward bias corresponds to SCL current injection. Diodes have been made from single crystals of CdS by Ruppel and Smith (1959), by Shao and Wright (1961), by Webb and Wright (1964), and by Page (1966), from evaporated CdS films by Dresner and Shallcross (1962), and by Zuleeg and Muller (1964), and from evaporated Cu-phthalocyanine films by Sussman (1967). The CdS-based diodes were n-type devices, the electrons being injected mostly from indium cathodes (Dresner and Shallcross used gold). The noninjecting electrodes were of tellurium. The phthalocyanine diodes were p-type with gold injecting, and chromium, tin oxide, and tin noninjecting contacts (Sussman, 1967). These diodes have yielded rectification ratios on the order of 10^6 and better.

The analog triode as first proposed by Shockley (1952) utilizes SCL majority carrier injection into an intrinsic region, the modulation of this current being achieved via a bias signal across a back-biased p–i or n–i junction. Analogue triode devices were first made from single crystals of CdS by Ruppel and Smith (1959) and from evaporated films of CdS by Weimer (1961) and by Zuleeg (1963). Since p-type CdS is not available, in none of this work were back-biased junctions used to apply the modulation signal. Rather, Ruppel and Smith used tellurium blocking contacts and Weimer used the field effect configuration with an evaporated SiO layer as the gate insulator. The single crystal device was high impedance (10^{11} Ω) with a transconductance of about 10 micromhos but with a rather poor frequency response (upper limit ~10 kHz). The thin film device was much more promising and served as a point of departure for the exploration of the evaporated CdS thin-film-transistor (TFT) (Weimer 1962) in which the equilibrium carrier concentration of a doped semiconductor, rather than an injected carrier concentration of an insulator, is modulated by a field electrode. Wright (1962, 1964) has also discussed the SCL triode. An important technical problem which must be overcome with these insulator SCL analogue devices is a relatively poor frequency response (bandwidth) due to low carrier mobility and severe trapping, both common in insulator materials, and especially in films.

6

TRANSIENT INJECTION CURRENTS

In the foregoing chapters, we have considered only steady-state injection currents. Before the steady state is reached, following, say, the application of voltage, a variety of transient response behaviors may be encountered whose relaxation times can vary from the nanosecond range to as long as several days or even much longer, depending on their physical origins. The long relaxation times are associated with the establishment of equilibrium between free and trapped charge, that is, with the establishment of a new quasi-Fermi level (Section 7.3). Consequently, these relaxation times are determined by the defect structure of the specimen. The short relaxation times, on the other hand, represent the motion of the initially injected charge through the crystal under the action of the applied field. They are related to the transit time of the injected carriers, which depends on the transport parameters and length of the specimen and on the applied voltage, and which may or may not be strongly influenced by the defect structure of the specimen. We shall see presently that the determination of the drift transit time has, in practice, been reduced to a time-of-flight measurement.

Until quite recently, only transient measurements of the first kind, with the long relaxation times, were investigated, as in the work of Rose and Smith on CdS and of Helfrich and Mark on organic insulators (see Chapter 5 for references). These are a useful adjunct to the steady-state measurements, primarily in confirming existence of SCL currents and in obtaining additional information about trapping phenomena.

As a result of the work of Brown (1955), Spear (1957), and Kepler (1960) with space-charge-free (SCF) conditions, and Many and his collaborators (1961, 1962) and Helfrich and Mark (1962a) with space-charge-limited (SCL) conditions, the time-of-flight injection-current measurements have come into their own as perhaps the best, if not the only, means of studying carrier drift mobilities in insulating materials where the more conventional techniques, e.g., Hall effect, fail completely.

The experimental studies of steady-state, one-carrier injection currents in insulators have generally been done with "good" injecting contacts, that is, contacts with a reservoir of free carriers sufficiently large to supply whatever SCL current is demanded at the given applied voltage. Accordingly, the theory presented in the preceding chapters has assumed the existence of just such a contact, and in fact, has idealized the contact into an infinite reservoir. The situation with transient, one-carrier injection currents is quite different. Here much useful experimental work has been done not only with "good" injecting contacts, but equally with "poor" injecting contacts, that is, contacts with a very limited reservoir of free carriers, in fact, so limited as to support only a SCF transient current through the insulator over the voltage range of interest. For example, if the physical cathode contact is noninjecting, but rather the carriers are injected into the insulator via a light flash which is strongly absorbed near one surface, the "strength" of the reservoir is determined by the intensity of the flash: At low light levels, there will be a "weak" reservoir and SCF current injection, and at sufficiently intense light levels there will be a "strong" reservoir and SCL current injection. (Of course, another convenient mode of injection, when practicable, is that via a voltage step-function applied across an insulator with a physical cathode which is already a good injecting contact.) Since both "strong" and "weak" contacts are used in the experimental work, the theory for the current transient presented in the following section will cover both extremes as well as the transition between them. The basic equations are the same for all cases; it is the boundary condition at the cathode that determines the "quality" of the contact, hence the nature of the current transient. Following the theoretical discussion, we present a review of the experimental information that has been obtained with a variety of materials through study of the injection transient.

6.1 Theory

The theory is presented here for the perfect, trap-free insulator for the situation in which current flow begins at $t = 0$ under constant applied voltage V. The influence of traps will be considered later. One-dimensional, planar, electron current flow is assumed.

The basic equations which must be solved to obtain the time-dependence of the injection current are the total current density equation

$$j(t) = j_{\text{cond}} + j_{\text{displ}}, \qquad\qquad (6.1)$$
$$j_{\text{cond}}(x, t) = en(x, t)\,\mu\mathscr{E}(x, t), \qquad j_{\text{displ}} = \epsilon\,\partial\mathscr{E}(x, t)/\partial t,$$

and the Poisson equation

$$\epsilon\,\partial\mathscr{E}(x, t)/\partial x = en(x, t). \qquad (6.2)$$

In writing (6.1) we have, as usual, neglected the diffusion current. This procedure is legitimate for applied voltages large compared with kT/e (Schilling and Schachter, 1967a, b). The fact that the total current is a function of time t only; that is, independent of position x, follows directly from the Maxwell equation, curl $\mathbf{H} = \mathbf{j}$, by taking the divergence of both sides.

The boundary conditions needed to determine a unique solution to (6.1) and (6.2) are, in addition to the constant voltage V, specification of $\mathscr{E}(0, t)$ and $\mathscr{E}(x, 0)$. The fields at the cathode and anode, $\mathscr{E}_c(t) = \mathscr{E}(0, t)$ and $\mathscr{E}_a(t) = \mathscr{E}(L, t)$, respectively, are related to each other through the total charge $Q(t)$ per unit area in the insulator at time t, namely

$$\mathscr{E}_c(t) = \mathscr{E}_a(t) - [Q(t)/\epsilon]. \qquad (6.3)$$

Equation (6.3) is Gauss' theorem which is just the integral of (6.2). It generalizes (3.25) which holds for $\mathscr{E}_0(t) \equiv 0$.

Substituting for $en(x, t)$ from (6.2) into (6.1), we obtain

$$j(t) = \epsilon\left\{\frac{\mu}{2}\frac{\partial\mathscr{E}^2(x, t)}{\partial x} + \frac{\partial\mathscr{E}(x, t)}{\partial t}\right\}. \qquad (6.4)$$

Integrating (6.4) from the cathode to the anode,

$$j(t) = (\epsilon\mu/2L)\{\mathscr{E}_a^2(t) - \mathscr{E}_c^2(t)\}, \qquad (6.5)$$

noting that $\int_0^L (\partial\mathscr{E}/\partial t)\,dx = (\partial/\partial t)\int_0^L \mathscr{E}\,dx = \partial V/\partial t = 0$, by assumption. (It is of interest to note that in cylindrical and spherical, one-dimensional current flows, the relevant current i explicitly involves the radial coordinate so that the step leading to (6.5) cannot be carried out. Consequently, transient injection current theory in these geometries is very much more difficult—in fact, it appears to be analytically intractable.)

Equation (6.5) can also be written, noting from (6.3) that

$$\mathscr{E}_a^2 - \mathscr{E}_c^2 = (\mathscr{E}_a - \mathscr{E}_c)(\mathscr{E}_a + \mathscr{E}_c) = Q\{2\mathscr{E}_a - Q/\epsilon\}/\epsilon,$$

as

$$j(t) = [\mu Q(t)/2L]\{2\mathscr{E}_a(t) - [Q(t)/\epsilon]\}. \tag{6.6}$$

6.1.1 *The Space-Charge-Free Transient*

We shall consider first the case where $Q(t)$ is small and independent of time t: $Q(t) = Q$, constant. This would correspond, for example, to the case of a blocking cathode and a weak light flash. Specifically, we assume that $Q \ll \epsilon\mathscr{E}_a$, that is, that Q is much less than the maximum charge the insulator can store according to its capacitance ϵ/L. Then, from (6.3), $\mathscr{E}_c(t) = \mathscr{E}_a(t) = \mathscr{E}_\Omega = V/L$, and (6.6) becomes

$$j(t) = j, \text{ constant} = \mu Q\mathscr{E}_\Omega/L = \mu QV/L^2. \tag{6.7}$$

Assuming that the charge Q is initially near the cathode, it flows out at the anode after the time t_0 given by the integral: $L = \int_0^{t_0} \mu\mathscr{E}\, dt$, that is,

$$t_0 = L/\mu\mathscr{E}_\Omega = L^2/\mu V, \tag{6.8}$$

which is, of course, the space-charge-free transit time. The constant current (6.7) flows only over the time interval 0 to t_0. An expanded version of this discussion, including the effects of a finite sample conductivity, is given in early work of Schumann (1933).

What happens physically in this small, constant Q approximation can be understood with reference to the photo-injection mechanism. The sample is initially in the dark between two noninjecting electrodes and subject to a constant applied voltage V. The field is everywhere constant, $\mathscr{E}(t) = \mathscr{E}_\Omega = V/L$; there are no excess carriers in the contact region. At $t = 0$, a light flash releases at or adjacent to the cathode a sheet of carriers of total number Q/e per unit area, an amount much smaller than that required to produce a potential minimum at the cathode. Thus, the field at the cathode is essentially unaltered from its ohmic value E_Ω. The carriers now drift in the applied field as a well-defined sheet. If there is no attenuation of Q, either by trapping or ohmic relaxation, the current rises suddenly at $t = 0$ to the constant value (6.7), persists for the time t_0 at which time the sheet arrives at the anode, and then suddenly decays to zero. If there is attenuation of Q during transit, the current will naturally decrease with time over the interval $0 \leqslant t \leqslant t_0$. (The concept of a sheet of charge is, of course, an idealization. The sheet, at its formation, will have finite thickness due either to the finite duration of the light flash or to the finite absorption length of the light in the insulator. The former will also cause a finite rise time for the current. The finite sheet thickness, which will be enhanced by diffusion during transit, will likewise cause a finite decay time for the current.) This method was applied, for example, by Kepler (1960), to an examination of electron and hole transport in anthracene.

A useful variant of this technique is applicable when t_0 is smaller than the resolving time of the electronic circuit used to measure the time-of-flight of the injected carriers. It consists of integrating the current and measuring the charge passed by the specimen. In the absence of diffusion and attenuation of Q, the time-integral of (6.7) yields a curve that rises from zero with a constant slope until $t = t_0$, after which the curve becomes horizontal. Inclusion of diffusion and attenuation rounds off the sharp corner at t_0. This technique was introduced by Brown (1955) and was later developed by Spear (1957) who used a very short duration, electron beam pulse impinging on the specimen to inject the excess carriers.

6.1.2 *The Space-Charge-Limited Transient*

When Q is large enough to create a charge reservoir at the cathode, either because a sufficiently intense flash is used or because the sample is provided with an injecting contact, a rather different transient response—the SCL current transient—is obtained. The main features of this transient are as follows (see Fig. 6.1): (i) an instantaneous current, at $t = 0$, which is slightly less than one-half of the steady-state, trap-free current, (ii) a monotonic rise to a maximum current overshooting the steady-state current by 21% and occurring after a time t, which is 20% shorter than the space-charge-free transit time t_0 at the same voltage, (iii) a discontinuous change of slope at the maximum with current now decreasing; i.e., a cusp-like behavior at the maximum, (iv) a slight undershoot of the steady-state current, and (v) approach to the steady-state after a time $t \simeq 2t_0$.

The behavior of the SCL current transient up to the cusp can be obtained, in an elementary manner, from the basic equations (6.1) and (6.2) and the boundary condition

$$\mathscr{E}_c(t) \equiv 0. \tag{6.9}$$

This is the familiar boundary condition at the cathode for a simplified theory of SCL currents neglecting diffusion. Its use is justified so long as the reservoir of free carriers at the contact is sufficiently large that there will be a potential minimum in the insulator at the voltage V.

The time required for establishment of the potential minimum, namely, a time comparable to the ohmic relaxation time for the carriers in the reservoir, is so short as compared to the time scale of even the initial portion of the transient, which is a time of the order of t_0, that it can be considered instantaneous.

Our further analysis of the planar problem is confined to the time domain $0 \leqslant t \leqslant t_1$, where t_1 is the time of arrival at the anode of the leading front of charge. Over this time interval the conduction current j_{cond} in (6.1), *evaluated at the anode*, is identically zero, since injected carriers have not yet arrived at the anode. Thus, Eq. (6.1), taken at the anode, combined with (6.5) and using (6.9), gives the differential equation

$$d\mathscr{E}_{\text{a}}/\mathscr{E}_{\text{a}}^2 = (\mu/2L)\,dt, \tag{6.10}$$

with solution

$$\frac{L}{V} - \frac{1}{\mathscr{E}_{\text{a}}} = \frac{\mu t}{2L}; \qquad \mathscr{E}_{\text{a}} = \frac{V}{L}\,\frac{1}{[1-(t/2t_0)]}, \tag{6.11}$$

since, at $t = 0$, $\mathscr{E}(x,0) \equiv V/L$, there being no injected charge in the bulk of the insulator. From (6.3), the total injected charge $Q(t)$ in the insulator at time t is given by

$$Q(t) = \epsilon\mathscr{E}_{\text{a}} = \frac{Q_0}{1-(t/2t_0)}; \qquad Q_0 = \frac{\epsilon V}{L}. \tag{6.12}$$

Now, using (6.11) in (6.1), taken at the anode, we obtain

$$j(t) = j_0\,\frac{1}{[1-(t/2t_0)]^2}; \qquad j_0 = j(0) = \frac{\epsilon\mu V^2}{2L^3}. \tag{6.13}$$

It remains to calculate t_1. This is easily done utilizing the fact that the field at the leading front of charge at time $t < t_1$ is the same as the field at the anode at time t, namely $\mathscr{E}_{\text{a}}(t)$. Thus,

$$L = \mu \int_0^{t_1} \mathscr{E}_a\,dt = (2\mu t_0 V/L)\ln[1 - (t_1/2t_0)]^{-1}$$

using (6.11). From the definition of t_0, this gives finally

$$t_1 = 2(1 - e_0^{-1/2})\,t_0 \simeq 0.786 t_0. \tag{6.14}$$

Letting $j_1 = j(t_1)$ and $j_\infty = j(\infty) = 9\epsilon\mu V^2/8L^3$, the trap-free, steady-state SCL current, Eq. (6.13) gives, using (6.14),

$$j_1/j_0 = e_0 \simeq 2.72; \qquad j_1/j_\infty = 4e_0/9 \simeq 1.21. \tag{6.15}$$

From (6.11) and (6.14), and the fact that $\mathscr{E}(L, \infty) = 3V/2L$, it follows that

$$\frac{Q_1}{Q_\infty} = \frac{\mathscr{E}_a(t_1)}{\mathscr{E}_a(\infty)} = \frac{2e_0^{1/2}}{3} \simeq 1.10; \qquad \left.\frac{dj}{dt}\right|_{t_{1,-}} = e_0^{1/2}\frac{j_1}{t_0} \simeq 1.65\frac{j_1}{t_0}, \quad (6.16)$$

where $Q_1 = Q(t_1)$, $Q_\infty = Q(\infty)$, and the notation $t_{1,-}$ means that the derivative is taken as t approaches t_1 from below.

The above development is due, independently, to Many *et al.* (1961) and Helfrich and Mark (1962a). In addition, the latter authors show, by some further algebra, that

$$(dj/dt)_{t_{1,+}} = (dj/dt)_{t_{1,-}} \times (1 - 2e_0^{-1/2}) \times (1 - e_0^{-1/2})^{-1} \simeq -0.90j_1/t_0,$$

where subscript $t_{1,+}$ indicates that the derivative is taken as t approaches t_1 from above. Thus the plot of current versus time exhibits a cusp, or discontinuity of slope, at time $t = t_1$, as shown in Fig. 6.1. Mathematically, a discontinuity in slope is expected because, differentiating (6.5) with $\mathscr{E}_c = 0$, $dj/dt \propto \mathscr{E}_a\, d\mathscr{E}_a/dt$. But, from (6.4), a discontinuity in $\partial\mathscr{E}/\partial x$ across the moving front must be accompanied by a "compensating" discontinuity in $\partial\mathscr{E}/\partial t$ across the moving front, and obviously $\partial\mathscr{E}/\partial x \propto n$ is discontinuous across this front. To close the argument, at time t_1, the leading front has just reached the anode, whence $\partial\mathscr{E}_a/\partial t$ has a discontinuity at time t_1. That the discontinuity is such that the slope of $\mathscr{E}_a$ changes sign is also to be expected, since at time t_1 the current has overshot its final, steady-state value by 21%, from (6.15). (Clearly, if diffusion were included in the problem, the cusp would be rounded off, since an infinitely sharp leading front of charge would not be maintained in the face of diffusive spreading.)

A complete solution of the problem, for all times t, has been given in the important theoretical paper of Many and Rakavy (1962). These authors also present analytical and numerical solutions for the problems of transient SCL currents in a conducting, trap-free crystal and in an insulating crystal with traps. Figure 6.1 is taken from their paper. In the dimensionless plots in this figure, the unit of current density is $2j_0$, the unit of time is t_0, and τ is the trapping time in units of t_0. Because of the mathematical intricacies of these problems, we do not discuss their paper further here. However we do note the very important theoretical prediction that the cusp-like, current overshoot persists in the presence of trapping provided that the trapping is not too fast, i.e., provided that $\tau > 1$, and, further, the peak in the current occurs at the same time t_1. Likewise the initial current density j_0 is independent of trapping time, with virtually no restrictions on τ.

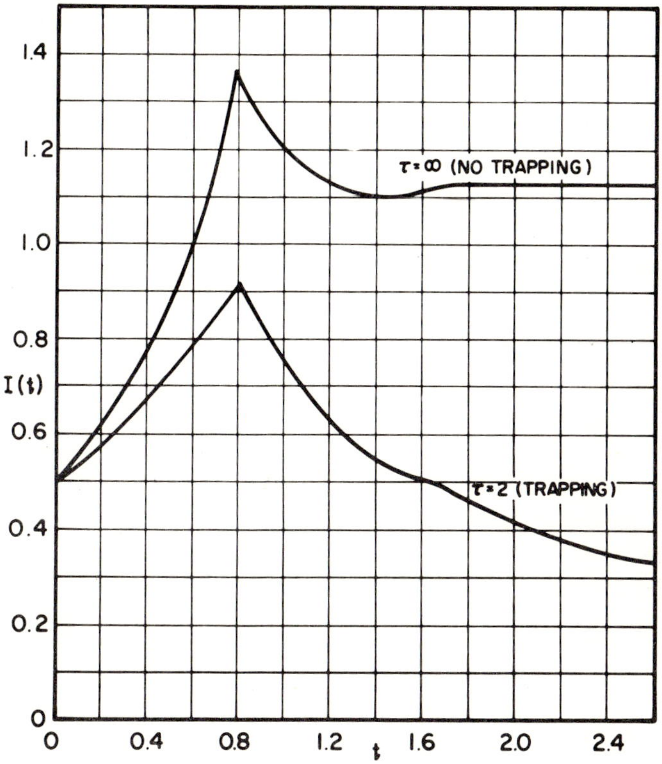

FIG. 6.1. Computed transient current response [current $I(t)$ versus time t] for one-carrier SCL current injection; linear plot with I and t measured in suitable normalized units. Upper curve is for a trap-free insulator, lower curve includes trapping effects with trapping time $\tau = 2$ in normalized time units. (From Many and Rakavy, 1962, Fig. 5.)

Before leaving the theoretical aspects of transient SCL currents, we should like to discuss the main features of the SCL current transient from a physical point of view. In the discussion of the SCF transient, we saw that the injected carriers traversed the specimen as a more or less well defined sheet; they are generated in this manner and maintain this property during transit because the total charge Q in the sheet, per unit area, is small compared to the maximum charge allowed by the capacitance and is therefore quickly removed from the contact regions. In the present SCL case, as the charge advances into the insulator behind the leading front at $x_1 = x_1(t)$, the entire space between the cathode and x_1 is filled with space charge, because now there is a reservoir at the cathode that can meet the demands for charge imposed by the capacitance of the specimen. As t increases, the capacitance C of the system keeps increasing with time as the "mean" displacement of the injected charge cloud from the anode keeps shrinking. Between the leading front and the anode there is no charge and so the field is constant: $\mathscr{E}(x) = \mathscr{E}_a$ for $x_1 \leqslant x \leqslant L$. Thus the field pulling on the leading front of charge is also the field at the anode at the same instant. These are the dominating influences on the behavior up to time t_1 of arrival of the leading front at the anode.

At very early times, $t \approx 0$, the leading front has advanced very little into the crystal, $x_1 \approx 0$, so that the capacitance has deviated negligibly from its geometric value C_0 and the total injected charge (per unit area) is $Q = C_0V = \epsilon V/L$. Each injected carrier contributes a current $ev/L = e\mu\mathscr{E}/L$ since v/L is the number of transits per second the carrier would make at drift velocity v. At very early times, the injected charge cloud is still quite uniform in density; therefore, the average field *inside* the cloud is $V/2L$, since $\mathscr{E}(0) = 0$ and $\mathscr{E}(x_1) = \mathscr{E}_a = V/L$. Thus the total current density j_0 at very early times is: $j_0 = Q\mu V/2L^2 = \epsilon\mu V^2/2L^3$. This explains the first conspicuous feature of the current transient, quantitatively as well as qualitatively.

As injection proceeds in time, and the capacitance concomitantly increases, the field ahead of the leading front of charge, namely $\mathscr{E}_a$, keeps growing in excess of the ohmic field $\mathscr{E}_{a,0} = V/L$. Since the transit time t_1 of the leading front is determined by $\mathscr{E}_a(t)$, t_1 is necessarily shorter than the space-charge-free transit time $t_0 = L^2/\mu V$. This explains qualitatively a second feature.

As for the current overshoot, this is easily understood as follows. The total injected charge Q_1 at time t_1 is some 10% larger than the steady-state injected charge, from (6.16). This circumstance derives from a more uniform concentration of charge throughout the crystal at time t_1 than in the steady state, namely relatively lower concentration near the cathode and higher concentration near the anode. As we have already noted, the charge concentration is very uniform, up to the leading front, at early times $t \ll t_1$. This uniformity of charge concentration up to the leading front, except very close to the cathode, persists up to time t_1. [See Fig. 4 of Many and Rakavy (1962).] On the other hand, the charge concentration at $t = \infty$ is monotonically decreasing between cathode and anode, namely $n \propto x^{-1/2}$. Greater uniformity of charge distribution in this case obviously is equivalent to higher capacitance: $C_1 = C(t_1) > C_\infty = C(\infty) = 3\epsilon\backslash 2L$ and $Q_1 = C_1V > C_\infty V = Q_\infty$. Further, although the *average* transit time $\bar{t}_1$ of *all* the injected charge Q_1 at time t_1 will certainly be longer than $t_1 \simeq 0.8t_0$, it is reasonable to expect that $\bar{t}_1 < t_\infty = 4t_0/3$, the transit time in the steady state from (4.14). Finally with $Q_1 > Q_\infty$ and $\bar{t}_1 < t_\infty$, it follows that $j_1 = Q_1/\bar{t}_1 > Q_\infty/t_\infty = j_\infty$.

The existence of the cusp; i.e., the discontinuity of slope at t_1, is an artifact associated with the unrealistic, infinitely sharp leading front of charge. If diffusion had been included in the theoretical treatment it would obviously have caused a "smearing out," somewhat, of the sharp leading front and a rounding of the cusp.

Following the arrival of the leading front, the current, which has overshot, must start decaying toward its steady-state value. The time scale of the decay will be the ohmic relaxation time for the charge injected at time t_1, namely $\epsilon/e\bar{n}_1\mu = \epsilon L/Q_1\mu \simeq \epsilon L/Q_0\mu = L^2/\mu V = t_0$. The slight undershoot seen in the predicted decay is due to the inhibition of charge injection around time t_1 because of the overshoot of charge already present at that time. The analytical problem of approach to the steady state following the cusp has been investigated by Schilling and Schachter (1967c) who demonstrate the existence of decaying current oscillations in this region of the transient. The related problem of the SCL current transient following a perturbing step voltage increase superposed on an already existing steady-state voltage has also been investigated by Baron *et al.* (1966) and by Schilling and Schachter (1967c). Here too, decaying oscillations in the SCL current transient are obtained.

6.1.3 *Other Cases*

We consider now a case intermediate between the two extremes examined above, namely one for which there is a fixed amount of charge Q initially available in the insulator and such that $QL < 2\epsilon V$ but is of comparable magnitude. This case is important, for example, if the injected carriers are produced by a light flash and either the flash intensity is not quite strong enough to satisfy the condition for SCL current flow: $QL > 2\epsilon V$, or the quantum yield for photogeneration of carriers is itself field dependent. In the latter case, investigated by Schwartz and Hornig (1965) for anthracene, Q at the cathode depends both on the flash intensity and on $\mathscr{E}_c$, and it may turn out that the condition for SCL current flow cannot be satisfied at any illumination level.

It is assumed that Q is initially created as a sheet of charge right at the cathode. The current transient is obtained by equating the general equation for the time-dependent current (6.6), in terms of the field at the anode, with the displacement part of the total current (6.1), also taken at the anode, and integrating, with Q taken as a constant, independent of time. This yields:

$0 \leqslant t \leqslant t_2$:

$$j(t) = \frac{Q\mu V}{L^2}\left[1 - \frac{QL}{2\epsilon V}\right]\exp\left(\frac{QL}{\epsilon V}\frac{t}{t_0}\right), \tag{6.17}$$

where t_2, the time of arrival of charge at the anode, is given implicitly by

$$\frac{QL}{2\epsilon V}\frac{t_2}{t_0} = 1 + \left[\frac{1}{2} - \frac{\epsilon V}{QL}\right]\left[\exp\left(\frac{QLt_2}{\epsilon V t_0}\right) - 1\right]. \tag{6.18}$$

The condition $QL/2\epsilon V < 1$ assures that the exponential in (6.17) is always less than 2 and that $t_1 \leqslant t_2 \leqslant t_0$, where t_1 is given by (6.14). It is clear that the total charge passing through the sample in the interval 0 to t_2 is always less than $2C_0V$. Further, Eq. (6.17) reduces to (6.7) and $t_2 \to t_0$ in the limit of SCF current flow: $QL/2\epsilon V \to 0$, while at the mathematically permissible maximum $QL/2\epsilon V = 1$, Eqs. (6.17) and (6.18) agree precisely with their SCL counterparts (6.13) and (6.14), respectively. When $QL/2\epsilon V$ is only slightly less than unity, the shapes of (6.17) and its SCL counterpart (6.13) are qualitatively very similar, as shown in Fig. 6.2, where (6.17) has been plotted with $QL/2\epsilon V = 1$. However, an important quantitative difference is that the total charge passed through the sample in the present case is always less than $2C_0V$. It is useful to note other quantitative differences. Although (6.13)

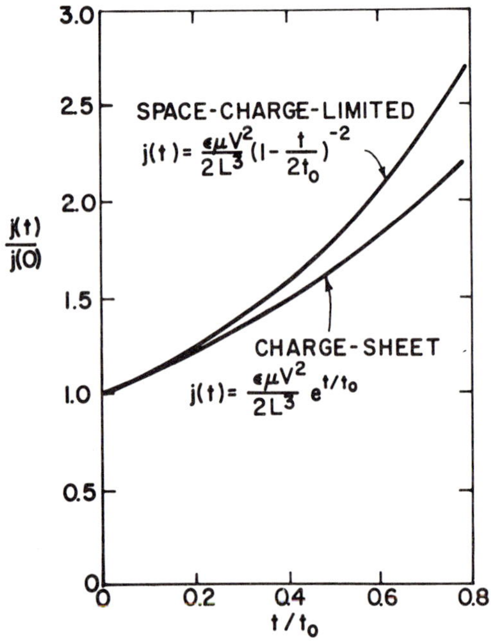

FIG. 6.2. Normalized current transient for the two cases: complete space-charge limitation (injected charge distributed) and injected charge sheet. (From Schwartz and Hornig, 1965, Fig. 1.)

requires that j_0 be simultaneously independent of the light flash intensity and proportional to V^2, Eq. (6.17) shows a more varied behavior: $j(0)$ will depend on the flash intensity if $QL/2\epsilon V < 1$ but this dependence becomes weaker as $QL/2\epsilon V \rightarrow 1$. Also, $j(0)$ will transform from its linear dependence on V to the quadratic dependence on V as the inequality $QL/2\epsilon V < 1$ approaches equality.

A very similar behavior results when the optically generated contact reservoir becomes exhausted either within the first transit time or shortly thereafter (Trester *et al.*, 1966; Papadakis, 1967; Weisz *et al.*, 1968). So long as the reservoir exists for a time interval $t_A \geqslant t_1$, the SCL current transient prior to the cusp, $0 < t < t_1$ is unaffected, but for $t > t_1$ the current falls to zero within 3 to 4 t_1-intervals rather than to the steady-state SCL current when $t_A = \infty$. For $t_A < t_1$, the initial current and the location of the cusp are unaffected but the cusp magnitude decreases and the subsequent decay of the current to zero is accelerated. In the limiting case $t_A = 0$, $J(t_1) = 2.25\, j_0$ and the transient vanishes at $t = 1.7\, t_0$.

It should perhaps be reiterated here that SCL current transients are always obtained if the sample is provided with a contact that is injecting in the steady state and if the transient is triggered by a step voltage.

It is worth noting that there are circumstances where it is possible to obtain a cusp-like transient with photo-injected carriers and still have $QL/2\epsilon V \ll 1$. This can occur if the sample is provided with injecting contacts and if the voltage is applied to the sample well before the light pulse. The action of the voltage prior to the light pulse is to establish a trap-limited, steady-state SCL current that may be many decades smaller in magnitude than the subsequent transient, and thus may escape detection. This steady-state SCL current, even though it may be imperceptible, will nevertheless establish the field distribution in the sample via the nonuniform spatial distribution of the injected trapped charge. Regardless of the energy distribution of the traps, the field will always increase towards the anode. For example, if the trapping is almost everywhere shallow, then $\mathscr{E}(x) \propto x^{1/2}$. If now a small amount of charge (much less than $\epsilon V/L$) is released by a light flash, the ensuing charge sheet will be accelerated, by the nonuniform field, on its way to the anode and the current will therefore increase with time. Thus, rather than obtaining the flat transient of the SCF current flow, which is flat because the field is constant, one obtains a rising current instead. The sheet arrives at the anode at the (transit) time t_r and leaves the crystal, following which the current rapidly decays toward its steady-state value. (Obviously we have assumed that $t_r \ll \tau$, the trapping time.) Thus, here one obtains a cusp-like transient because of the nonuniform field distribution enforced by the space charge that was trapped prior to the release of charge by the incident light pulse. An "artifact" cusp, such as this, can be avoided by working with noninjecting contacts, if photoinjection is contemplated.

6.2 Information Obtainable from Transient Injection-Current Measurements

Since the SCL current transient has more structure than the SCF current transient, it naturally has a higher information content. All the information disclosed by the latter is also revealed in the former. Therefore, in what follows, we shall confine our attention almost entirely to the SCL current transient and shall comment only briefly on the SCF current transient, where appropriate. We shall also assume that the numerous criteria, that were mentioned in the last section, for recognizing the various types of current transients are borne in mind.

If an oscilloscope trace of current versus time exhibits a current overshoot, then the location-in-time t_1 of the current peak directly measures the free-carrier drift mobility, namely, from (6.8) and (6.14),

$$\mu = \frac{0.786\, L^2}{t_1 V} \quad \frac{\text{cm}^2}{\text{V sec}}, \tag{6.19}$$

with L in centimeters, t_1 in seconds, and V in volts. This determination of μ has the same overriding virtue as the SCF transient technique. It requires no knowledge whatever of any material parameters of the crystal. Further, the current overshoot is a relatively large, easily identifiable effect which lends fair precision to the determination of μ, say to within about 10%. As we shall see shortly, this technique has been successfully used to determine hole mobilities in iodine, anthracene, and sulfur.

The proportionality of t_1 to $1/V$ presents a good check on the interpretation of the measurement. So likewise does the proportionality of the magnitude of the entire transient to V^2. For the SCF case, the transient is proportional to V. Thickness dependence measurements of the initial current amplitude j_0, at fixed voltage, also furnish a good check. Where the injecting contact is photogenerated by an incident light flash, the independence of j_0 of the light intensity is strong evidence that a sufficient reservoir of free carriers has been generated to support SCL current flow into the insulator.

Having obtained μ from (6.19), the initial current density j_0 gives a determination of the relative static dielectric constant, namely, from

$$\kappa = 2.25 \times 10^{13}(J_0 L^3/\mu V^2), \tag{6.20}$$

with j_0 in amperes per centimeter squared, L in centimeters, μ in cm²/V sec, and V in volts. Actually, because of some sensitivity of j_0 to the detailed nature of the contact, only a factor-of-2 accuracy is available from this measurement. Since κ may be measured with much better accuracy, say by capacitance techniques, Eq. (6.20) may better be regarded as an additional check of μ where κ is known.

If the insulator is anisotropic, then the components of the diagonalized mobility (and dielectric-constant tensors) can be determined from (6.19) and (6.20) by measuring the transient SCL currents along the principal axes. Such orientation-dependent measurements have been made on anthracene and iodine.

Where the observed SCL current transient resembles the "$\tau = 2$" curve of Fig. 6.1, curve-fitting with τ as an adjustable (dimensionless) parameter can determine the trapping time $t_t = \tau t_0$. Self-consistency can be achieved by fitting several current transients at different voltages: Since $t_0 \propto 1/V$, τ will be different for each curve, but the product $\tau t_1 = t_t t_1/t_0$ should be the same for all curves. If the concentration of traps N_t can be obtained from steady-state, dc SCL current measurements on the same sample, as discussed in Section 5.2, then this information can be combined with the transient SCL current determination of τ to obtain the cross section for capture of the free carriers by the traps.

For long trapping times, $\tau \gg 1$, the current transient will appear to fit the no-trapping ($\tau = \infty$) curve on a short time scale, say $t < 5t_1$. However the current decay will then be apparent on a time scale comparable with τ. This is seen in measurements on iodine as shown in Fig. 6.10, (a) representing the transient on the short time scale, (b) on a much longer scale and at several voltages. Two trapping times are recovered from the latter plots by curve fitting.

Criteria for the observation of transient SCL currents in insulators, based on the above considerations, are given by Silver *et al.* (1962a).

It is also possible to apply the techniques of transient injection currents to study photoconductivity processes. When the injected carriers are photogenerated, there are two processes that compete for the photocarriers: extraction by the field into the bulk of the specimen and recombination within the photoexcited contact region. For a given illumination level (absorbed photon flux density), Q, in the contact region, depends on the recombination rate. The nature of the transient, whether it is SCF or SCL, also depends on Q. It is clear that photoconductivity recombination rates can be investigated by observing the transition from SCL to SCF transient current flow as a function of the illumination. Such measurements have been performed by Simhony and Gorelik (1965) on iodine crystals.

6.3 Materials Studied with the Transient Injection-Current Technique

The first unambiguous application of these techniques to the study of charge transport in solids is, apparently, Brown's work (1955) with AgCl, which made use of the SCF method described at the end of Section 6.1.1, in which the current transient is integrated. Since then, as the various other transient techniques were developed, a variety of materials have come under investigation. In the following review, we shall limit detailed consideration to the original earlier work and shall mention only briefly more recent applications.

6.3.1 *Silver Chloride*

As mentioned, the first application of injection currents to the study of carrier drift mobilities in solids appears to be the measurements of Brown (1955) on electrons in AgCl, using the SCF transient method discussed at the end of Section 6.1.1. Brown measured the transit time, through crystals approximately 5 mm thick, of secondary electron pulses released by individual 2 MeV β-particles incident on the surface of the crystals. The orientation of the samples was not specified. At $-187°C$, the electron drift mobility $\mu_n = 274\ cm^2/V\ sec$ and varied as $\mu_n = 2.54 \times 10^5 T^{-3/2}$ from -187 to 77°C, indicating acoustic scattering.

6.3.2 *Selenium*

The next step in the development of this technique is due to Spear (1957) who improved upon Brown's method of carrier injection by replacing the β-source with an electron gun. Spear's experimental arrangement is schematically illustrated in Fig. 6.3. The electron gun gave pulses of 0.1 μsec duration which were incident on the specimen, in this case evaporated vitreous Se films 2–12 μ thick, through an evaporated gold electrode. The sample was in series with a dc voltage source and an amplifier, the latter being shunted by an integrating capacitor. Spear measured the transport of both holes and electrons by suitably choosing the polarity of the applied voltage. From a direct observation of the transit time of the injected carriers, Spear found that the drift mobility of electrons, μ_n, ranged from 4.7×10^{-3} to 5.5×10^{-3} $cm^2/V\ sec$ at room temperature and that the temperature variation was described by an exponential, $\exp(-E_a/kT)$, with an activation energy E_a of 0.25 eV. Similarly, for holes, μ_p was between 0.1 and 0.2 $cm^2/V\ sec$ with an activation energy of 0.16 eV. The temperature ranged from -30 to 42°C. Spear attributed the low values of μ_n and μ_p, as well as

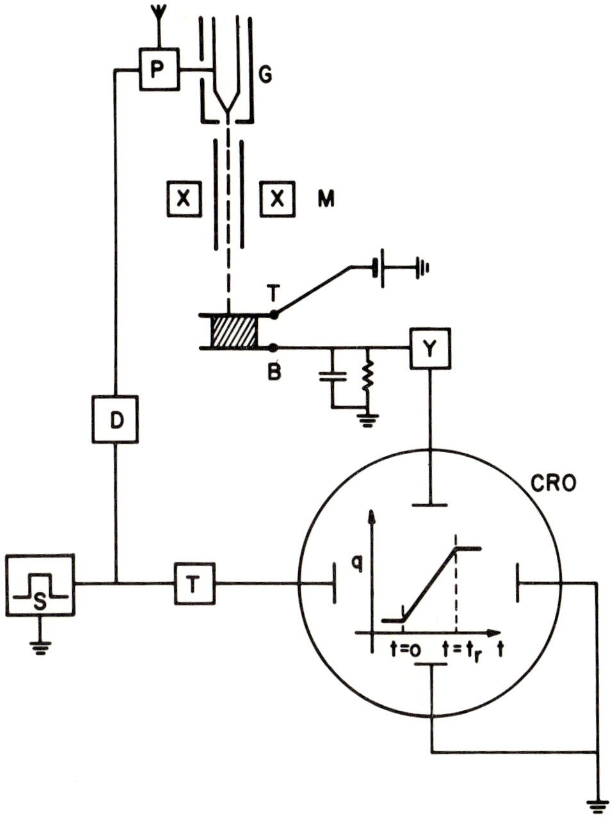

FIG. 6.3. Experimental arrangement used by Spear (1957) for study of carrier injection into selenium using an electron gun.

their temperature dependences, to a trap dominated drift mechanism suggested earlier by Rose (1951). In a later paper, Spear (1960) improved the accuracy of μ_p and quotes a range between 0.13 and 0.14 cm^2/V sec and an activation energy of 0.14 eV.

Spear (1961) has also studied charge transport in crystals of α-monoclinic selenium with the same technique. The transport of both electrons and holes were investigated perpendicular to the (101) plane. These transit time measurements led to an electron mobility $\mu_n \simeq 2$ cm^2/V sec at 21°C. Above 0°C, $\mu_n \propto T^{-3/2}$; but with decreasing T, a sharp transition was evident from the lattice-dominated mobility to a trap-controlled mobility with an activation energy of 0.25 eV, the same value obtained for vitreous Se. The hole mobility was found to be trap-controlled from -120 to 110°C.

6.3.3 *Cadmium Sulfide*

Mort and Spear (1962) and Spear and Mort (1963) have also studied μ_n and μ_p in single crystal CdS over the temperature range -190 to 230°C with the same technique used to study Se. Above -110°C, μ_n agrees well with the Hall mobility. The mobility is governed by acoustic scattering and obeys the relation $\mu_n = 1.28 \times 10^6 T^{-3/2}$. Below -110°C, they observed a transition to a transport mechanism predominantly controlled by a level 0.049 eV below the conduction band. The hole drift mobility μ_p was also determined directly, by the same technique, for the first time in this material. Hole lifetime was also measured. At room temperature, μ_p ranged from 10 to 18 cm²/V sec with a mean value of about 15 cm²/V sec. The hole lifetime in their crystals ranged from 0.1 to 0.3 μsec at room temperature. The temperature dependence of μ_p was measured in the range -190 to 230°C. Coming down in temperature, the mobility reached a maximum at about -63°C, below which it decreased exponentially with temperature with an activation energy of 0.019 eV.

6.3.4 *Anthracene: SCF Transient Currents*

The first thorough investigation of the charge transport properties of an organic molecular crystal by transient injection currents is that of Kepler (1960, 1962). He studied both electron and hole transport in anthracene by the SCF technique, observing the current transient directly. Anthracene is a molecular-type (van der Waals binding) crystal in which the unit cell is monoclinic and contains two molecules. It has a well defined cleavage plane, the *ab*-plane, which is the base plane of the monoclinic cell. Kepler's measurements were made with the carriers moving either parallel or perpendicular to this plane.

A schematic diagram of the measuring circuit used by Kepler is shown in Fig. 6.4; it is self-explanatory. The crystal itself was isolated from the conducting glass electrodes by means of thin (1.4 mm) Mylar sheets. The transient was triggered by a light pulse 1–2-μsec long, incident on the crystal through one of the transparent electrodes. Light of wavelength shorter than the absorption edge was used and it did not penetrate the crystal more than 1 μ. The crystals were all 1 to 2 mm thick. Normally, anthracene exhibits only photoconduction by holes since electrons are rapidly trapped by impurities. By taking great pains in purifying his material, Kepler was able to study the transport of both electrons and holes. The sign of the carriers comprising the current pulse corresponds to the polarity of the illuminated electrode.

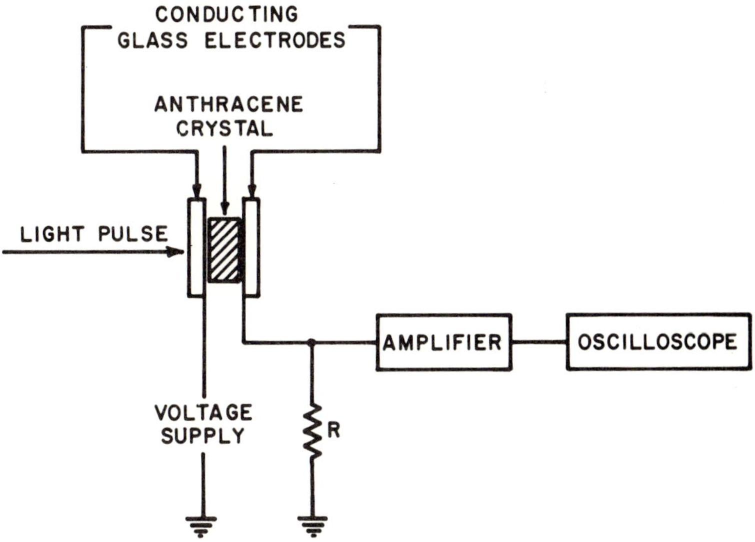

FIG. 6.4. Schematic diagram of circuit used by Kepler (1960) to study the SCF current transient in anthracene.

Figure 6.5 shows a typical SCF transient current pulse. Each horizontal division is 50 μsec so that the pulse is over 300 μsec long. The slow rise of the current at the beginning of the pulse is caused by the time constant of the measuring circuit while the long tail is a measure of the spreading of the carrier sheet owing to diffusion. Initially, the carrier sheet is rather well defined. The ratio of the pulse duration to the transit time is about 5×10^{-3} so that the initial sheet width is about 7 μ.

Kepler (1962) examined the temperature, pressure, and orientation

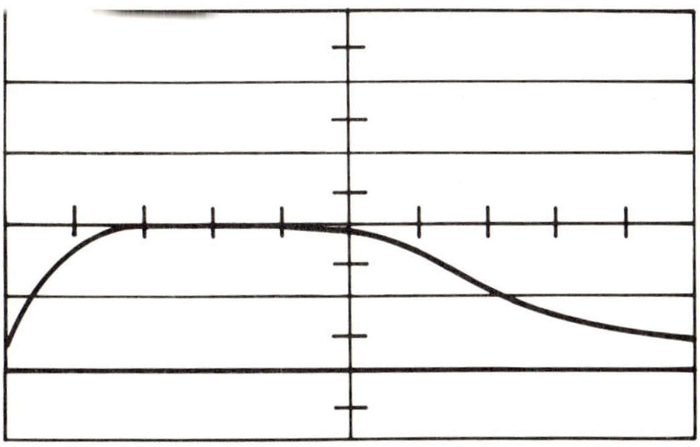

FIG. 6.5. Typical SCF current transient observed by Kepler (1962, Fig. 4) in anthracene.

dependence of the mobility for both carrier types. Altogether, this work provides a fine demonstration of the power of the injection-current technique in obtaining information about the electronic properties of a material. Figure 6.6 shows the temperature dependence of the drift mobility for both electrons and holes, in a direction perpendicular to

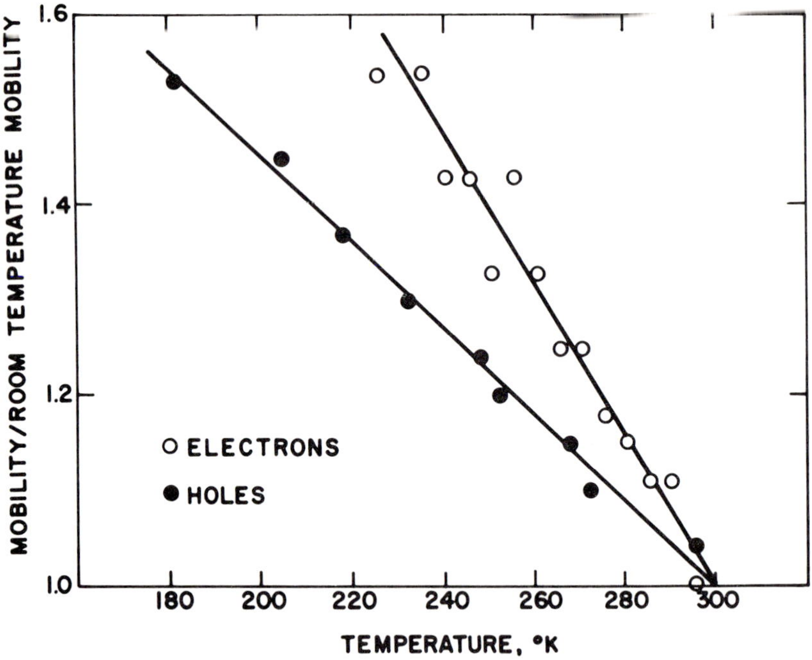

FIG. 6.6. Temperature dependence of the normalized drift mobilities for electrons and holes in anthracene, obtained by SCF transient current measurements. (Taken from Kepler, 1962, Fig. 5.)

the *ab*-plane and normalized to the room temperature mobilities which are 0.3 and 0.4 cm^2/V sec for electrons and holes, respectively. Similar dependences are observed along other orientations. The orientational anisotropies of the electron and hole drift mobilities are shown on Fig. 6.7 in the *ab*-plane.

6.3.5 *Anthracene: SCL Transient Currents*

Measurements of SCL current transients in anthracene have been made by Helfrich and Mark (1962a) using steady-state ohmic contacts, and by Silver *et al.* (1962b) using optically generated carriers. In both

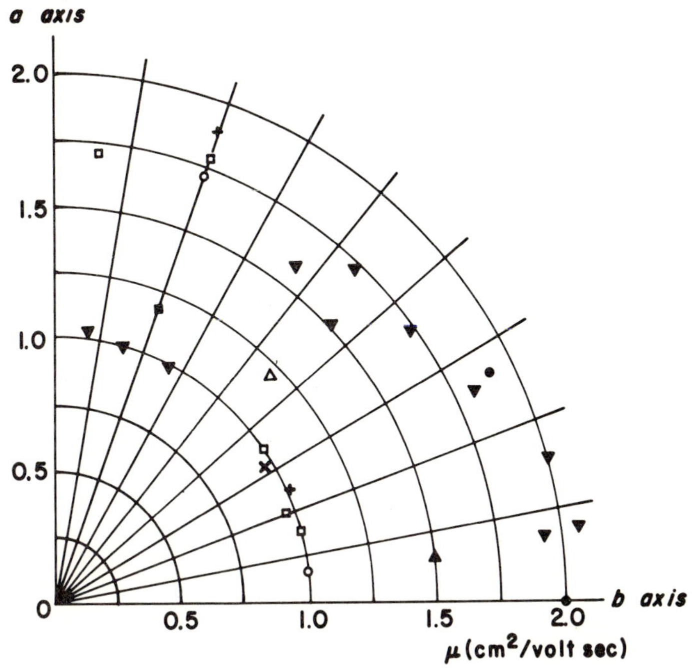

FIG. 6.7. Anisotropy of the electron and hole drift mobilities in anthracene, obtained by SCF transient current measurements. (Taken from Kepler, 1962, Fig. 17.)

instances, only hole transport was studied. We discuss the former measurements first.

The problem of resolving the transient SCL current injected from a steady-state contact by a step voltage is complicated by the fact that the measuring circuit must have a time constant appreciably smaller than t_0, and that a substantial capacitive current pulse accompanies the application of the voltage to the insulating specimen. These problems were overcome by using the balanced circuit shown in Fig. 6.8. On closing the mercury relay S, the voltage $+V/2$ appears at P_2 and $-V/2$ appears at P_1. The point P_2 is connected to the injecting contact I, which was a saturated solution of iodine in 1 M KI, and to one beam of a double beam oscilloscope S_1. The other electrode NI of the specimen was a 1 M KI solution which is noninjecting for both carrier types. The point P_1 is connected to a variable capacitor. The internal resistance of the battery is shunted by the large parallel capacitance so that the full voltage develops at P_1 and P_2 in less than 10^{-7} sec. The current through the anthracene $[C_P]$ is obtained by measuring the voltage developed

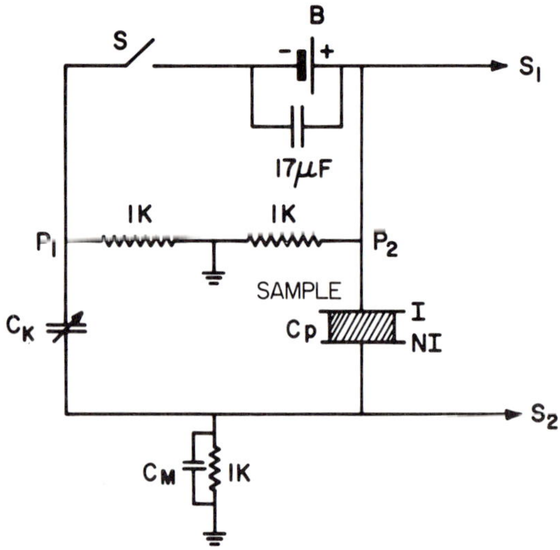

FIG. 6.8. Schematic diagram of circuit used by Helfrich and Mark (1962a) for the study of SCL current transients in anthracene.

across the load resistance with the second beam of the oscilloscope connected at S_2. C_M represents the unavoidable stray capacitance of the cables. The capacitance transient due to C_P is compensated by the equal and opposite transient through C_K which is adjusted to equal C_P.

A typical transient is shown in Fig. 6.9a for a 102-μ thick anthracene crystal with current flow perpendicular to the *ab*-plane. The sweep is 5 μsec/div. The upper trace is the voltage. It is practically horizontal and shows that the voltage rises in less than 10^{-7} sec. The lowest trace is the current with reverse bias and represents the residual capacitor-charging transient that cannot be bucked out because of stray inductances in the circuit. The middle trace is the SCL current transient obtained with forward bias. Trapping causes the current to decay below its steady-state trap-free value. The same measurement was repeated at different voltages and the transit time was plotted against the reciprocal voltage. This is shown in Fig. 6.9b for a 10-μ thick crystal. From the slope of the curve, one obtains $\mu_n = 0.38$ cm^2/V sec while the magnitude of the current, which varied as V^2, yields $\mu_n = 0.4$ cm^2/V sec. These values are in excellent agreement with Kepler's results. Steady-state measurements were also made that confirmed the existence of trap-limited SCL current flow. In the later work of Helfrich and Schneider (1965), SCL current transients for both electrons and holes were observed with steady-state ohmic contacts for both carrier types. These were nonaqueous solutions containing anthracene ions. The ohmic anode contained positive anthracene ions and was obtained by adding $AlCl_3$ to a solution of anthracene in nitromethane. The ohmic cathode, which contained negative anthracene ions, was prepared by interacting sodium with a solution of anthracene in tetrahydrofuran. Similar results have been obtained for holes in anthracene by Silver *et al.* (1962a) using noninjecting electrolytic electrodes and a light flash to produce the free holes.

The trapping mechanism can be examined by studying the decay of the transient after the cusp. Such an investigation has been performed on anthracene, with flash-generated holes, by Weisz *et al.* (1964). Two trapping processes were found, a fast trapping with $\tau = 150\ \mu$sec and a slower process for which $\tau \simeq 10$ msec. The authors demonstrated that the trapping with the short time constant can be reduced by annealing the crystal and consequently they associate this trapping with defects.

Hoesterey and Letson (1963) have studied the trapping of electrons and holes in anthracene by impurities using the SCF transient technique of Kepler. The impurities examined were anthraquinone, anthrone, and naphthacene in concentrations of 10 ppm or less. Anthraquinone and anthrone are more effective at trapping electrons than holes, possibly as a consequence of their higher electron affinities. Naphthacene acts as a hole trap located 0.43 eV above the hole band and has a capture cross section exceeding 10^{-15} cm^2 (Mark and Helfrich, 1962).

6.3.6 *Iodine*

Iodine single crystals were intensively investigated by Many and his collaborators. Iodine is a low melting point (114°C) molecular crystal belonging to the base-centered orthorhombic system. It has a resistivity of about $10^{11}\ \Omega$ cm at room temperature, which varies exponentially with temperature with an activation energy of about 1 eV. Only SCL hole currents have been observed in this material. Presumably there are too many electron traps with large electron capture cross section (hence fast trapping times) to permit the observation of even transient electron SCL currents under the experimental conditions employed up to the present. The hole current was released by a flash of strongly absorbed light from either a xenon or a mercury discharge tube.

In the earlier work (Many *et al.*, 1961, 1962), all measurements were made perpendicular to the ac crystallographic plane. With some crystals, the slow-trapping type of current transient was seen, and with others, the fast-trapping type, resembling the "$\tau = 2$" curve of Fig. 6.1. An example

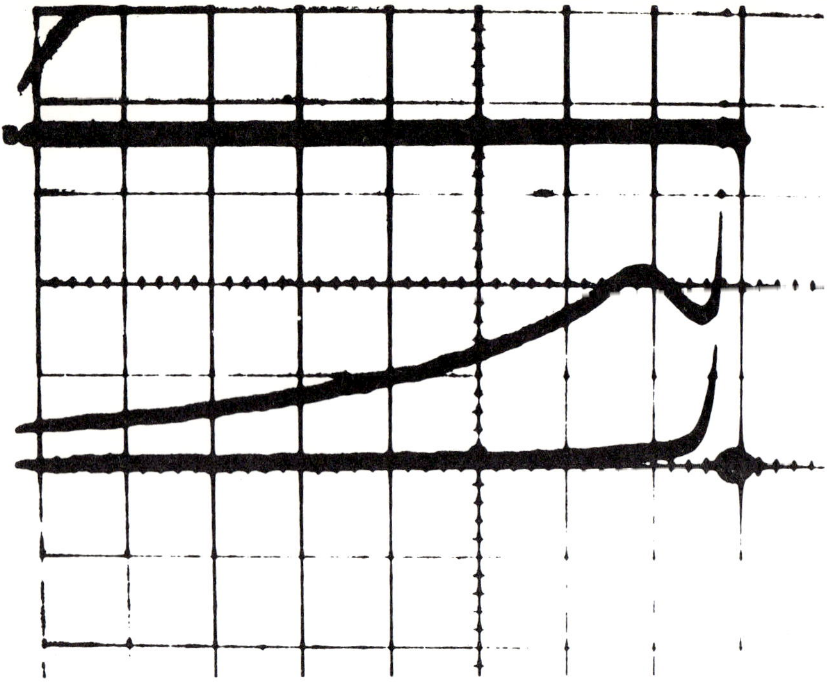

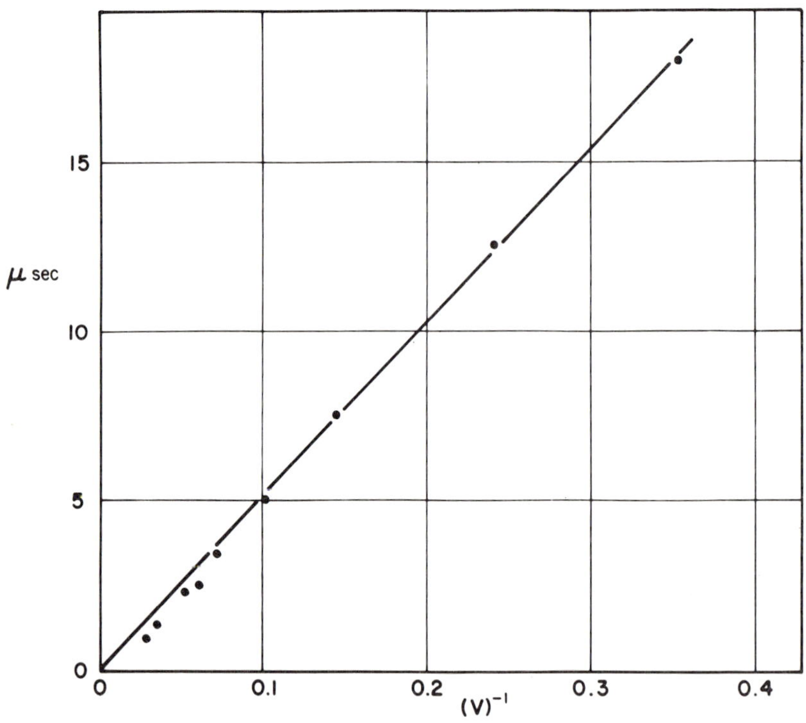

μ sec
15
10
5
0
0
0.1
0.2
0.3
0.4
(V)$^{-1}$

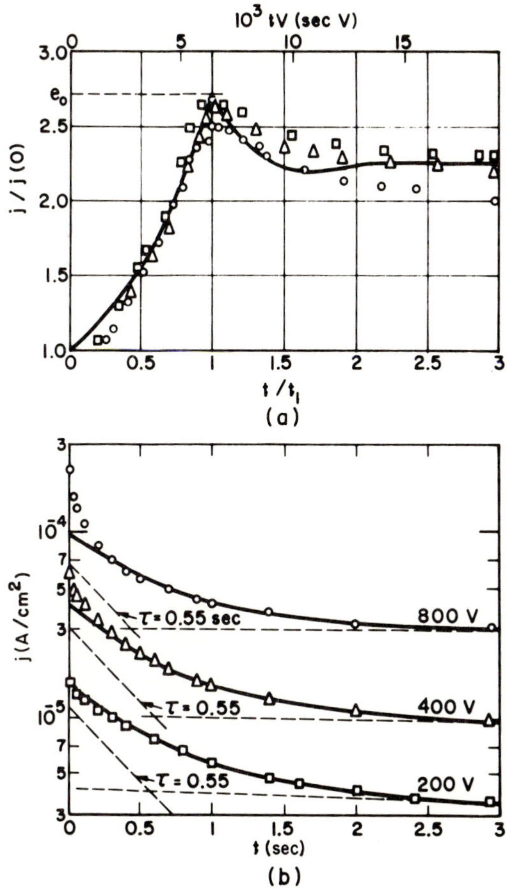

FIG. 6.10. SCL current transients in an iodine crystal ($L = 7.2 \times 10^{-2}$ cm): (a) transient on a "short" time scale; open squares are experimental points taken at 200 V, open triangles at 400 V, and open circles at 800 V; current is normalized to $J_0 = \epsilon\mu V^2/2L^3$, time to $t_1 = 0.786L^2/\mu V$; the solid curve is the theoretical, trap-free transient; (b) transients on a "long" time scale at three voltages, with data points as in (a); solid curves are theoretical fits to the data the dashed lines are the individual decay components making up the solid curves. (From Many *et al.*, 1962, Fig. 8.)

FIG. 6.9. (a) (top) Typical SCL current transient observed by Helfrich and Mark (1962a, Fig. 4) in anthracene. Note that the origin for time is at the right-hand end of the tracing and that time increases to the left; (b) (bottom) transit time versus reciprocal voltage obtained from SCL current transient measurements in anthracene ($L = 10\ \mu$). (From Helfrich and Mark, 1962a, Fig. 6.)

of the former is seen in Fig. 6.10a where data from one sample, taken at three different voltages, are plotted in normalized form, namely as j/j_0 vs t/t_1. It is seen that the theoretical no-trapping ($\tau = \infty$) curve, the solid line, gives a rather good fit to the data. The room-temperature hole mobility obtained from the measured t_1's for this, and other iodine, samples, using (6.19), was $\mu_p = 0.7$ cm^2/V sec. The temperature dependence of the hole mobility was found to be $\mu_p \propto T^{-1/2}$.

On a much longer time scale, this same sample of iodine did exhibit slow-trapping decay, as shown in Fig. 6.10b. Here current density is directly plotted against time for three voltages. The data are well represented by theoretical curves, the solid lines, which are compounded out of two decay components, the dashed lines, one corresponding to a trapping time of 0.55 sec and the other to a trapping time on the order of minutes. Examples of experimental, fast-trapping current transients are given in Fig. 9 of the original (1962) paper. The corresponding trapping time for the sample studied there was 50 μsec.

Steady-state SCL current measurements were also reported (1962). As a check on the SCL current interpretation of their data, both steady state and transient, the authors studied the thickness dependence of the current density, at fixed voltage (100 V), with results shown in the log–log plot of Fig. 5.5. (Other checks of the transient SCL current data were also made.) For the transient data, the ordinate is the initial current density j_0. The solid line is a plot of the theoretical expression for j_0, namely (6.13), taking $\mu_p = 0.7$ cm^2/V sec and $\kappa = 2$, as independently measured in their earlier work (1961). The fit is seen to be excellent, though it will be changed slightly with the revised value of $\kappa_b = 3$ for this crystallographic direction. Because the observed steady-state SCL current follows a shallow-trap square law, Eq. (4.45) with an unknown θ, a theoretical curve cannot be drawn through the experimental points. Rather a best-fit line corresponding to $J \propto L^{-3}$, namely the dashed line, is drawn through the points. The large scatter of the data about this line is attributed to a variation of trap content among the different samples used for the measurements. The corresponding scatter is absent from the fit of the transient data to the theoretical (solid) line, because j_0 is not affected by trapping, which takes place later in time. From (4.45) and (6.13), the vertical displacement of the solid and dashed lines is precisely $0.44\theta^{-1}$, yielding a value for θ at room temperature, averaged over the samples studied: $\theta \simeq 0.02$. From the temperature dependence of the shallow-trap, square-law current at fixed voltage (essentially a plot of $\ln \theta$ vs $1/T$), measured on one crystal, a trap depth $E_t - E_v \simeq 0.6$ was determined. The corresponding trap concentration N_t was 5×10^{13} cm^{-3}, taking $N_v \simeq 10^{22}$ cm^{-3}. Finally combining these data with transient SCL current data, the hole capture cross section of this set of traps was estimated to be 10^{-20} cm^2 or lower.

In more recent work (Simhony, 1963; Many *et al.*, 1964), transient SCL currents were measured along the three principal crystallographic directions, *a*, *b*, and *c*, of the orthorhombic crystal. The mobility data obtained from these studies are summarized in Table 6.1. The data

TABLE 6.1

HOLE MOBILITIES IN IODINE[a]

Crystallographic direction	Room temperature values (cm^2/volt sec)	Temperature dependence in ranges		
		190–230°K	230–260°K	260–320°K
a	$\mu_a = 2.3 \pm 0.2$	$T^{2.2}$	Constant	$T^{-1.5}$
b	$\mu_b = 0.72 \pm 0.05$	$T^{-0.2}$	$T^{-0.2}$	$T^{-1.3}$
c	$\mu_c = 2.1 \pm 0.2$	$T^{0.5}$	Constant	$T^{-2.1}$

[a] From Many *et al.* (1964).

corresponding to the *b*-direction confirm the earlier (1962) work. The room temperature relative dielectric constants, as determined by capacitance measurements with a bridge circuit (Many *et al.*, 1964) are: $\kappa_a = 6 \pm 0.5$, $\kappa_b = 3 \pm 0.3$, and $\kappa_c = 40 \pm 2$. In the temperature range between 180 and 330°K, these values vary between 0.9 and 1.3 times the room temperature values. A plot of the temperature variation of κ_a, κ_b, and κ_c is given in Fig. 5 of the publication.

The lifetime of the holes generated by the light flash has been measured in iodine by observing the attenuation of the SCL current transient due to recombination in the photoexcited region (Simhony and Gorelik, 1965). In the samples studied, the lifetime ranged from 1 to 3 μsec in a surface layer parallel to the ac plane.

6.3.7 *Sulfur*

Charge transport in sulfur (S_8) crystals has been examined by Adams and Spear (1964) and Adams *et al.* (1964), using the integrated current SCF technique and by Many *et al.* (1965) utilizing SCL current transients released by a light flash.

Adams and Spear (1964) measured the drift mobility of holes perpendicular to the (111) plane of the orthorhombic sulfur crystal over a temperature range from 180 to 370°K. Their results showed that holes drift at room temperature with a lattice mobility of about

10 cm^2/V sec and possess a recombination lifetime of 10 μsec. At lower temperatures, the transport of holes is an activated process, dominated by trapping centers 0.19 eV above the valence band. Electron transport (Adams *et al.*, 1964) was also studied by the same method; at 19°C, the drift mobility is 5.5×10^{-4} cm^2/V sec and increases exponentially with an activation energy of 0.94 eV. Many *et al.* (1965) confirmed the room temperature electron drift mobility (6×10^{-4} cm^2/V sec) but found a rather different activation energy (0.2 eV). Thus, the charge transport properties of sulfur are quite similar to those of vitreous selenium. The electron drift mobility is very low and therefore characterized by an intermolecular hopping mechanism (with a hopping frequency of 4×10^9 sec^{-1}), while the hole drift mobility is sufficiently large to suggest transport via a (narrow) band.

6.3.8 *Silicon*

The mobility of holes in uncompensated (trap-free), 6×10^4 Ω cm Si has been measured by Lemke and Müller (1967) using the transient SCL current technique. The holes were injected either from a p$^+$–p contact or from an evaporated Pt contact. Cusp-like transients were observed, with a sampling technique, that agree very well with the theory up to two transit times. The hole mobility obtained was 450 cm^2/V sec, for an average of four samples. Note that this is some 40 times higher than the highest mobility previously measured by the cusp technique, namely 10 cm^2/V sec for the holes in sulfur.

7

SPECIAL TOPICS

In the present chapter we shall discuss some pertinent, miscellaneous topics in one-carrier current injection which, for a variety of reasons, have not been included in the previous chapters. We consider first some effects of photoexcitation on one-carrier SCL current flow. We then review the various methods that have been developed for measuring directly the amount of space charge in a solid owing to one-carrier injection. Such a determination is certainly the most direct verification of SCL current flow. We next discuss the question of thermal equilibration times, which has crucial bearing on the possibility of doing experiments starting from thermal equilibrium conditions. Then we discuss the mathematical difficulties of dealing with high-field phenomena (field-dependent mobilities). Here we suggest a method of approximation which preserves the analytic tractability, which is such an attractive feature of the simplified, low-field theory. We then discuss space-charge-free, quasi-steady-state injection currents with trapping, as studied, for example, with wide-bandgap insulators using the internal photoemission effect as a source of injection currents. Next we discuss performance limitations, imposed by majority-carrier injection, in the operation of devices based on the phenomenon of photoconductivity —a subject of obvious, practical importance. Finally we conclude with the presentation of recently published evidence for SCL ionic currents in a variety of materials, namely oxide layers on semiconductors, ionic and liquid crystals.

7.1 Optical Effects

Illumination of a sample sustaining SCL current flow can lead to an increase in the SCL current provided that part of the space charge is trapped and that the carriers in the traps can in some way receive energy from the incident light. The latter can be accomplished in two ways. A trapped carrier may directly absorb a photon and thereby undergo a transition to one of the bands. The other possibility is an indirect release mechanism in which an absorbed photon first produces an exciton that eventually transfer its energy to a trapped carrier. In either case, the light has the effect of redistributing the space charge among the conduction and trapping states without altering the total amount of space charge in the solid. The space charge, of course, is determined by the applied voltage and by geometric considerations.

Let us consider first the case of a solid with a single, discrete trap level of concentration N_t, degeneracy g, and energy E_t. The rates of capture and thermal re-emission of free carriers by the traps are given by (3.12) and (3.13) respectively, rewritten here for convenience

$$r = n\langle v\sigma\rangle(N_t - n_t), \tag{7.1}$$

where σ is the capture cross section for the dominant capture process (usually nonradiative),

$$g_{th} = n_t\langle e\rangle N_c; \qquad \langle e\rangle = (\langle v\sigma\rangle/g)\exp[(E_t - E_c)/kT], \tag{7.2}$$

where (3.15) has also been used. With photons incident on the solid, the optical release rate may be written as

$$g_{op} = n_t\sigma_{op}f, \tag{7.3}$$

where σ_{op} is the total photoionization cross section at the given photon frequency and f is the photon flux density. It is clear that, in the context of SCL current theory, Eq. (7.3) is the mathematical equivalent of (7.2); we have merely replaced the "thermal" quantity $\langle e\rangle N_c$ by the "optical" quantity $\sigma_{op}f$. Therefore, we may, in analogy with (3.15), equate (7.1) and (7.3) and define the *equivalent* energy E_t^* by the relation

$$\sigma_{op}f/\langle v\sigma\rangle = (N_c/g)\exp[(E_t^* - E_c)/kT]. \tag{7.4}$$

We see that the *entire* effect of the impurity excitation light on a discrete level at energy E_t may be subsumed by assigning the energy E_t^* to the trap. Note that E_t^* is a function of the excitation level f, as well as of σ_{op}. Using $E_t{}^*$ instead of E_t, all known results of SCL current theory may be taken over. However, since the *total* excitation rate is the sum of the thermal rate (7.2) and the optical rate (7.3), the SCL current is increased by optical energy absorbed in the traps only when the condition

$$f > (\langle e\rangle N_c/\sigma_{op}) = (N_c\langle v\sigma\rangle/g\sigma_{op}) \exp[(E_t - E_c)/kT] \tag{7.5}$$

is satisfied; that is, there is a threshold condition on the illumination level. This mode of optically stimulated SCL current flow has been observed in single crystals of CdS by Driedonks and Zijlstra (1966).

The situation is somewhat more complicated when the traps are distributed in energy. As an example, consider the exponential trap distribution defined by (2.36) in Chapter 2. Following Helfrich (1964), we define a demarcation energy E_D as that energy where the thermal and optical release rates are equal. Traps below E_D are emptied mainly by optical excitation while carriers in traps above E_D are released primarily through thermal excitation. This is shown in Fig. 7.1a. Since

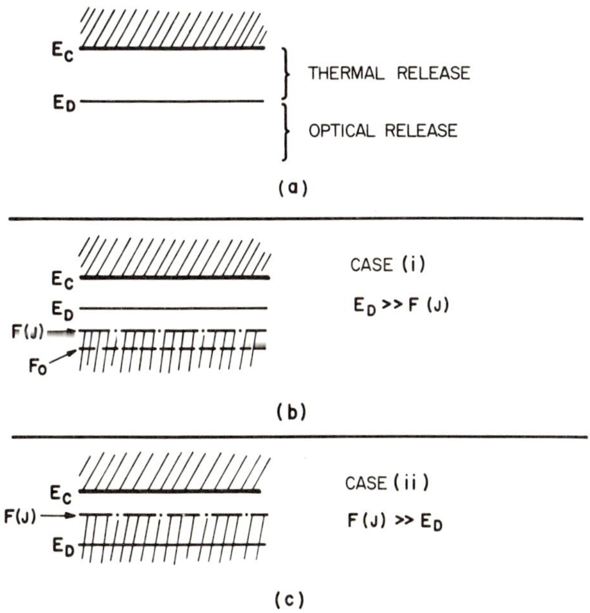

FIG. 7.1. Schematic energy-band diagrams for consideration of thermal versus optical release of trapped carriers. E_D is the demarcation level and $F(J)$ the quasi-Fermi level at current level J in the dark. Case (a) general; (b) $E_D \gg F(J)$; (c) $F(J) \gg E_D$.

the traps are distributed in energy, the thermal release rate $g_{\mathrm{th}}(E)$ for traps located at the energy E is given by (7.2), rewritten as

$$g_{\mathrm{th}}(E) = n_{\mathrm{t}}(E)\langle e\rangle N_{\mathrm{c}} = [n_{\mathrm{t}}(E)\, N_{\mathrm{c}}\langle v\sigma\rangle/g] \exp[(E - E_{\mathrm{c}})/kT]. \tag{7.6}$$

The optical release rate is given by (7.3). The demarcation energy is obtained from the relation: $g_{\mathrm{th}}(E_{\mathrm{D}}) = g_{\mathrm{op}}$:

$$E_{\mathrm{D}} = E_{\mathrm{c}} - kT \ln(N_{\mathrm{c}}\langle v\sigma\rangle/g\sigma_{\mathrm{op}} f). \tag{7.7}$$

Here we are neglecting the energy variation of σ_{op} for light with the given $\hbar\omega$; in particular $\hbar\omega$ must be large enough to produce photoionization over the entire energy range of the traps under discussion.

We must now distinguish two cases: (i) $E_{\mathrm{D}} \gg F(J)$ and (ii) $E_{\mathrm{D}} \ll F(J)$, where $F(J)$ is the quasi-Fermi level at current level J in the dark. (Note that we are here using the phenomenological approach of Chapter 2.) Only the first alternative, shown in Fig. 7.1b, is important because in case (ii), illustrated by Fig. 7.1c, only the lower portion of the traps that are occupied are available for optical release and these traps constitute only a small fraction of the total number of occupied traps. The SCL current flowing under condition (ii) will be virtually unaffected by light. For case (i), however, all the traps that would have been occupied by injection in the absence of light, namely those below $F(J)$, are capable of being emptied by light, leading to a higher current in the presence of the light.

In formulating case (i) quantitatively, we neglect the charges in the traps between E_{c} and E_{D}. Even in the absence of light, these are emptied too rapidly by thermal excitation to play a significant role. The effective trap concentration $\mathcal{N}_{\mathrm{t}}'$ is then obtained by integrating over the exponential distribution (4.115) between the limits F_0 and E_{D} :

$$\mathcal{N}_{\mathrm{t}}' = \int_{F_0}^{E_{\mathrm{D}}} n_{\mathrm{t}}(E)\, dE = N_{\mathrm{t}}\{\exp[(E_0 - E_{\mathrm{c}})/kT_{\mathrm{t}}] - (n_0/N_{\mathrm{c}})^{1/l}\}.$$

Since the second term is negligible,

$$\mathcal{N}_{\mathrm{t}}' \simeq N_{\mathrm{t}} \exp[(E_0 - E_{\mathrm{c}})/kT_{\mathrm{t}}], \tag{7.8}$$

and finally, using (7.7),

$$\mathcal{N}_{\mathrm{t}}' = N_{\mathrm{t}}(\sigma_{\mathrm{op}} f/N_{\mathrm{c}}\langle v\sigma\rangle)^{1/l}, \qquad l = (T_{\mathrm{t}}/T) > 1. \tag{7.9}$$

The ratio θ of free to trapped charge is obtained by equating the optical release rate from the effective traps with the capture rate of free carriers by these same traps, giving

$$\theta = (\sigma_{\mathrm{op}} f/\mathcal{N}_{\mathrm{t}}'\langle v\sigma\rangle). \tag{7.10}$$

To obtain the corresponding SCL photocurrent, we merely multiply the trap-free current (4.9) by this θ and eliminate N_t' using (7.9):

$$J_{op} = \frac{9}{8} \frac{\epsilon\mu N_c^{(1/l)}}{N_t} \left[\frac{\sigma_{op} f}{\langle v\sigma\rangle}\right]^{[1-(1/l)]} \frac{V^2}{L^3}. \tag{7.11}$$

The SCL photocurrent varies as V^2/L^3, as contrasted with the SCL dark current that varies as $V^{(l+1)}/L^{(2l+1)}$ [see (4.139)]. In addition, there is an intensity dependence $J_{op} \propto f^{[1-(1/l)]}$ involving the same parameter l that appears in the dark current. The schematic behavior of the SCL dark current and of several SCL photocurrents corresponding to different intensities is illustrated by Fig. 7.2a. The exact way in which

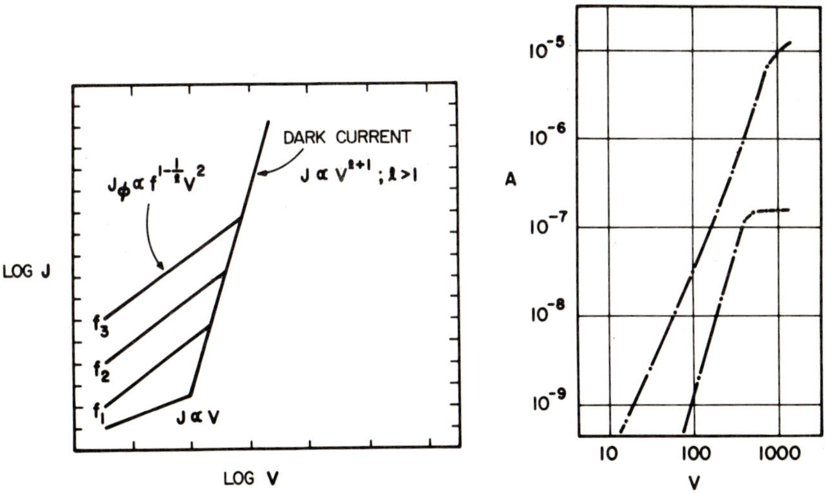

FIG. 7.2. SCL photocurrent-voltage characteristics: (a) schematic characteristic for exponential trap distribution, with $T_t/T = l$; (b) experimental curves for anthracene: upper curve, with illumination, lower curve, in the dark. (From Mark and Helfrich, 1962a, Fig. 7.)

the SCL photocurrent and the SCL dark current merge at higher injection levels is schematically drawn on this figure. At any given light level f, which from (7.7) fixes $E_D = E_D(f)$, the point of merger of the two curves is that current level J at which $F(J)$ coincides with E_D.

SCL photocurrents corresponding to an exponential trap distribution were first observed by Mark and Helfrich (1962) (see also Helfrich and Mark, 1962a) with anthracene. One of the original curves is shown in Fig. 7.2b. At low injection levels, the photocurrent varies very nearly as V^2 while the dark current varies as a higher power of V. A detailed examination of the effect was later undertaken by Jansen *et al.* (1964), again working with anthracene. These authors confirmed not only the V^2-dependence of the photocurrent but also the interrelated intensity dependence of the photocurrent and voltage dependence of the dark current as predicted by (7.11) and (4.139), respectively.

By a similar analysis, Helfrich (1964) also derived the SCL photocurrent appropriate to a uniform trap distribution (see Section 4.7.2):

$$J_{\mathrm{op}} = \frac{9}{8} \frac{f\sigma_{\mathrm{op}}\epsilon\mu}{[F_0 + kT\ln(f\sigma_{\mathrm{op}}/N_{\mathrm{c}}\langle v\sigma\rangle)]\langle v\sigma\rangle\mathcal{N}_0} \frac{V^2}{L^3}, \tag{7.12}$$

where $\mathcal{N}_0$ is the (constant) concentration of traps per unit energy range [see (4.143)]. This is to be contrasted with the nearly exponential current-voltage characteristic of the dark current, given by (4.159).

The situation can be much more complicated when light is absorbed in the fundamental absorption band while the solid carries a SCL current. This type of absorption does more than merely redistribute the injected charge between the band states and the defect states of the lattice. It produces minority carriers which, until trapped, can respond to the nonuniform electrical forces produced by the majority-carrier space charge and thereby tend to neutralize that space charge. It is, by definition, a two-carrier effect. We shall not pursue the matter further except to mention that the ultimate effect of such illumination is to wash out the SCL dark current in favor of a larger, ohmic photocurrent, as demonstrated by the early experiments of Smith (1955) and Smith and Rose (1955) as illustrated by Fig. 5.1, and later again by Bräunlich (1967).

7.2 Direct Determination of the Space Charge

In the earliest attempts to observe SCL currents in solids, it was very important to pin down experimentally as many of the theoretically predicted properties of SCL currents as possible. Perhaps the most obvious of all such properties is the very existence of the space charge itself. If that can be measured and correlated with the amount predicted by the applied voltage and the geometry of the solid, one has a powerful argument indeed for the existence of SCL current flow. If all the injected space charge in a conventional solid, say CdS, were free, it would be quite difficult to measure it. The reason for this is, as we shall see momentarily, that the technique for accomplishing this measurement relies on determining the space charge in the sample *immediately after removal of the voltage* by which it was originally injected. With conventional geometries ($L \simeq 1$ mm), mobilities ($\mu \simeq 100$ cm^2/V sec), and applied voltages (V $\simeq$ 100 V), the injected free space charge would leave the crystal in an interval of about one transit time ($L^2/\mu V$) or about 10^{-6} sec after the injecting voltage is removed. This is a very short time indeed. Fortunately, in the present instance at least, most of the space charge injected into a real material resides in traps so that the flow of space charge from the crystal upon removal of the applied voltage is governed not by the transit time but by the thermal release time of the injected carriers from the traps. The latter is usually sufficiently large to allow for the measurement of the injected space charge with little difficulty. (In fact, with wide bandgap materials, the likelier problems will arise from unduly long trapping times, as discussed in the following section.)

The first direct observation of injected space charge was reported by Smith and Rose (1955) working with CdS. In their apparatus, the crystal was clamped between two indium (ohmic) pressure contacts, as shown in Fig. 7.3. The apparatus was also supplied with an arrangement with which the crystal could be mechanically released by spreading the pressure contacts apart, thus allowing the crystal to fall into an electrometer pan situated underneath the crystal. A SCL current was caused to flow by applying a voltage to the contacts. Subsequently, the contacts were spread, and the crystal fell onto the electrometer pan for a direct measurement of the charge in the crystal. The values so obtained coincided, within a factor of 2, with the expected value, the uncertainty being ascribed to a comparable error in the determination of the

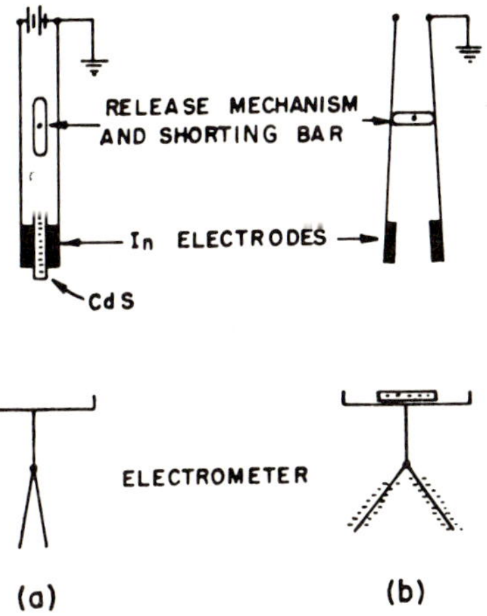

FIG. 7.3. Schematic drawing of experiment to measure charge injection in an insulator. (From Smith and Rose, 1955b, Fig. 2.)

electrode areas. The same technique was later applied by Ruppel (1958) to measure the injected space charge in ZnS crystals.

A more elaborate technique is necessary when ohmic pressure contacts cannot be made, as is the case when electrolytic electrodes are used to inject currents into anthracene, for example. The sample is supplied with one injecting and one noninjecting contact. After SCL current flow is established at the voltage V_0, a reverse bias V_r is suddenly switched on that sweeps the excess space charge from the crystal through the ohmic contact without additional carriers entering through the noninjecting electrode. When $V_r < V_0$, not all the injected space charge is recorded leaving the crystal, but for $V_r > V_0$, essentially all the space charge is extracted. Thus, a measurement of the extracted charge Q vs V_r should give Q increasing with V_r until $V_r = V_0$, beyond which Q should remain constant. This effect was observed by Helfrich and Mark (1962b), working with anthracene. The magnitude of the space charge so measured agreed to within 20% of the computed value.

7.3 Thermal Equilibration Times

Throughout all of this book the basic assumption has been made that the solid is in thermal equilibrium prior to the injection of current. The experimentalist cannot, of course, merely assume this but should attempt to ascertain it by measurements or at least plausibility arguments. In order to indicate the difficulties that may be encountered in realizing the hypothesized thermal equilibrium, particularly with wide bandgap insulators, we present here an estimate of the thermally activated trap-release time, a quantity which governs the recovery to thermal equilibrium following a disturbance from that condition, e.g., current injection.

Letting R_c and R_e be respectively the electron capture rate to, and thermal emission rate from, a given trap level, we may write

$$R_c = n/\tau_c, \qquad 1/\tau_c = n\langle v\sigma\rangle, \tag{7.13}$$

$$R_e = n_t/\tau_e, \qquad 1/\tau_e = \langle e\rangle N_c, \tag{7.14}$$

where τ_c, τ_e are the capture and emission lifetimes, respectively. Equations (7.13) and (7.14) are restatements of (3.12) and (3.13), respectively. Detailed balance relates $\langle e\rangle$ and $\langle v\sigma\rangle$, namely the result (3.15). Using (3.15) in (7.14), we obtain

$$\tau_e = (g/\langle v\sigma\rangle N_c) \exp[(E_c - E_t)/kT]. \tag{7.15}$$

Taking as "typical" numbers at room temperature, $g = 2$, $N_c = 2 \times 10^{19}\ \text{cm}^{-3}$, $\langle v\sigma \rangle = 10^{-8}$, corresponding to $\langle \sigma \rangle \approx 10^{-15}\ \text{cm}^2$, Eq. (7.15) then becomes $\tau_e \simeq 10^{-11} \exp[(E_c - E_t)/kT]$ with $kT = 1/40$ eV. In Table 7.1 we have listed the values of τ_e corresponding to different

TABLE 7.1

THERMAL RELEASE TIMES FROM TRAPS AT ROOM TEMPERATURE

$\tau_e = 10^{-11} \exp[(E_c - E_t)/kT]$, $kT = (1/40)$ eV

$E_c - E_t$	τ_e
1.6 (1.61)	1 ae[a]
1.1 (1.09)	3 yr
1.0 (0.98)	10 days
0.9 (0.92)	1 day
0.8 (0.81)	$\frac{1}{4}$ hr
0.75 (0.75)	2 min
0.7 (0.69)	10 sec
0.6 (0.58)	10^{-1} sec
0.5 (0.52)	10^{-2} sec
0.4 (0.40)	10^{-4} sec

[a] 1 ae = 3×10^9 yr $\approx$ the age of the earth.

$(E_c - E_t)$ as computed from this last formula. It is seen that when the trap depth exceeds 0.75 eV, the thermal release time at room temperature becomes exceedingly long. In dealing with such deep trap levels, the approach to thermal equilibrium must obviously be stimulated by auxiliary means, for example heating followed by slow cooling down to room temperature, or shining light, in a suitable wavelength range, on the sample.

7.4 HIGH-FIELD PHENOMENA

The effect of field-dependent mobility on the SCL current-voltage characteristic has already been examined phenomenologically in Section 2.10. We saw there that if the functional dependence $v(\mathscr{E})$ of the drift velocity v on the field $\mathscr{E}$ is known, then the I–V characteristic is readily derived by replacing $\mathscr{E}$ by its average, V/L: $v(\mathscr{E}) \to v(V/L)$. If more detailed information about the properties of the current flow is required, then the governing equations must be studied. These equations are (3.7), (3.8), and (3.11), and the boundary condition (3.21). We now follow closely the discussion of the simplified theory in Section 3.4. In place of the field parameters $\mathscr{E}_J$ and $\mathscr{E}_{J,j}$ of (3.18), it is convenient to define analogous velocity parameters v_J and $v_{J,j}$, respectively, as follows

$$v_J = J/en_0, \qquad v_{J,j} = g_j J/eN_j. \tag{7.16}$$

Note that $v_J = \mu\mathscr{E}_J$ and $v_{J,j} = \mu\mathscr{E}_{J,j}$. Then the following relations obtain:

$$n = n_0(v_J/v), \qquad n_{tj} = N_{tj}/[1 + (v/v_{J,j})], \tag{7.17}$$

these being analogous to (3.17).

Corresponding to (3.19) we have here

$$dx/d\mathscr{E} = a(v; J), \tag{7.18}$$

$$1/a(v; J) = (e/\epsilon)(v_J - v)\left[(n_0/v) + \sum_j n_{tj,0}/(v_{J,j} + v)\right]. \tag{7.19}$$

In contrast to the phenomenological approach of Section 2.10, it is here more useful to consider the field as functionally dependent on the drift velocity: $\mathscr{E} = \mathscr{E}(v)$. Then writing $d\mathscr{E} = \mathscr{E}'(v)\,dv$, with $\mathscr{E}'(v) = d\mathscr{E}/dv$, Eq. (7.18) becomes

$$dx = a(v; J)\,\mathscr{E}'(v)\,dv. \tag{7.20}$$

Integrating (7.20) from the cathode to the anode, we obtain

$$L = \int_0^{v_a} a(v; J)\,\mathscr{E}'(v)\,dv = b(v_a; J), \tag{7.21}$$

where $v_a = v(L)$ is the drift velocity at the anode.

Finally,

$$V = \int_0^L \mathscr{E}\,dx = \int_0^{v_a} a(v; J)\,\mathscr{E}'(v)\,\mathscr{E}(v)\,dv = c(v_a; J). \tag{7.22}$$

The point of this formal development is the following: In the low-field case, $\mathscr{E}(v) = v/\mu$, the above is identical to the formal development in Section 2.10—note that $a(v; J)$ is precisely the same function of v that $f(\mathscr{E}; J)$ is of $\mathscr{E}$. In the low-field case, for many problems of interest, the integrations in (3.22) and (3.23) [(7.21) and (7.22)] can be carried out analytically, via an expansion of $f(\mathscr{E}; J)$ or $a(v; J)$ in simple partial fractions, yielding explicit analytic expressions for the functions $g(\mathscr{E}_a; J)$ and $h(\mathscr{E}_a; J)$ or $b(v_a; J)$ and $c(v_a; J)$. For these problems, the purely algebraic equations (3.24), or here,

$$L = b(v_a; J), \qquad V = c(v_a; J) \tag{7.23}$$

define implicitly the J–V relationship, via the intermediate variable $\mathscr{E}_a(v_a)$. The key observation is that, in the high-field case, there are quite useful functional forms $\mathscr{E}(v)$ which do not impair the integrability of (7.21) and (7.22), if present in the low-field case. In general, for example, a polynomial function

$$\mathscr{E}(v) = \frac{v}{\mu} + \frac{1}{\mathscr{E}_1}\left(\frac{v}{\mu}\right)^2 + \frac{1}{\mathscr{E}_2{}^2}\left(\frac{v}{\mu}\right)^3 + \cdots, \tag{7.24}$$

where μ is the low-field mobility, does not impair analytic tractability.

The simplest example of (7.24) is the quadratic representation:

$$\mathscr{E} = \frac{v}{\mu} + \frac{1}{\mathscr{E}_1}\left(\frac{v}{\mu}\right)^2. \tag{7.25}$$

It is instructive to solve (7.25) for $\tilde{\mu} = v/\mathscr{E}$:

$$\tilde{\mu} = \tfrac{1}{2}\mu(\mathscr{E}_1/\mathscr{E})\{[1 + (4\mathscr{E}/\mathscr{E}_1)]^{1/2} - 1\}. \tag{7.26}$$

At low fields, Eq. (7.26) approximates to

$\mathscr{E} \ll \mathscr{E}_1$:

$$\tilde{\mu} = \mu[1 - (\mathscr{E}/\mathscr{E}_1)]. \tag{7.27}$$

At high fields, Eq. (7.26) approximates to

$\mathscr{E} \gg \mathscr{E}_1$:

$$\tilde{\mu} = \mu(\mathscr{E}_1/\mathscr{E})^{1/2}[1 - \tfrac{1}{2}(\mathscr{E}_1/\mathscr{E})^{1/2}]. \tag{7.28}$$

Fixing on the leading terms of (7.27) and (7.28), we see that the mobility function (7.26) has the following useful properties:

(i) At low fields, it approaches a constant.

(ii) At high fields, it has a field dependence appropriate to constant-mean-free-path problems [e.g., deformation potential, acoustic mode scattering (Shockley, 1951)]. Accordingly, the leading term of (7.28) was used by Dacey (1953) to discuss one-carrier SCL currents in Ge under punch-through conditions. Further,

(iii) It smoothly connects the low and high-field regions.

(iv) It will generally not impair the analytic tractability of one-carrier injection problems which are tractable under low-field conditions.

The properties (i)–(iv) recommend (7.26) as a useful mathematical representation of the mobility for constant-mean-free-path problems.

Another useful functional dependence of $\mathscr{E}$ on v is the hyperbola

$$(v - v_s)[\mathscr{E} - (v/\mu) + \mathscr{E}_0] = -\mathscr{E}_0 v_s \rightarrow \mathscr{E} = (v/\mu) - \mathscr{E}_0 - [\mathscr{E}_0 v_s/(v - v_s)], \tag{7.29}$$

where $\mathscr{E}_0$ and v_s are parameters. This hyperbola in the v–$\mathscr{E}$ plane is asymptotic to the straight lines $v = v_s$ and $v = \mu\mathscr{E} + \mu\mathscr{E}_0$; further, as $\mathscr{E}_0 \rightarrow 0$ the hyperbola degenerates into the two asymptotes. Thus (7.29) gives the normal, v–$\mathscr{E}$ dependence, $v = \mu\mathscr{E}$, at low fields, and v saturates at high fields: $v \rightarrow v_s$. The sharpness of the transition from the ohmic to the saturation behavior of v is determined by the magnitude of $\mathscr{E}_0$: $\mathscr{E}_0 \ll v_s/\mu$ corresponds to a sharp transition, $\mathscr{E}_0 \gtrsim v_s/\mu$ to a very gradual transition. As with (7.24), Eq. (7.29) will generally not impair the analytic tractability of one-carrier injection problems which are tractable under low-field conditions. Therefore (7.29) appears to be a useful mathematical representation of the v–$\mathscr{E}$ relation in problems where a transition is made directly from the ohmic, low-field relation to velocity saturation.

Much of the above discussion is taken from a paper on one-carrier injection theory with field-dependent mobilities by Lampert (1958).

7.5 Space-Charge-Free, "Steady-State" Injection Currents

The steady-state, one-carrier current-voltage characteristic for a sample with a good injecting contact is a space-charge limited characteristic such as discussed in Chapter 4 for numerous cases. There are experimental conditions of interest under which quasi-steady-state current (that is, steady for a "long," but not infinite, time) can flow that is essentially space charge free; these conditions are: (i) extremely low injection level (corresponding to a poor injecting contact), (ii) extremely small thermal equilibrium current, and (iii) relatively strong trapping of injected carriers (the case of negligible trapping is without interest for the quasi-steady-state, though it is useful for transient measurements, as discussed in Section 6.11). The current levels satisfying these conditions and remaining essentially space-charge-free for adequate lengths of time are small but, in practice, sufficiently large to measure (if necessary, using integrating techniques). A practical example of this mode of injection is internal photoemission, from a metal or semiconductor, into a wide bandgap insulator.

In carrying through the analysis of the injected current, we completely neglect the thermal free carriers and we assume the space charge small enough that it does not perturb the ohmic field distribution $\mathscr{E}_\Omega = V/L$, where V is the applied voltage and L the cathode–anode spacing. For the sake of definiteness, assume the current carriers are electrons. After their release from the cathode, at $x = 0$, the electrons drift in the constant field $\mathscr{E}_\Omega$ toward the anode. They are able to move in the insulator only an average distance λ_s, called the mean range or *schubweg*, before they are captured by the deep traps. If τ_s is the mean time against capture (τ_s is a constant, independent of field, in the absence of hot-carrier effects),

$$\lambda_s = \mu \mathscr{E}_\Omega \tau_s = \mu \tau_s (V/L). \tag{7.30}$$

The free-electron concentration inside the insulator is given by

$$n = n_0 \exp -(t/\tau_s) = n_0 \exp -(x/\lambda_s), \tag{7.31}$$

where n_0 is the concentration at the cathode (but inside the insulator). This follows since the probability of capture in time dt is $dt/\tau_s = dx/\lambda_s$.

The mean distance $\bar{x}$ traveled by an injected electron is the sum of two terms:

(i) The fraction of the injected electrons that do not reach the anode times their mean distance of travel:

$$\begin{aligned} \frac{(n_0 - n_a) \int_{n_0}^{n_a} x(-dn)}{n_0 \int_{n_0}^{n_a} (-dn)} &= \frac{1}{n_0} \int_0^L x \left(-\frac{dn}{dx}\right) dx \\ &= \lambda_s \int_0^L \frac{x}{\lambda_s} \exp\left(-\frac{x}{\lambda_s}\right) \frac{dx}{\lambda_s} \\ &= \lambda_s[1 - \exp(-L/\lambda_s)] - L \exp(-L/\lambda_s), \end{aligned}$$

and

(ii) The fraction of the injected electrons that reach the anode times L:

$$(n_a/n_0)L = L \exp(-L/\lambda_s),$$

Thus,

$$\bar{x} = \lambda_s[1 - \exp(-L/\lambda_s)] \tag{7.32}$$

In the limit $L/\lambda_s \gg 1$, $\bar{x} \simeq \lambda_s$; in the opposite limit, $L/\lambda_s \ll 1$, $\bar{x} \simeq L$. Equation (7.32) was first derived by Hecht (1932).

The current in the external circuit is $\bar{x}/L$ times the rate j_ϕ of electron injection at the cathode:

$$j = j_\phi \frac{\lambda_s}{L}\left[1 - \exp\left(-\frac{L}{\lambda_s}\right)\right] = j_\phi \frac{\mu \tau_s V}{L^2}\left[1 - \exp\left(-\frac{L^2}{\mu \tau_s V}\right)\right], \tag{7.33}$$

where (7.30) has been used. Note that (7.33) can be obtained more directly using $j = Q/T$ with $Q = e\int_0^L n\,dx$ and n given by (7.31), $t = L^2/\mu V$, and $j_\phi = e\mu n_0 V/L$. This simple calculation works because of the uniform field distribution.

As mentioned, this mode of injection current flow has found useful application in the field of photoemission from metals and from semiconductors into insulators. When this process is studied as a function of the wavelength of the light incident on the metal-insulator junction, one should observe a threshold for electron injection at the optical energy corresponding to the gap between the Fermi level of the metal (or the valence band edge of the semiconductor) and the conduction band edge of the insulator when the junction has a negative bias, and another threshold for hole injection at the optical energy corresponding to the gap between the metal Fermi level (or the conduction band edge of the semiconductor) and the valence band edge of the insulator when the junction has a positive bias. A more accurate treatment of the problem includes a correction to these energy intervals arising from the Schottky effect, owing to the usually high fields at the junction. This procedure was first applied by Gilleo (1953) to study the work function of Ag relative to KBr and NaCl (4.3 eV in both cases) and also relative to AgCl (1.1 eV). The same technique was later applied by Williams (1965, 1966) and by Goodman (1966) to study the photoemission of electrons from Si into a thermally grown silicon oxide layer in order to obtain the important energy band levels at the silicon-oxide interface. This work revealed, for example, that the Si valence band lies 4.25 eV below the conduction band of the oxide. The equation (7.33) yields the range $\bar{r}$ per unit field; $\bar{r} = \mu\tau_s$. Measurements by Goodman (1966) of electrons photoemitted into silicon oxide, either from a Au contact or from the silicon substrate, yielded values for $\bar{r}$ of the order of 10^{-9} cm^2/V. A systematic review of the internal photoemission technique has been given by Williams (1970).

7.6 Some Device Limitations

Much of the early interest in SCL current flow in solids derived from the desire to understand the fundamental limitations on the operating characteristics of devices based on the photoconductivity phenomenon. Since the latter is an extensive subject in its own right (see, e.g., Bube, 1960; Rose, 1963), we shall confine our brief discussion here to the specific question as to how injected space charge sets an upper limit on the performance of photoconductive devices.

At applied fields less than the breakdown field, the sensitivity of a photoconductor is defined by the gain G. This is the number of electrons passed through the photoconductor per absorbed photon and is equal to the ratio of the lifetime τ of a free electron (the time spent in the conduction band) and the transit time t_0 [see (2.5)] of a free electron from the cathode to the anode:

$$G = \tau/t_0, \qquad t_0 = L^2/\mu V. \tag{7.34}$$

In the absence of significant trapping, the response time is also given by τ and since the response time is equal to the reciprocal bandwidth, the gain-bandwidth product for the trap-free photoconductor is

$$\text{gain} \times \text{bandwidth} \rightarrow \frac{G}{\tau} = \frac{1}{t_0} = \frac{\mu V}{L^2}. \tag{7.35}$$

Ideal performance is defined by this relation.

The chief reason why real materials fall well below this ideal limit in their performance is attributable to trapping. In that event, τ must be replaced by the observed response time τ_0, the measured rise or decay time of photoconductivity, using the relation

$$\tau_0 = \tau(n_t/n), \tag{7.36}$$

where n and n_t are, respectively, the free and trapped carrier concentrations: $n \ll n_t$, generally. Thus the gain–bandwidth product is reduced by the factor n/n_t:

$$\text{gain} \times \text{bandwith} = \frac{G}{\tau_0} = \frac{1}{t_{0,\text{eff}}}, \qquad t_{0,\text{eff}} = t_0\frac{n_t}{n} = \frac{L^2 n_t}{\mu V n}. \tag{7.37}$$

Regardless of whether or not trapping takes place, space-charge injection sets an upper limit on performance. At sufficiently low voltages, where Ohm's law is observed, the gain-bandwidth product increases linearly with voltage, according either to (7.35) or to (7.37). However, this increase with voltage cannot continue indefinitely for, as the effective transit time $t_{0,\text{eff}}$ decreases with increasing voltage, it will eventually become less than the ohmic relaxation time, $t_\Omega = \epsilon/e\mu n_0$, which is a material constant, and space-charge injection commences as discussed in Section 2.2. At the onset of space-charge injection, the gain-bandwidth product reaches its maximum value

$$\text{maximum gain} \times \text{bandwidth} = 1/t_\Omega \tag{7.38}$$

and at higher voltages, the photoconductor suffers a reduction in its performance characteristics. This result specifically precludes the avalanche multiplication process which permits higher gain and, concomitantly, higher performance.

For a deeper discussion of how space-charge injection limits the performance of various photoconductive devices, including possible refinements of (7.37), the reader is referred to the original literature (Rose, 1957a, b; Redington, 1958; Rose and Lampert, 1959a, b; *RCA Rev.* **20**, Dec., 1959).

7.7 SCL Ion Currents

The great bulk of the literature on SCL currents, both experimental and theoretical, refers to electronic currents; i.e., SCL currents of electrons or holes. Since ion currents certainly exist in solids, and since "defect" sites for the trapping of ions likewise exist, we would expect comparable SCL injection-current effects for ions. Indeed, in recent years SCL ion current observations have been reported for several materials: steady-state proton currents in ice by Engelhardt and Riehl (1965), steady-state cation currents in KCl by Tödheide-Haupt *et al.* (1966), transient proton currents in silicon dioxide films by Hofstein (1967a, b), and transient ion (unidentified) currents in liquid crystals (*p*-azoxyamisole) by Heilmeier and Heyman (1967). The SCL proton currents in ice, at 77°K, were injected from powdered Pd electrodes that had been saturated with hydrogen. A steep rise in current, shown in Fig. 7.4, associated with trap filling, corresponds to 3.5×10^{12} traps/cm^3. The trapping was confirmed by the measurement of the thermally stimulated current of protons released from traps following the SCL current injection. The nature of the proton traps was not established. The measurements with KCl were made on 6-μ-thick crystals at 100°C. A V^2/L^3 dependence was observed which was attributed to a SCL cation current because its temperature dependence coincided with that of the ionic conductivity. The transient proton SCL currents in silicon dioxide films were measured on standard MOS (metal-oxide-semiconductor) structures in which the oxide is grown on a single-crystal silicon substrate. The protons were injected from the positively biased gate electrode after contamination of this surface with ethanol. A transient current cusp was observed with the substrate biased negative, as shown in Fig. 7.5. The SCL current interpretation of the cusp was substantiated by noting that its location on the time axis varied as $1/V$ and its magnitude varied as V^2. The protons were identified by using tritium-substituted ethanol and by monitoring the radioactivity

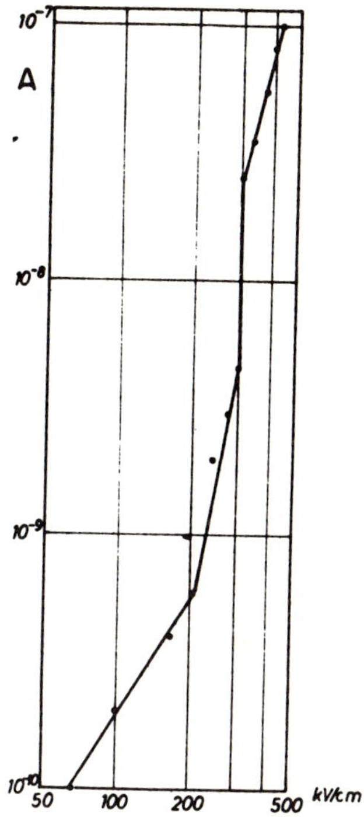

FIG. 7.4. Typical current-voltage characteristic for a pure ice single crystal at 77°K. (From Engelhardt and Riehl, 1965, Fig. 1.)

accumulated at the oxide–semiconductor interface following the passage of current. The mobility of the ions at 40°C was measured to be 5×10^{-11} cm^2/V sec.

In the liquid crystal measurements, a current cusp was observed whose magnitude varied as $V^n/L^{2.3}$, where $1.5 < n < 2$, and whose location on the time scale varied as $L^{1.3}/V$. An interpretation in terms of SCL current transients yielded an ionic mobility of approximately 2.5×10^{-4} cm^2/V sec for negative ions in *p*-azoxyanisole, a rather typical value for ions in solution.

With regard to the theory of SCL ion currents, there is one particular point that we wish to consider here, and this is the relation of the theory to the particle statistics. In working out the theory of SCL electronic currents, trap occupancies are calculated using the Fermi–Dirac occupation function (2.15). This procedure would not be questioned because electrons and holes are fermions. Since some ions are bosons (e.g., the deuteron, singly ionized ^{6}Li, doubly ionized ^{4}He, etc.), one might naively expect a rather drastically different theory for the effects of trapping on SCL currents of such ions. However, on examination it comes out that such an expectation is unjustified. The critique that

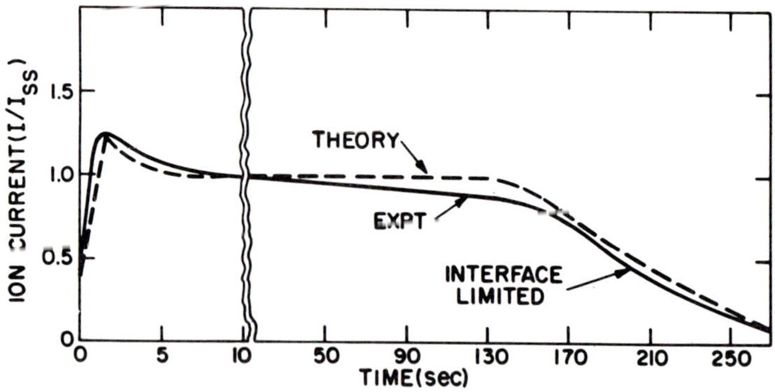

FIG. 7.5. SCL transient current of protons through a silicon dioxide film grown on a single-crystal silicon substrate. (From Hofstein, 1967, Fig. 1.)

validates the use of the Fermi–Dirac occupation function (2.15) in calculating trap occupancies is the assertion that a localized trap state can hold one, and only one, electron (or hole). In fact, this circumstance normally obtains, not because of the statistics of the electron (or hole), but because of simple electrostatic considerations: The electrostatic repulsive energy would simply be too great if two electrons (or holes) were held in the state. But precisely the same circumstance will hold in the trapping of ions, whether bosons or fermions: Normally, a localized trap will be able to hold only one ion because of nonstatistical (e.g., electrostatic) considerations. Therefore, the Fermi–Dirac occupation function is appropriate for the calculation of ion trapping whether the ions are bosons or fermions. We conclude then that the SCL current injection theory developed for the electronic case is applicable, without drastic modification, also to the ionic case.

8

INJECTION FROM A POINT AND OTHER GEOMETRIES

In the preceding chapters, we have confined our discussion entirely to the properties of one-dimensional, planar current flow. In practice, of course, other current-flow configurations may be present, either by design or by uncontrollable properties of the contact, an outstanding example being the point contact configuration. Under suitable conditions, injection from a point will closely approximate spherical, one-dimensional current flow; therefore the bulk of this chapter is directed to a study of some of the interesting properties of this latter type of flow. In addition, some attention is paid to cylindrical, one-dimensional flow and to planar flow in a sample long compared to at least one of its cross-sectional dimensions.

We proceed to discussion of majority-carrier current injection at a point contact. The analytical problem involves radial current flow in a spherical geometry, between two concentric spherical electrodes of radius r_c and r_a as shown in Fig. 8.1a for the case that $r_c \ll r_a$. Further, we shall assume that the smaller-radius electrode, at r_c, is the injecting cathode. The reverse situation will be discussed later. We obtain the point-contact configuration, Fig. 8.1b, simply by cutting Fig. 8.1a in half. There is now the flat surface, labeled SS, in the problem. If this surface does not itself have special properties that influence the current flow, such as conducting channels or an unusually high concentration of surface traps, the current-flow properties of the hemisphere will closely

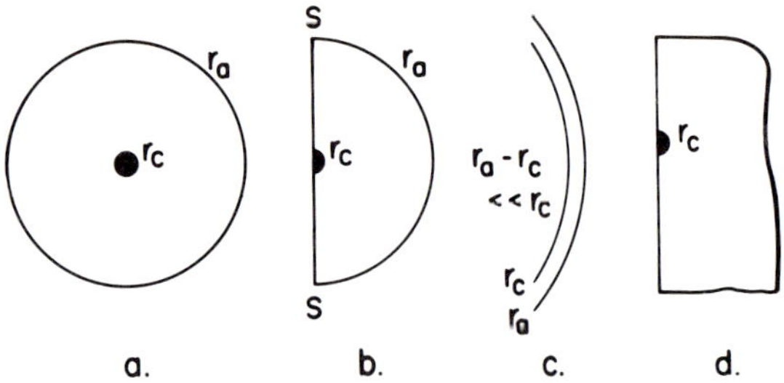

FIG. 8.1. Spherical, radial-flow geometries: (a) complete spherical case, $r_c \ll r_a$; (b) point-contact configuration (hemispherical case), $r_c \ll r_a$; (c) quasi-planar flow case, $r_a - r_c \ll r_c$; (d) practical point-contact configuration.

approximate those of the sphere except for the obvious factor of two. If $r_a \simeq r_c$, namely $(r_a - r_c) \ll r_c$, as in Fig. 8.1c, then it is obvious by inspection that current-flow properties under this almost-planar condition will necessarily resemble those of the conventional, one-dimensional, plane-parallel, current flow treated in Chapter 4. Clearly then, anything novel and unusual in spherical current flow (and indeed in nonplanar flow generally) must originate with the situation in which $r_c \ll r_a$. Accordingly, we make this assumption throughout the discussion.

Before proceeding to the analysis of injection currents in spherical configurations, it is instructive to review beforehand its major features. At first glance, it might be expected from the comparable problem in the vacuum that nothing fundamentally new should emerge from the change in geometry, planar to spherical. It is well known, for example, that for SCL current flow in vacuum diodes, the planar-, spherical-, and cylindrical-flow geometries all yield the same functional dependence of current on voltage, namely, $I \propto V^{3/2}$. In fact, it has been demonstrated, through a dimensional analysis (Spangenberg, 1948) that this same functional dependence obtains, in the vacuum problem, for electrodes of arbitrary (but not too pathological) shape. A like result, but with $I \propto V^2$, can be proved, in like manner, for pure SCL current flow in a perfect insulator (no thermal carriers) with isotropic transport and dielectric properties.

The outstanding result of current injection from a point contact into a real solid derives from the existence of compensated conduction charge in thermal equilibrium, a fundamental impossibility in the vacuum case. In the case of the solid, the initial space-charge controlled regime of the I–V characteristic, following the ohmic regime ($I \propto V$) that always obtains at sufficiently low voltages (but not in the vacuum), is generally of the form: $I = KV^{3/2}$. The constant K depends on the concentration and energy distribution of the traps in the forbidden gap that may capture and immobilize the injected carriers. *However, K is independent of the point radius r_c as well as of the anode radius r_a*, provided $r_a \gg r_c$. Thus, the practical point-contact configuration of Fig. 8.1d is well approximated by the hemispherical electrode configuration of Fig. 8.1b.

We shall show in Section 8.4 that the 3/2-power law is a consequence of the particular way in which the injected space charge increasingly spreads into the solid, beyond the immediate vicinity of the injecting contact, as the applied voltage is increased. Over the range of voltages for which Ohm's law is valid, the injected space charge is confined essentially inside a radius $r_x < 2r_c$. Following the departure from ohmic behavior; that is, following the onset of the space-charge-controlled characteristic, the confining radius r_x for the injected space charge increases as $V^{1/2}$, and this relationship leads to the above-cited 3/2-power law. Similarly, only the traps in the neighborhood of the injecting point are initially important for the determination of K. Thus, the 3/2-power law is a regime in which $r_c \ll r_x \ll r_a$ and this suggests the Regional Approximation for the analysis.

The above-described features of the 3/2-power law suggest immediately the application of this kind of measurement to the determination of trap concentrations *locally* in the solid, namely, in the vicinity of the injecting point. It is an important experimental convenience in this measurement that K is independent of both radii. Thus, only a reasonably small point is needed; its exact radius is unimportant. Furthermore, the anode need not be actually spherical; it need only be a broad area contact, and can then be quite flat without impairing the measurements. This is particularly convenient in single-crystal studies since a spherical or hemispherical anode would be a generally impractical configuration.

In the above discussion, and in much of the succeeding discussions, we have followed closely the paper by Lampert *et al.* (1964).

8.1 Analytical Formulation of the Point-Contact (Spherical Flow) Problem

The equations governing steady-state one-carrier injection current flow in the spherical geometry are the particle conservation equation

$$I = 4\pi e\mu n r^2 \mathscr{E} = \text{constant}, \tag{8.1}$$

where, in keeping with our overall approach, the diffusive contribution to the current has been neglected, and the Poisson equation

$$(\epsilon/er^2)(d/dr)(r^2\mathscr{E}) = n_1 + n_{t,1}; \qquad n_1 = n - n_0\,, \quad n_{t,1} = \sum_j (n_{tj} - n_{tj,0}), \tag{8.2}$$

where the relationship between n_{tj} and n is the usual one specified by (3.11). Equations (8.1) and (8.2) are the spherical-flow analogs of Eqs. (4.1) and (4.2), respectively. In (8.1), I is the total current, not the current density; further, the solid is assumed isotropic in its transport and dielectric properties so that both μ and ϵ are scalar quantities. The one-dimensional radial coordinate r extends from the cathode radius r_c to the anode radius $r_a \gg r_c$.

Also, the usual assumption is made that the free and trapped carriers in the solid are in quasi-thermal equilibrium; that is, that the quasi-Fermi level $F(x)$ determines the occupancy of the traps at position x through Eq. (4.3). The usual boundary condition for an ohmic contact, appropriate to the simplified theory which neglects diffusion, obtains:

$$\mathscr{E}(r_c) = 0. \tag{8.3}$$

Finally, the applied voltage V is

$$V_a = \int_{r_c}^{r_a} \mathscr{E}(r)\, dr. \tag{8.4}$$

8.2 Pure SCL Current Flow in the Point-Contact Geometry

The simplified theory for spherical flow, formulated above, is not, in general, analytically tractable. (This is in contrast to the simplified theory for planar flow which is analytically tractable for most cases of interest, as shown in Chapter 4.) Before we develop the Regional Approximation technique for obtaining solutions, it is instructive to examine briefly the extreme limits of solution, that of ohmic current flow at very low injection levels and that of pure SCL current flow at very high injection levels, and to compare these limiting cases with their counterparts in the plane-parallel electrode configuration. Ohmic current flow in the planar case yields a uniform field, shown by the solid line of Fig. 8.2a, except for the usual correction at the injecting contact, namely the dashed line, where the field is driven to zero by the boundary condition. For the spherical case, the ohmic field is given by (8.1) with $n = n_0$, namely

$$\mathscr{E}_\Omega(r) = I/4\pi e \mu n_0 r^2. \tag{8.5}$$

The field drops off as $1/r^2$, as shown by the solid line of Fig. 8.2b, and again the field must be brought to zero at the contact—the dashed line—so that the field maximum is extremely close to, but not precisely at, the cathode. A comparable difference for these two geometries persists for

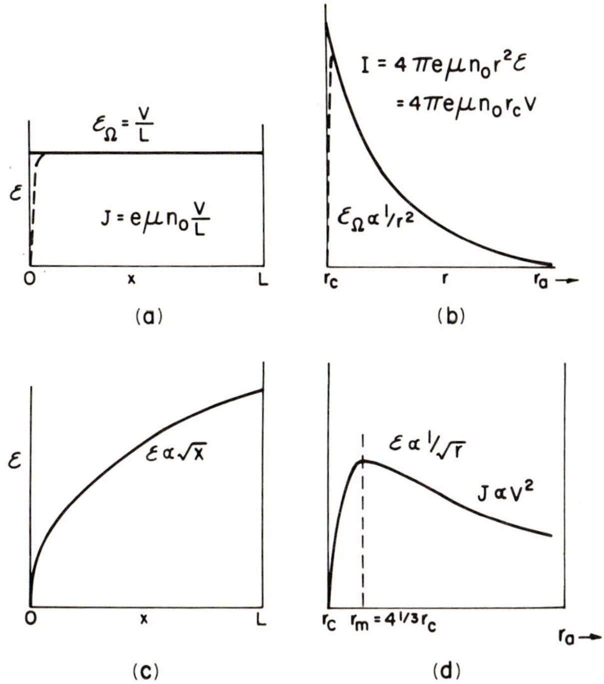

FIG. 8.2. Comparative field distributions: (a) Ohmic distribution, planar-flow case; (b) ohmic distribution, spherical radial-flow case; (c) trap-free SCL distribution, planar-flow case; (d) trap-free SCL distribution, spherical radial-flow case.

the case of pure SCL current flow in a trap-free, perfect insulator. For planar current flow, the field increases monotonically as $x^{1/2}$ [see Eq. (4.7)] on going from the cathode to the anode as in Fig. 8.2c. The field for the spherical case is obtained from the simultaneous solution of (8.1) and (8.2), with n_0, n_{tj}, and $n_{tj,0}$ all set equal to zero, with the boundary condition (8.3),

$$\mathscr{E}(r) = (I/6\pi\epsilon\mu)^{1/2}[(r^3 - r_c{}^3)/r^4]^{1/2}. \tag{8.6}$$

The field increases very rapidly from zero at the cathode, reaches a maximum at the radius r_m, given by

$$\mathscr{E}_m \equiv \mathscr{E}(r_m) = (I/6\pi\epsilon\mu)^{1/2}(3/4^{4/3}r_c)^{1/2}, \qquad r_m = 4^{1/3}r_c, \tag{8.7}$$

and thereafter decreases as $1/r^{1/2}$, as indicated in Fig. 8.2d. The pure SCL current is obtained by inserting (8.6) into (8.4), namely, in the limit $r_a \gg r_c$,

$$I = (3\pi/2)\,\epsilon\mu(V^2/r_a). \tag{8.8}$$

As previously mentioned, $I \propto V^2$, just as in the planar case. Meltzer (1960) has carried out numerical calculations for the situation where r_a does not greatly exceed r_c.

We compute also the reverse current I_r; that is, the pure SCL current that flows when the contact at r_a, assumed ohmic, is biased for injection. In this case, the field is (8.6) with r_c replaced by r_a:

$$\mathscr{E}(r) = (I_r/6\pi\epsilon\mu)^{1/2}[(r_a^3 - r^3)/r^4]^{1/2}, \tag{8.9}$$

since the ohmic contact boundary condition is applied at r_a : $\mathscr{E}(r_a) = 0$. The current I_r is given by inserting (8.9) into (8.4) and integrating the latter with the limits reversed, again in the limit of $r_a \gg r_c$:

$$I_r = I(2r_c/r_a)^2. \tag{8.10}$$

The rectification ratio $(r_a/2r_c)^2$, by which the forward current I exceeds the reverse current I_r, can obviously be exceedingly large, should the requisite trap-free conditions be realizable. This rectification is of purely geometric origin; it does not rely on any asymmetry of the contacts other than the disparity of radii.

These results for the case of pure SCL current flow are summarized in Table 8.1 for the trap-free case as well as for the case of an exponential trap distribution (see Section 4.7.1). The case of the single shallow trap level (see Section 4.3) is obtained by multiplying the trap-free currents by the θ factor [see Eq. (4.45)]. The solutions for the planar configuration are also given for ease of comparison. As mentioned in the introduction, for pure SCL current flow, the functional dependence on voltage is not contingent on the electrode geometry but is determined by the trap configuration in the forbidden gap. Also, the nonplanar pure SCL current depends only on the outer (larger) radius. As an example of the current magnitudes to be expected, consider the typical insulating material CdS, with $\epsilon/\epsilon_0 = 11$ and, say, $\mu = 200$ cm^2/V sec. For $r_a = 0.5$ cm, the *trap-free*, pure SCL currents for the cylindrical and spherical cases at $V = 100$ V are 4×10^{-5} A/cm of cylinder length and 1.5×10^{-5} A, respectively. These currents are to be compared with 1.5×10^{-5} A/cm^2 of electrode area, the planar, *trap-free*, pure SCL current for an electrode spacing of 0.5 cm. Generally, the traps present in real insulators reduce these currents by many decades.

We turn next to the spherical-flow case mentioned in the introduction, that of injection from a point where the injected space charge does not extend all the way through the solid. This is the situation characterized by the 3/2-power law between current and voltage, which has been observed with semiconducting CdS (see Section 8.6).

TABLE 8.1

I–V CHARACTERISTICS FOR PURE SPACE-CHARGE REGIMES (n_0 NEGLIGIBLE)[a,b]

Current-flow geometry	No traps[f]	Exponential distribution of traps: $N_t(E) = \mathcal{N}_0 \exp(E - E_c)/kT_t$; $l = T_t/T > 1$
Plane-parallel[c]	$I = \frac{9}{8}\epsilon\mu A \frac{V^2}{(r_a - r_c)^3}$	$I = \left[\frac{\epsilon l}{e\mathcal{N}_0 kT_c(l+1)}\right]^l (e\mu N_c A)\left[\frac{2l+1}{l+1}\right]^{l+1} \frac{V^{l+1}}{(r_a - r_c)^{2l+1}}$
Cylindrical[d]	$I = 2\pi\epsilon\mu L \frac{V^2}{r_a{}^2}$	$I = \left[\frac{2\epsilon l}{e\mathcal{N}_0 kT_c(l+1)}\right]^l (2\pi e\mu N_c L)\left[\frac{2l}{l+1}\right]^{l+1} \frac{V^{l+1}}{r_a^{2l}}$
	$I_r = I\left[\ln\left(\frac{r_a}{r_c}\right)\right]^{-2}$	$I_r = I\left[\frac{2l}{l+1}\ln\left(\frac{r_a}{r_c}\right)\right]^{-(l+1)}$
Spherical[e]	$I = \frac{3\pi}{2}\epsilon\mu \frac{V^2}{r_a}$	$I = \left[\frac{3\epsilon l}{e\mathcal{N}_0 kT_c(l+1)}\right]^l (4\pi e\mu N_c)\left[\frac{2l-1}{l+1}\right]^{l+1} \frac{V^{l+1}}{r_a^{2l-1}}$
	$I_r = I\left[\frac{r_a}{2r_c}\right]^{-2}$	$I_r = I\left[\frac{2l-1}{l+1}\frac{r_a}{r_c}\right]^{-(l+1)}$

[a] From Lampert *et al.* (1964).

[b] I corresponds to current injection at r_c, collection at r_a ; I_r corresponds to the reverse polarity.

[c] A is the electrode area; $(r_a - r_c)$ is the cathode-anode spacing; $I_r \equiv I$.

[d] L is the length of the cylinder; it is assumed that $r_a \gg r_c$.

[e] It is assumed that $r_a \gg r_c$.

[f] The shallow-trap result is given simply by multiplying the corresponding expressions by θ, the ratio of free to trapped charge.

8.3 Point-Contact Injection into an Insulator with Traps and Thermal Free Carriers; the Universal 3/2-Power Law

Our main interest in this chapter is the initial space-charge dominated regime immediately following the Ohm's law regime. This is the regime in which the injected space charge, with increasing voltage, is pushed out into the solid beyond the immediate vicinity of the point-contact cathode, but has not yet reached the anode. In the planar-flow problems discussed in Chapter 4, the nature of this regime depended strongly on specific properties of the traps and, accordingly, a separate discussion was given for each trap model. For the spherical-flow problems, the situation is quite otherwise; there is a common feature in this regime independent of the trap model, namely $I = KV^{3/2}$, in which only the coefficient K is trap-dependent. We establish this remarkable result using the Regional Approximation Method. Since the regime in question is a transition regime between two "pure" regimes, namely the Ohm's law regime and a pure SCL current regime (such as discussed in Section 8.2), this method is particularly well suited for treatment of the problem.

At the outset our discussion has much in common with the Regional Approximation treatment of the planar-flow problem for the trap-free insulator with thermal free carriers, Section 4.2.2. As previously, we note that two components contribute additively to the total current flow, the thermal free carrier n_0 and the injected carriers $n_\mathrm{i} = n - n_0$. Again, $n_\mathrm{i}(r)$ is a monotonic decreasing function of r and, over the current range of interest here, there is a spherical surface at radius r_x inside the insulator where $n_\mathrm{i}(r_x) = n_0$. For $r < r_x$, $n_\mathrm{i} > n_0$ and for $r > r_x$, $n_0 > n_\mathrm{i}$. Accordingly, the basic approximations are to neglect n_0 for $r < r_x$ and n_i for $r > r_x$. Thus, there are two regions in the insulator characterized as follows:

Region I ($r_\mathrm{c} \leqslant r \leqslant r_x$):

$$I = 4\pi e\mu n_\mathrm{i} r^2 \mathscr{E}. \tag{8.11}$$

The boundary condition (8.3) obtains in this region.

Region II ($r_x \leqslant r \leqslant r_\mathrm{a}$):

$$I = 4\pi e\mu n_0 r^2 \mathscr{E}. \tag{8.12}$$

The radius r_x is characterized by

$$n(r_x) \simeq n_\mathrm{i}(r_x) = n_0. \tag{8.13}$$

The specification for joining the solutions in Regions I and II is the usual continuity condition at r_x :

$$\mathscr{E}(r_x^-) = \mathscr{E}(r_x^+), \tag{8.14}$$

where r_x^-, r_x^+ denote the limits obtained approaching r_x from below and from above, respectively.

The total voltage (8.4) is, of course, the sum of the voltages across the two regions:

$$V = V_{c,x} + V_{x,a}; \qquad V_{c,x} = \int_{r_c}^{r_x} \mathscr{E}\, dr, \quad V_{x,a} = \int_{r_x}^{r_a} \mathscr{E}\, dr. \tag{8.15}$$

A straightforward, conventional analytical development of the solution would, at this point, require us to deal with the Poisson Equation (8.2), hence to specify the trap configuration. Since we are here deriving a result which is independent of the trap configuration, we circumvent the Poisson equation in favor of an elementary dimensional analysis, namely we assume that the voltage $V_{c,x}$ across Region I, and the total injected charge,

$$Q_x = 4\pi e \int_{r_c}^{r_x} (n_i + n_{t,i}) r^2\, dr$$

contained in Region I, can be written, respectively, as:

$$V_{c,x} = c_1 r_x \mathscr{E}_x\,, \tag{8.16}$$

Region I:

$$Q_x = c_2(4\pi/3)\, r_x^3 e(n_0 + n_{t,0}), \tag{8.17}$$

where $\mathscr{E}_x = \mathscr{E}(r_x)$, $n_{t,0}$ is the excess, trapped-electron concentration in quasi-thermal equilibrium with the excess free electron concentration $n_i = n_0$ (that is, $n_{t,0}$ is the trapped-electron excess corresponding to upward motion of the Fermi level by $0.7kT$), and c_1 , c_2 are constants of order unity. These assumptions concerning c_1 and c_2 give significant content to the relations (8.16) and (8.17).

The relation (8.16) states that the main contribution to the voltage $V_{c,x}$ comes from the region of the effective anode for Region I, namely, at $r = r_x$, rather than from near the cathode. Similarly, since $n_i(r_x) = n_0$ and $n_{t,i}(r_x) = n_{t,0}$, Eq. (8.17) stipulates that Q_x in Region I is adequately estimated by assuming that the excess charge density at the effective anode, $e(n_0 + n_{t,0})$, is uniformly distributed throughout Region I. Because of the monotonic decreasing character of $n_i(r)$, $c_2 \geqslant 1$ necessarily.

A discussion as to why relations (8.16) and (8.17) are expected to hold generally is given at the end of this section.

The 3/2-power law is derived by first noting that Region II is an "ohmic" region, characterized by

Region II:

$$V_{x,\mathrm{a}} = r_x \mathscr{E}_x , \tag{8.18}$$

$$I = 4\pi e\mu n_0 r_x V_{x,\mathrm{a}} , \tag{8.19}$$

as can be seen by solving (8.12) for $\mathscr{E}$, inserting this into (8.15), and evaluating the integral in the limit $r_x \ll r_\mathrm{a}$. Then, from (8.15), (8.16), and (8.18) it follows that

$$r_x \mathscr{E}_x = V/(1 + c_1), \tag{8.20}$$

so that the current (8.1) may be written as

$$I = 4\pi e\mu n_0 r_x [V/(1 + c_1)]. \tag{8.21}$$

To obtain the current-voltage dependence, we need only eliminate r_x from (8.21). This is accomplished, first by integrating (8.2) between the limits r_c and r_x and using boundary condition (8.3) to give

$$r_x{}^2 \mathscr{E}_x = Q_x/4\pi\epsilon. \tag{8.22}$$

Then, using (8.20), (8.22), and (8.17), we obtain

$$r_x = [3\epsilon V/(1 + c_1)\, c_2\, e(n_0 + n_{\mathrm{t},0})]^{1/2}. \tag{8.23}$$

The universal 3/2-power law follows immediately from the substitution of (8.23) into the current-voltage relation (8.21):

$$I = 4\pi e\mu n_0 [3\epsilon/c_2 e(n_0 + n_{\mathrm{t},0})]^{1/2} [V/(1 + c_1)]^{3/2}. \tag{8.24}$$

Note that (8.24) refers to the full spherical geometry. For the hemispherical geometry, used to approximate the point contact, the right-hand side of (8.24) must be divided by 2.

It is argued in Section 8.3.1 that the possible ranges of c_1 and c_2 are $\frac{1}{2} \leqslant c_1 \leqslant 2$ and $1 \leqslant c_2 \leqslant 2$, respectively, where the lower limits obtain in the deep-trap situation and the upper limits in the trap-free and shallow-trapping situations. The corresponding variation in the factor $1/c_2^{1/2}(1 + c_1)^{3/2}$ in (8.24) covers the range from $(\frac{2}{3})^{3/2} = 0.545$ at the lower limits to $(3\sqrt{6})^{-1} = 0.136$ at the upper limits,a factor of four between the two extremes. Explicit values for c_1 and c_2 are calculated for the trap-free, shallow-trap, deep-trap, and exponential-distribution trap configurations in following sections.

It remains to determine the onset voltage for the 3/2-power law regime and the range of voltages over which the law is valid. The 3/2-power law emerges from Ohm's law when the current given by (8.24) exceeds the current given by (8.19) with $r_x = r_c$ and $V_{x,a} = V$. Letting the subscript "cr" denote the quantity evaluated at this transition, we readily obtain

$$r_{x,\mathrm{cr}} = (1 + c_1)r_c\,, \tag{8.25}$$

$$V_{\mathrm{cr}} = r_c^2(1 + c_1)^3 c_2 e(n_0 + n_{t,0})/3\epsilon, \tag{8.26}$$

and

$$I_{\mathrm{cr}} = 4\pi e^2 \mu n_0 c_2 (1 + c_1)^3 (n_0 + n_{t,0}) r_c^3/3\epsilon. \tag{8.27}$$

Further, using (8.20) and (8.22), we establish the capacitance relationship:

$$Q_x \equiv CV, \qquad C = 4\pi\epsilon r_x/(1 + c_1). \tag{8.28}$$

This physically plausible relationship establishes, from (8.23), that $C \propto V^{1/2}$. Note also, from (8.25), that $C_{\mathrm{cr}} = C(V_{\mathrm{cr}}) = 4\pi\epsilon r_c$.

It is evident from (8.23) and (8.24) that the boundary radius r_x between the two regions increases only as the $\frac{1}{3}$ power of the injection level (current), as opposed to the general linear dependence of region boundaries on injection level in the planar case [see, for example, Eqs. (4.33) and (4.75)]. It is precisely because the region boundary in the spherical case moves so slowly with current that the transitional regime between the purely ohmic and purely SCL regimes is so much more significant than it is in planar flow. The span of the transitional 3/2-power law region is obtained by noting that the transition from this initial space-charge-controlled regime of the I–V characteristic to the succeeding SCL regime obviously takes place at the voltage V'_{cr} at which $r_x = r'_{x,\mathrm{cr}} \simeq r_a$. [From (8.25), $r_{x,\mathrm{cr}} \simeq r_c$, whence it follows from (8.23) that $V'_{\mathrm{cr}}/V_{\mathrm{cr}} \simeq (r_a/r_c)^2$. Since it has been assumed that $r_a \gg r_c$, it is seen that the 3/2-power law prevails over an enormous range of voltages.]

All succeeding regimes of the one-carrier SCL I–V characteristic, following the 3/2-power law regime, depend explicitly on the anode radius r_a as seen in Table 8.1. Therefore the theoretical results for these succeeding regimes are applicable only if the physical shape of the anode bears a reasonably close resemblance to a spherical or hemispherical surface. For single-crystal materials, these shapes are certainly inconvenient, if not altogether impractical. On the other hand, for the study of polymers or vitreous materials, these shapes may indeed prove practical.

Before turning to the computation of the I–V characteristics and the electric field distribution peculiar to various trap configurations, we note two useful properties of the constants c_1 and c_2, namely:

$$\text{If} \quad \mathscr{E} \propto r^n, \quad \text{with} \quad n > -1, \quad \text{then} \quad c_1 = 1/(n+1). \tag{8.29}$$

$$\text{If} \quad (n_i + n_{t,i}) \propto r^m, \quad \text{with} \quad 0 \geqslant m \geqslant -3, \quad \text{then} \quad c_2 = 3/(m+3). \tag{8.30}$$

The exponent m cannot be positive since $(n_i + n_{t,i})$ must be non-increasing with r. These relations are proved below. The conditions $n > -1$ and $m > -3$ enable us to neglect $r_c^{(n+1)}$ relative to $r_x^{(n+1)}$, and $r_c^{(m+3)}$ relative to $r_x^{(m+3)}$, respectively, thereby validating (8.16) and (8.17). Thus, in the analysis of specific problems as in the following sections, the main steps are simply the determination of the exponents n and m.

The basic step in the argument verifying that $n > -1$ and $m > -3$ and establishing the ranges of c_1 and c_2, respectively, is the representation of the functional relationship between $n_{t,i}$ and n_i by the simple form: $n_{t,i} \propto n_i{}^p$. To the extent that such a representation is valid, it is easy to see that $0 < p < 1$. For a given shift of the quasi-Fermi level, $F \to F + \delta F$, the maximum change in trap occupancy will occur if the traps are shallow, $n_{t,i} \propto n_i$, so that $p = 1$, and the minimum change will occur if the traps are deep, i.e., no change at all, $n_{t,i} =$ constant, so that $p = 0$. More formally, $n_{t,i} \propto n_i{}^p$ is equivalent to $dn_{t,i}/dn_i = pn_{t,i}/n_i$. Now, for a single trap-level, the Fermi–Dirac occupancy relationship (3.29) [noting that in Region I, where these results are pertinent, we are neglecting n_0 so that $n = n_i$ and $n_t = n_{t,i}$] gives $dn_t/dn = qn_t/n$ with $q = N/(N + gn)$, so that $0 \leqslant q \leqslant 1$ where the left- and right-hand equality signs hold in the respective limits $n = \infty$ and $n = 0$.

Taking $n_{t,i} \gg n_i$ in (8.2) (the trap-free case is explicitly calculated in Section 8.4.1), and noting that $n_{t,i} \propto n_i{}^p \propto (r^2\mathscr{E})^{-p}$ from (8.11), Eq. (8.2) can be written: $(r^2\mathscr{E})^p \, d(r^2\mathscr{E}) \propto r^2 \, dr$ so that $(r^2\mathscr{E})^{p+1} \propto r^3$. Thus, $\mathscr{E} \propto r^{(1-2p)/(1+p)}$ and $n_{t,i} \propto r^{-3p/(p+1)}$. As p varies from 0 to 1, $(1 - 2p)/(1 + p)$ decreases monotonically from 1 to $-\frac{1}{2}$, and $-3p/(p + 1)$ decreases monotonically from 0 to $-\frac{3}{2}$. Thus, referring to (8.29) and (8.30), we see that the 3/2-power law is indeed validated. Further, the corresponding ranges of c_1 and c_2 are from $\frac{1}{2}$ to 2 and from 1 to 2, respectively.

The above argument is rigorous only to the extent that the representation $n_{t,i} \propto n_i{}^p$ is valid. We have seen above, in discussing a single, discrete set of traps that this representation is indeed valid only in the limit of deep trapping or shallow trapping. For intermediate positions of the Fermi level, the equivalent p is a function of the position, hence of n_i. However since $0 < p < 1$ (in the exponential trap distribution, $p = 1/l$) and since the corresponding ranges of c_1 and c_2, for this range of p, are relatively narrow, it would appear plausible that the 3/2-power law (8.24) is indeed generally valid, with the coefficients c_1 and c_2 having possibly a very weak voltage dependence, both being bounded within the above-specified, relatively narrow limits. The general validity of the 3/2-power law is argued further in the following sections.

8.4 Some Specific Trap Configurations

We have seen above that the 3/2-power law (8.24) is given in terms of a trap-concentration parameter $n_{t,0}$ and two dimensionless constants, c_1 and c_2, that depend "weakly" on the trap configuration, being of order unity. The three quantities $n_{t,0}$, c_1, and c_2 are defined in (8.16) and (8.17). In the following sections, we discuss four specific trap configurations and compute the relevant quantities for each of them. The four configurations are: trap-free case; shallow traps; deep traps; and an exponential distribution of traps. In establishing the dependence of current on voltage, in each case it is only the Region I: $r_c \leqslant r \leqslant r_x$, with which we need be concerned. Here again we shall assume that $r_x \gg r_c$. Where $r_x < 3r_c$ we are dealing with the transition from Ohm's law to the 3/2-power law, and we cannot expect a simplified treatment to describe accurately such a transition. The electric field distributions for the above cases are also derived. In particular, for the deep trap and exponential distribution cases, this requires a further subdivision of Region I according to whether the free or the trapped space charge dominates in Poisson's equation (8.2).

8.4.1 *The Trap-Free Case*

The defining equations are (8.11) and (8.2) with $n_{t,i} = 0$ from which we obtain (8.6) for the field in Region I. The field in Region II is given by (8.5). In the limit $r \gg r_c$, the field in Region I can be written:

Region I ($r \geqslant r_c$):

$$\mathscr{E} \simeq (I/6\pi\epsilon\mu)(1/r^{1/2}). \tag{8.31}$$

The electric field distribution derived from (8.6) and (8.5) is shown in Fig. 8.3. Since $r_x \gg r_c$ in our present approximations, the maximum of the field is given by (8.7). Further, since $n_i \propto 1/r^2\mathscr{E}$ from (8.11), it follows from (8.31) that $n_i \propto r^{-3/2}$. Thus, referring to (8.29) and (8.30), $n = -\frac{1}{2}$ and $m = -\frac{3}{2}$ so that $c_1 = 2$ and $c_2 = 2$. The current-voltage characteristic is therefore, from (8.24),

$$I = 0.94\pi e n_0 \mu(\epsilon/e n_0)^{1/2} V^{3/2}. \tag{8.32}$$

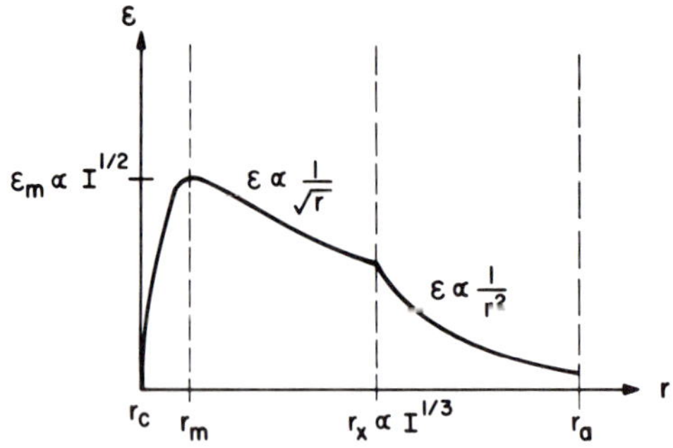

FIG. 8.3. Electric field distribution for the trap-free case in the 3/2-power law regime.

The exact solution, discussed in Section 8.5.1, contains the numerical factor 1.06 in place of 0.94.

In view of the presence of the field maximum so close to the cathode, one might well question our use of the approximate form of $\mathscr{E}$, given in (8.31), to evaluate c_1 and c_2 above, since the approximate form is badly in error near the cathode. One way of seeing that the use of the approximate form of $\mathscr{E}$ does not produce significant error is to show that its use in the derivation of the current-voltage characteristic leads to the correct result. Direct integration of (8.31) to obtain the voltage $V_{c,x}$ gives $I = 3\pi\epsilon\mu V_{c,x}^2/2r_x$, which is precisely the pure SCL current characteristic (8.8), obtained using the exact form for $\mathscr{E}$, namely (8.6), and taking $r_x \gg r_c$.

8.4.2 *The Shallow-Trap Case*

We turn next to the case in which the only effective traps are a single set of shallow traps, i.e., traps lying at an energy E_t above the thermodynamic Fermi level F_0, of concentration N_t and statistical weight g. As injection proceeds, the quasi-Fermi level F rises in the forbidden gap. So long as $(E_t - F) > kT$, the ratio of free-to-trapped carrier concentrations is a constant given by $n_i/n_{t,i} = \theta = N/gN_t$, as in (3.32). This condition will be realized throughout most of Region I, except close to the cathode, where $F > E_t$ necessarily. For the sake of simplicity, we assume that this shallow-trap condition holds everywhere in Region I. In that case, $(n_i + n_{t,i})$ in (8.2) can be replaced by n_i/θ, assuming $\theta \ll 1$, which is the interesting case. It is now evident that mathematically the shallow-trap problem is formally identical with the trap-free problem if ϵ is everywhere replaced by $\theta\epsilon$, e.g., in (8.8), (8.29), and (8.24), a result already noted for planar flow in Section 4.3. Since trapping effects are completely subsumed in this substitution, $n_{t,0}$ is taken equal to zero in (8.17), (8.23), and (8.24) and Q_x refers only to free charge. Finally, $c_1 = 2$ and $c_2 = 2$, as in the trap-free case, since these quantities are unaffected by the above substitution.

If one wants to treat the cathode region more rigorously, it is necessary to divide Region I into three separate subregions: Region $\mathrm{I_a}$, $r_c \leqslant r < r_{x''}$, with $n_i(r_{x''}) = N_t$; Region $\mathrm{I_b}$, $r_{x''} < r < r_{x'}$, with $n_i(r_{x'}) = \theta N_t$; and Region $\mathrm{I_c}$, $r_{x'} < r < r_x$, with $n(r_x) = n_0$. (Regions $\mathrm{I_a}$, $\mathrm{I_b}$, and $\mathrm{I_c}$ correspond to Regions I, II, and III, respectively, in the corresponding planar flow problem, Section 4.6.1. Likewise, $r_{x''}$, $r_{x'}$ and r_x correspond to x_1, x_2, and x_3, respectively.) In Region $\mathrm{I_a}$, $n_{t,i}$ is neglected in (8.2); in Region $\mathrm{I_b}$, n_i is neglected in (8.2) and $n_{t,i} = N_t =$ constant; and in Region $\mathrm{I_c}$, n_i is neglected in (8.2) and $n_{t,i} = n_i/\theta$. With $r_x \gg r_c$, the corrections introduced by this more rigorous treatment are indeed small and the above-derived result holds as stated. This more careful treatment of Region I is, however, useful for the study of succeeding regimes in the space-charge-controlled current-voltage characteristic, a matter which is not discussed further here.

8.4.3 *The Deep-Trap Case*

We next discuss the problem where the only effective traps constitute a single, discrete set lying at an energy below the thermodynamic Fermi level F_0. Let $p_{t,0}$ be the thermal-equilibrium concentration of unoccupied traps. Where the quasi-Fermi level F has been raised above F_0, through injection, by more than kT, namely over Region I, the previously unoccupied traps are filled with electrons, and $n_{t,i} \simeq p_{t,0}$. Assuming that $p_{t,0} \gg n_0$, which is the interesting case, it follows that $n_{t,i} > n_i$ over most of Region I, namely everywhere except very close to the cathode. It simplifies matters considerably in the derivation of the I–V characteristic (although not in obtaining the field distribution) to take $n_{t,i} > n_i$ *everywhere* in Region I, so that n_i can then be neglected in (8.2), and $n_{t,i}$ is taken equal to $p_{t,0}$. It is then trivial to integrate (8.2), obtaining

$$\mathscr{E} = \frac{ep_{t,0}}{3\epsilon}\left(\frac{r^3 - r_c^3}{r^2}\right) \simeq \frac{ep_{t,0}}{3\epsilon} r \qquad \text{for} \quad r \gg r_c, \tag{8.33}$$

where the first relationship is the exact one satisfying the boundary condition $\mathscr{E}(r_c) = 0$. Further, since $(n_i + n_{t,i}) \simeq n_{t,i} = p_{t,0} \propto r^0$ (constant), referring to (8.29) and (8.30), we see that $n = 1$ and $m = 0$ so that $c_1 = \frac{1}{2}$ and $c_2 = 1$. The current voltage characteristic is

$$I = 3.86\pi e n_0 \mu (\epsilon/ep_{t,0})^{1/2} V^{3/2}. \tag{8.34}$$

The exact solution, which differs only by a numerical constant, is discussed in Section 8.5.2.

Here again, as with the previous problem, one might well raise the question as to the magnitude of error introduced by the incorrect treatment of the cathode region. Strictly speaking, the Region I should be divided into two separate subregions: Region I_a, $r_c \leqslant r < r_{x'}$, with $n_i(r_{x'}) = p_{t,0}$ and Region I_b, $r_{x'} \leqslant r < r_x$ with $n_i(r_x) = n_0$ as previously. Over Region I_a, $n_i > p_{t,0}$ so that $n_{t,i}$ can be neglected in (8.2), and over Region I_b, $n_i < p_{t,0}$ so that n_i can be neglected in (8.2) and the right side of the Poisson equation is constant. (In the above argument, we have assumed that Region I_b covers all of Region I.) If this three-region problem is now analyzed, one finds, using obvious notation, that $V_{x',x} = (V_{c,x'} + V_{x,a})/2$, so that $V = 3(V_{c,x'} + V_{x,a})/2$. Further $V_{x,a}/V_{c,x'} = (p_{t,0}/4n_0)^{2/3}$, so that for $p_{t,0} \gg n_0$, $V_{c,x'}$ can be neglected and the above result is shown to be correct.

It is also instructive to examine the field distribution that derives from this more exact three-region treatment. The detailed analysis is presented below; here we summarize the outstanding results. These are illustrated schematically in Fig. 8.4. We note that $\mathscr{E}$ as a function of r has three extrema—two maxima and one minimum. The first maximum lies in the perfect-insulator domain characterized by Region I_a; it is identical to the maximum exhibited in Fig. 8.2d. We have already seen that in Region I_b, where the space charge in Poisson's equation is the constant quantity $n_{t,i} = p_{t,0}$, the field accordingly increases approximately linearly with r, namely as in (8.33). This is possible only if the field has a minimum near $r_{x'}$, the boundary between Regions I_a and I_b. Likewise, since $\mathscr{E}$ decreases

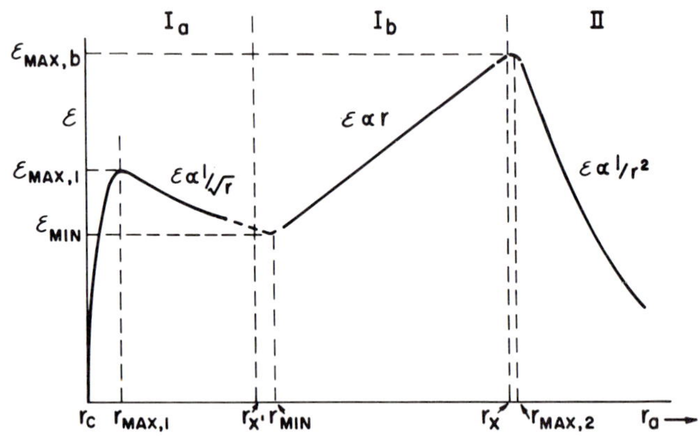

FIG. 8.4. Electric field distribution for the deep-trap case in the 3/2-power law regime (Regional Approximation diagram).

as $1/r^2$ in the ohmic Region II beyond r_x, the field must have a second maximum near r_x. The transition boundaries $r_{x'}$ and r_x both recede from the injecting point as $I^{1/3}$; for r_x this is already expected from (8.23) and (8.24). Also, the first field maximum, in Region I_a nearest the cathode, is initially *lower* than the second field maximum in the vicinity of the boundary between Regions I_b and II, but the first maximum increases faster with injection level than the second maximum, namely, as $I^{1/2}$ contrasted with $I^{1/3}$, and so the former finally overtakes the latter.

Since, from (8.33), $V_{c,x} \simeq ep_{t,0}r_x^2/6\epsilon$, the upper end of the 3/2-power regime occurs at the voltage $V_{TFL} \simeq ep_{t,0}r_a^2/6\epsilon$. At this voltage, all the traps throughout the volume are filled—hence the subscript TFL, denoting trap-filled-limit. Beyond this point, the current rises very steeply with voltage in a manner similar to the planar-flow TFL problem (see Section 4.4). Details of this regime and its transition to the trap-free regime (8.8) can be readily worked out in terms of the two-region picture, where the two regions are now I_a, $r_c \leqslant r < r_{x'}$ with $n_i(r_{x'}) = p_{t,0}$ as previously, and I_b, $r_{x'} \leqslant r \leqslant r_a$.

Finally we note that although we have discussed the deep-trap case in terms of one set of traps, at a single, discrete energy level, it is obvious that the argument holds for an arbitrary distribution of traps provided only that they all lie below F_0 in energy. In this case, $p_{t,0}$ is the total concentration of unoccupied traps at all energies.

We now provide somewhat more detail in the Regional Approximation treatment of the electric field distribution. In Region I_a, $\mathscr{E}(r)$ is given by (8.6). At the boundary $r_{x'}$, $n_i = p_{t,0}$, so that (8.6) and (8.1) give

$$r_{x'} = r_c(1 + w)^{1/3}, \qquad w = 3\epsilon I/8\pi e^2\mu p_{t,0}^2 r_c^3. \tag{8.35}$$

The maximum field in Region I_a is given by (8.7), and clearly this region is fully developed only if $r_{x'} > r_m$.

In Region I_b, the precise distribution for $\mathscr{E}$ is given by (8.33) which we rewrite here as

Region I_b:

$$\mathscr{E}(r) = Cr + (D/r^2); \qquad C = ep_{t,0}/3\epsilon, \quad D = r_c^3C(w - 1). \tag{8.36}$$

The transition boundary r_x is obtained by evaluating (8.36) at r_x and eliminating $\mathscr{E}_x$ using (8.1) with $n = n_0$:

$$r_x = r_c[2w(p_{t,0}/n_0) - (w - 1)]^{1/3}. \tag{8.37}$$

Now $D \gtreqless 0$ when $w \gtreqless 1$ or $(r_{x'}/r_c)^3 \gtreqless 2$. Further,

$$d\mathscr{E}/dr = C - (2D/r^3). \tag{8.38}$$

For $D < 0$, there is clearly no extremum of $\mathscr{E}$ inside Region I_b, and $d\mathscr{E}/dr$ has its maximum value at $r_{x'}$; $(d\mathscr{E}/dr)_{max} = C + 2|D|/r_{x'}^3$. For $D = 0$, $d\mathscr{E}/dr = C$, a constant. For $D > 0$, $d\mathscr{E}/dr = 0$ at r_{min} given by

$$r_{min} \equiv (2D/C)^{1/3} = r_c[2(w-1)]^{1/3}. \tag{8.39}$$

Note that $r_{min} > r_{x'}$ so long as $w > 3$. Thus $r_{x'} > (4)^{1/3} r_c$ is required for the existence of the maximum (8.7) in Region I_a as well as for the existence of the minimum in Region I_b. Also, $r_x \gg r_{x'}$ because we have assumed that $(p_{t,0}/n_0) \gg 1$, and further $r_x > r_{min}$ because

$$(p_{t,0}/n_0) > [(\tfrac{3}{2}) - (\tfrac{3}{2})w].$$

The minimum field $\mathscr{E}_{min}$ at r_{min} in Region I_b is

$$\mathscr{E}_{min} = \tfrac{3}{2}Cr_{min} = (ep_{t,0}r_c/2\epsilon)[2(w-1)]^{1/3}. \tag{8.40}$$

For $w \gg 1$, that is, for Region I_a extensively developed, $\mathscr{E}_{min} \propto w^{1/3} \propto I^{1/3}$ [through (8.35)] as shown in Fig. 8.4. From (8.7) and (8.40), the ratio of the maximum field in Region I_a and the minimum field in Region I_b is

$$\frac{\mathscr{E}_{max,a}}{\mathscr{E}_{min}} = \left(\frac{4w}{3}\right)^{1/2}\left[\frac{1}{4(w-1)}\right]^{1/3}, \tag{8.41}$$

which is unity for $w = 3$ and is greater than unity for $w > 3$; $\mathscr{E}_{max,a} > \mathscr{E}_{min}$ always, when Region I_a is fully developed. Further, $\mathscr{E}_{max,a}/\mathscr{E}_{min} \propto w^{1/6}$ for $w \gg 1$ so that the minimum in Region I_b is always less than the maximum in Region I_a as the injection level is raised.

Finally, the field at r_x is

$$\mathscr{E}_{r_x} = Cr_x\left[\frac{2w(p_{t,0}/n_0)}{2w(p_{t,0}/n_0) - (w-1)}\right], \tag{8.42}$$

and, for $w \gg 1$, we have that

$$\mathscr{E}_{r_x} \simeq Cr_x = (ep_{t,0}r_c/3\epsilon)[2w(p_{t,0}/n_0)]^{1/2} \propto I^{1/3}; \qquad w \gg 1, \tag{8.43}$$

where the final proportionality follows from (8.35). Since $\mathscr{E} \propto r$ in Region I_b for $r \gg r_{x'}$ from (8.33), the maximum field $\mathscr{E}_{max,b}$ in this region occurs at r_x : $\mathscr{E}_{max,b} = \mathscr{E}_{r_x}$. The ratio of $\mathscr{E}_{max,b}$ to $\mathscr{E}_{max,a}$ is given by

$$\frac{\mathscr{E}_{max,b}}{\mathscr{E}_{max,a}} = \left(\frac{16}{27}\right)^{1/6}\left(\frac{1}{w}\right)^{1/6}\left(\frac{p_{t,0}}{n_0}\right)^{1/3} \tag{8.44}$$

Since we assume $(p_{t,0}/n_0) \gg 1$, Eq. (8.44) exceeds unity at low injection levels (small w). This is the situation depicted in Fig. 8.4. However, w increases linearly with injection level so that ultimately the maximum field in Region I_a will exceed the maximum field in Region I_b provided, of course, that no new regime (e.g., $r_x > r_a$) has developed.

8.4.4 *The Exponential Trap Distribution*

As a final example of a trap configuration, we consider the exponential trap distribution that was previously treated in the planar-flow analysis in Section 4.7.1. The trap distribution is given in (4.115), rewritten here for convenience: $\mathscr{N}(E) = (N_t/kT_t) \exp(E - E_c)/kT_t$. As before, we note that for $l = T_t/T > 1$ the integrated trap concentration between the quasi-Fermi level F and F_0 is also approximately the concentration of trapped electrons: $n_{ti} \simeq N_t \exp(F - F_0)/kT_t$. In the context of the universal 3/2-power law, Section 8.3, we consider the problem as a two-region problem: Region I extending out from the cathode and dominated by the trapped, injected charge, and Region II extending inward from the anode and dominated by the thermal free carriers (an ohmic region). In Region I the thermal free carriers are neglected so that the relationship between trapped and free injected charge is given by (4.118), taking $n_0 = 0$, namely

$l > 1$:

$$n_{ti} = N_t(n_i/N_c)^{1/l}. \tag{8.45}$$

We have already seen in Section 8.3 that $n_{t,i} \propto n_i{}^p$ implies

$$n_{ti} \propto r^{-3p/(p+1)} \qquad \text{and} \qquad \mathscr{E} \propto r^{(1-2p)/(1+p)}.$$

Here $p = 1/l$ so that, from (8.29), $c_1 = (l+1)/(2l-1)$, and, from (8.30), $c_2 = (l+1)/l$. Substituting these results into (8.24), we obtain for the I–V characteristic:

$$I = 4\pi e\mu n_0[3\epsilon l/eN_t(l+1)]^{1/2}[(2l-1)/3l]^{3/2}V^{3/2}. \tag{8.46}$$

Just as in the deep-trap case, a rigorous treatment of the present problem necessitates splitting Region I into two subregions, I_a and I_b, where the injected free carriers dominate Region I_a, adjacent to the cathode, and the injected trapped carriers dominate the adjoining Region I_b. Although this does not influence the current-voltage dependence, again as in the deep-trap case, it markedly influences the electric field distribution which, naturally, depends on l. Since, in Region I_b, $\mathscr{E} \propto r^{(l-2)/(l+1)}$, $d\mathscr{E}/dr \propto [(l-2)/(l+1)]\, r^{-3/(l+1)}$. Thus, for $1 < l < 2$,

$d\mathscr{E}/dr$ is always negative, the field in Region $\mathrm{I_b}$ is always monotonically decreasing with r, and there is only one maximum in the field distribution, namely that given by (8.7), in Region $\mathrm{I_a}$. For $l = 2$, $d\mathscr{E}/dr$ vanishes in Region $\mathrm{I_b}$ so that the field there is constant. Finally, for $l > 2$, the field in Region $\mathrm{I_b}$ has a positive slope and the three field extrema that characterize the deep-trap problem appear here as well. Clearly, as $l \to \infty$, the field in Region $\mathrm{I_b}$ becomes proportional to r, this being mathematically equivalent to the deep-trap case. Similarly, setting l equal to unity conforms with the mathematics of the trap-free case. Thus, although the analysis of Section 8.3, establishing the existence of three extrema in the field for the deep-trap case, would correspond to setting $l = \infty$ in the present case, it holds more generally for the cases $l > 2$.

8.5 Further Mathematical Discussion

The one-carrier injection equations are given in (8.1) and (8.2). For the discussion in this section it is convenient to define dimensionless variables:

$$u = \frac{n_0}{n} = \frac{4\pi\epsilon\mu n_0 r^2 \mathscr{E}}{I}, \qquad i = \frac{3\epsilon I}{4\pi e^2 n_0{}^2 \mu r_\mathrm{c}{}^3},$$
$$v = \frac{12\pi e n_0 \mu r_\mathrm{c} V}{I}, \qquad z = \left(\frac{r}{r_\mathrm{c}}\right)^3. \tag{8.47}$$

In terms of these variables, (8.1), (8.2), and (3.11) combine to give the equation:

$$i\frac{du}{dz} = \frac{i-u}{u} + \sum_j B_j \frac{1-u}{1+C_j u}; \qquad B_j = \frac{N_{\mathrm{t},j}}{n_0}\frac{C_j}{(1+C_j)}, \qquad C_j = \frac{N_j}{g_j n_0} \tag{8.48}$$

and the voltage relationship (8.4) becomes

$$v = \int_1^{z_\mathrm{a}} u\, dz/z^{4/3}; \qquad z_\mathrm{a} = (r_\mathrm{a}/r_\mathrm{c})^3. \tag{8.49}$$

It is obvious that the integration in (8.49) cannot be performed analytically, even for the simplest case, namely with all $B_j = 0$ in (8.48); i.e., the case of no traps.

For further discussion, it is convenient to consider individual cases separately.

8.5.1 *The Trap-Free Case*

Since there are no traps, all of the $B_j = 0$, so that (8.48) can be written:

$$u\,du/(1-u) = \{-1 + 1/(1-u)\}\,du = (1/i)\,dz, \tag{8.50}$$

with solution

$$-u - \ln(1-u) = (1/i)(z-1) = y, \tag{8.51}$$

satisfying the boundary condition (8.3) which is $u = 0$ at $z = 1$ in the dimensionless notation. In terms of the variable y, Eq. (8.49) becomes

$$v = i^{-1/3}\int_0^{w_a} u\,dy/[y + (1/i)]^{4/3}, \qquad y_a = (1/i)(z_a - 1) \simeq z_a/i. \tag{8.52}$$

The range of the variable u is $0 \leqslant u < 1$. From (8.51) it follows that for $w > 6$, $0.999 < u < 1$. Thus, for $z > 1 + 6i$ or $y > 6$, there is negligible error in taking $u = 1$ in (8.49) and (8.52).

(a) Ohm's Law Regime

Rewrite (3.49) as $v = I_1 + I_2$ with

$$I_1 = \int_1^{1+6i} u\,dz/z^{4/3}$$

and

$$I_2 = \int_{1+6i}^{z_a} dz/z^{4/3} = 3\{(1+6i)^{-1/3} - z_a^{-1/3}\}.$$

Now, for $i \ll \frac{1}{6}$, $I_1 < 6i \ll 1$ and $I_2 \simeq 3$. Thus $v \simeq 3$, which corresponds, from (8.47), to Ohm's law: $I = 4\pi e n_0 \mu r_c V$.

(b) 3/2-Power Regime

Consider currents i in the range $1 \lll i \lll z_a$. There will be an enormous range of currents satisfying these inequalities even for r_a/r_c as small as 100, since then $z_a = 10^6$. Rewrite (8.52) as $v = i^{-1/3}K$, $K = K_1 + K_2$ with

$$K_1 = \int_0^6 u\,dy/[y + (1/i)]^{4/3}$$

and

$$K_2 = \int_6^{y_a} dy/[y + (1/i)]^{4/3} = 3\{[6 + (1/i)]^{-1/3} - [y_a + (1/i)]^{-1/3}\}.$$

For the range of i's considered,

$$K_2 \simeq 3/6^{1/3} \qquad \text{and} \qquad K_1 \simeq \int_0^6 u\, dy/y^{4/3} = 8.84,$$

as determined with the aid of a digital computer. Thus, $v = 10.49 i^{-1/3}$, or from (8.47),

$$I = 1.06 \pi e n_0 \mu (\epsilon / e n_0)^{1/2} V^{3/2}. \tag{8.53}$$

The approximate result (8.24), taking $n_{t,0} = 0$ and $c_1 = c_2 = 2$ as determined in Section 8.4.1, is the same as (8.53) except that the numeric 1.06 is replaced by $4/3\sqrt{2} = 0.94$.

In the above discussion, the approximation of K_1 by $\int_0^6 u\, dy/y^{4/3}$ involves a small error (an overestimate) in the range (approximately) $0 \leqslant w \leqslant 1/i$. The integral $\int_0^{1/i} u\, dy/[y + (1/i)]^{4/3} \simeq (1/i)^{-4/3} \int_0^{1/i} u\, dy \simeq (2\sqrt{2}/3)(1/i)^{1/6}$, since for small y, $u \simeq (2y)^{1/2}$ from (8.51). In the above we have used $\int_0^{1/i} u\, dy/y^{4/3} \simeq 6\sqrt{2}(1/i)^{1/6}$. Thus, for $1/i$ small enough, the error is negligible.

8.5.2 *The Deep-Trap Case*

Consider a single set of traps lying below F_0 so that $C = N/gn \ll 1$ in (8.48), where we have suppressed the subscript j. Thus the single trap term on the right-hand side of (8.48) is $B(1-u)/(1+Cu) \simeq B(1-u)$. With this approximation, Eq. (8.48) can be written as

$$\{[1/(1-u)] - [1/(1+Bu)]\}\, du = [(1+B)/i]\, dz = dy, \tag{8.54}$$

with solution

$$-\ln(1-u) - (1/B)\ln(1+Bu) = [(1+B)/i](z-1) = y. \tag{8.55}$$

The voltage integral (8.49) can now be written as

$$v = \left(\frac{i}{1+B}\right)^{-1/3} \int_0^{y_a} \frac{u\, dy}{[y + [(1+B)/i]\}^{4/3}}, \qquad y_a \simeq \frac{1+B}{i} z_a\,. \tag{8.56}$$

The 3/2-power law corresponds to the current range:

$$1 \ll i/(1+A) \ll z_a\,.$$

Following the same procedure as in the preceding problem, we write (8.56) as: $v = [i/(1+B)]^{-1/3} K$, $K = K_1 + K_2$ with

$$K_1 = \int_0^6 u\, dy/\{y + [(1+B)i]\}^{4/3}$$

and

$$K_2 = \int_6^{y_a} dy/\{y + [(1+B)/i]\}^{4/3}$$

$$= 3\,\{[6 + (1+B)/i]^{-1/3} - [y_a + (1+B)/i]^{-1/3}\}.$$

Over the range of i's considered,

$$K_2 \simeq 3/6^{1/3} = 1.65 \qquad \text{and} \qquad K_1 \simeq \int_0^6 u\, dy/y^{4/3}.$$

As with the preceding problem, K_1 can only be evaluated, using (8.54) or (8.55), with the aid of a computer, which has not been done.

Noting that for deep traps, $C \ll 1$, then $B \simeq p_{t,0}/n_0$, and the final result can be written, assuming $p_{t,0} \gg n_0$ or $B \gg 1$, as

$$I = \frac{9}{K^{3/2}}\, 4\pi e n_0 \mu \left(\frac{\epsilon}{e p_{t,0}}\right)^{1/2} V^{3/2}, \tag{8.57}$$

where K remains to be evaluated. The approximate result (8.24), taking $n_{t,0} = p_{t,0} \gg n_0$ and $c_1 = \frac{1}{2}$, $c_2 = 1$ as determined in Section 8.4.3, is the same as (8.57) except that the quantity $9/K^{3/2}$ is replaced by $2^{3/2}/3 = 0.94$.

As with the preceding problem, the approximation of K_1 by $\int_0^6 u\, dy/y^{4/3}$ is in error over the range $0 \leqslant y \leqslant (1+B)/i$. However, it is easily shown that $\int_0^{(1+B)/i} u\, dy/y^{4/3}$ is in any case negligibly small for $(1+B)/i \ll 1$.

8.5.3 *The General Case*

This case is characterized by (8.48). An analysis along the same lines as followed in Sections 8.5.1 and 8.5.2 above gives the result

$$v = \left(\frac{i}{1+B}\right)^{-1/3} \int_0^{y_a} \frac{u\, dy}{\{y + [(1+B)/i]\}^{4/3}},$$
$$B = \sum_j \frac{B_j}{1+C_j}, \qquad y_a = \left(\frac{1+B}{i}\right) z_a. \tag{8.58}$$

Again, in the current range $1 \ll i(1+B) \ll z_a$, $v = [i/(1+B)]^{-1/3} K$, with $K = K_1 + K_2$, $K_1 \simeq \int_0^6 u\, dy/y^{4/3}$ and $K_2 \simeq 3/6^{1/3}$, where K_1 can only be evaluated with the aid of a computer; and again the 3/2-power law is obtained.

8.6 Experimental Verification of the 3/2-Power Law and Further Applications

There is, in the published literature, only a single experiment bearing on the 3/2-power law known to the authors. It is reproduced in Fig. 8.5 in which the measured current-voltage characteristic is shown for

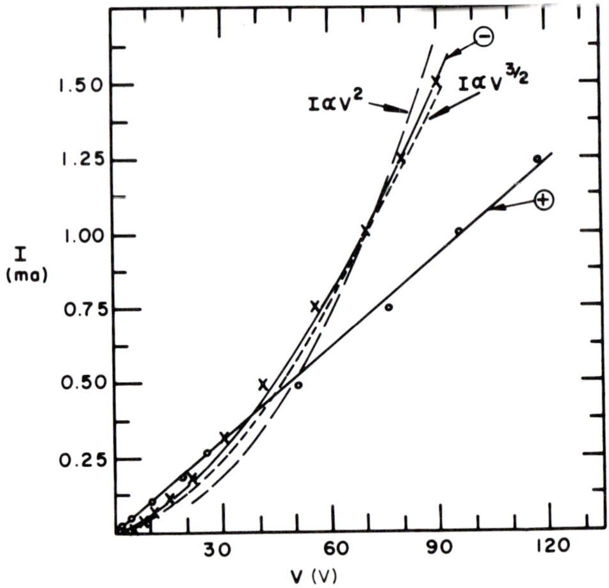

Fig. 8.5. Current-voltage characteristics obtained with a point-contact electrode on a single crystal of CdS, resistivity 20 Ω-cm. All measurements were made at room temperature. The point radius was approximately 1 μ. The crosses represent data taken with the point injecting, i.e., biased negatively; the solid line labeled ⊖ is a smooth fit to the data. The dashed curve is a 3/2 power-law curve fitted to the data at one point, $I = 1$ mA. The open circles represent data taken with the point biased positively. The straight line labeled ⊕ is an Ohm's-law fit to the data. (From Lampert *et al.*, 1964, Fig. 1.)

injection from a point contact of radius $r_c \simeq 1\ \mu$ into a single crystal of CdS at room temperature. It is seen that $I \propto V^{3/2}$ gives a good fit to the data whereas $I \propto V^2$ gives a rather poor fit. (With the polarity such that the point is positive, Ohm's law is observed over the entire voltage range, as expected.) Since the room-temperature resistivity of the sample was 20 Ω cm, the corresponding trap-free, 3/2-power law, Eq. (8.32), is $I = 4.77 \times 10^{-6} V^{3/2}$ A, for V in volts, taking $\epsilon/\epsilon_0 = 11$ and $\mu = 200\ \text{cm}^2/\text{V sec}$, and dividing by two to take the hemispherical geometry into account. The experimental curve in Fig. 8.5 is fitted by

$I = 1.75 \times 10^{-6} V^{3/2}$ A. From (8.24) it can then be concluded that trapping is taking place with $n_{t,0}/n_0 = 7$. Since 20 Ω cm corresponds to $n_0 \simeq 10^{15}$ cm^{-3}, the thermodynamic Fermi level F_0 was $\simeq$0.2 eV below the conduction band. The observation of trapping effects in this circumstance is not unexpected for CdS.

Regarding future applications of the point-contact injection-current technique to the analysis of defect states in solids, the following remarks are offered in a somewhat speculative vein. From the physical analysis of the 3/2-power law regime it follows, since the injected space charge spreads out from the point gradually with increasing voltage, that the anode configuration is irrelevant at voltages V such that $r_x \ll (A_a/\pi)^{1/2}$, where A_a is the effective anode area. Furthermore, even if the injecting point contact, of effective area A_c, cannot be adequately represented by a hemispherical electrode, one has the intuitive expectation that the one-dimensional, radial-flow analysis is still applicable to obtain the I–V characteristic over the range of voltages for which

$$(A_c/\pi)^{1/2} \ll r_x \ll (A_a/\pi)^{1/2}.$$

Therein lies the potential usefulness of the injecting point as a "probe" to study trapping locally.

The injecting-point "probe" and the planar-flow measurement technique appear to have rather complementary virtues for the study of defect states in insulators. The former studies traps locally, near the injecting point, and the latter necessarily studies an average over all traps between cathode and anode. The point-contact method is experimentally more convenient in at least two respects. First, where contact difficulties are present, which is frequently the case, they can be more easily overcome with a point contact than with a wide-area contact; e.g., tunneling to achieve injection is more easily promoted with a small-radius point. Also if injection has to be achieved by electronic means, e.g., by electron-bombardment-induced or photo-induced conductivity, a higher density of free carriers can be achieved locally (equivalent point contact). In this connection, the use of the naturally narrow-beam laser light may prove exceptionally useful. Thickness requirements with the planar-flow geometry are usually quite severe and often very thin samples have to be used. Certainly, thicker samples can be used with the point-contact technique. On the other hand, for very high-resistivity insulators, e.g., polystyrene, some background, light-stimulated conductivity may be necessary to obtain an n_0 high enough to use the point-contact method conveniently.

In favor of the planar-flow measurements, they can yield both more information, e.g., the trap energy level, from the richer structure in the

SCL I–V curve, and more accurate information, e.g., a remarkably accurate determination of the average trap density from the trap-filled-limit voltage V_{TFL}. (To be sure, in some cases difficulties of inconsistencies in the derived data have appeared.) The injecting-point measurement, at a single temperature, yields directly no information about the trap energy level and, further, yields the quantity $n_{t,0}$ in (8.24) with a maximum uncertainty of a factor of 16. (Actually the latter uncertainty can be narrowed by an order of magnitude with only minimal, qualitative information about the traps, such as whether they are shallow or deep or distributed in energy. The temperature-dependent behavior of $n_{t,0}$ can be expected to yield such information.)

A further uncertainty relates to our idealization of the injecting-point geometry. We have represented the injecting point mounted against a flat surface by a hemispherical flow geometry and then neglected possible surface effects. If the total number of surface states out to radius r_x is less than Q_x/e, presumably only small errors are introduced by our treatment. However, if these surface states exceed Q_x/e, the magnitude of the error is unknown and remains to be investigated. Here an experimental approach may prove more fruitful or convenient than an attempted generalization of the theory.

8.7 Pure SCL Current Flow in the Cylindrical Geometry

For the sake of completeness, we consider here the analog, for cylindrical geometry, of the problem discussed in Section 8.2, namely SCL current flow in a trap-free insulator with no thermal free carriers. The electrode configuration is that of concentric cylinders of radii r_c and r_a, respectively. The basic equations determining the current flow are

$$I' = 2\pi e\mu n_i r\mathscr{E} = \text{constant}, \tag{8.59}$$

where I' is the current per unit length of cylinder, and

$$(\epsilon/r)(d/dr)(r\mathscr{E}) = en_i , \tag{8.60}$$

subject to the usual ohmic contact boundary condition (8.3) for diffusion-free current flow. The field for the case that r_c is the injecting electrode is

$$\mathscr{E}(r) = (I'/2\pi\epsilon\mu)^{1/2}[(r^2 - r_c^2)/r^2]^{1/2}, \tag{8.61}$$

which differs basically from the field for injection at a point. In the present cylindrical case, the field increases monotonically from the injecting inner contact. Thus, it is much more like the field in the planar configuration. The forward and reverse currents I' and I_r', analogous to the spherical case, can be readily computed. They are

$$I' = 2\pi\epsilon\mu(V^2/r_a^2) \tag{8.62}$$

and

$$I_r' = I[\ln(r_a/r_c)]^{-1}. \tag{8.63}$$

There is geometric rectification here also, but it is considerably less than in the spherical case.

8.8 Planar SCL Current Flow in a Sample of Finite Cross Section

In the remainder of this chapter, we shall deal in a qualitative way with planar SCL current flow in a sample of finite cross section. The type of current flow that concerns us here is realized experimentally with a geometry in which the cathode–anode spacing L is larger than at least one of the transverse dimensions of the sample; that is, with typical "gap" electrode structures. Such measurements, on evaporated layers of crystalline hexagonal Se, have been reported by Polke *et al.* (1963) and by Polke (1964). The general feature of this type of current flow is ohmic flow at low applied voltages followed by a transition to a new regime, as the voltage is raised, in which $I \propto V^2/L^2$, in contrast to the L^{-3} dependence expected from conventional one-dimensional planar SCL current flow.

The crux of the theoretical problem in analyzing SCL current injection into, say, a long, needle-like or plate-like sample lies in the fact that the lines of force emanating from the distributed, injected charge are not confined to the sample. Concomitantly, even though there is a potential minimum near the cathode, the total injected charge to the right of the minimum need not be compensated entirely by positive charge on the anode. Lines of force of injected charge near the cathode will leave the insulator and return to the cathode structure through the external surroundings, i.e., some of the compensating charge will be on the cathode structure. Thus, the actual geometric configurations of the cathode and anode must be taken into account in formulating the boundary conditions. Further, if there are other objects near the sample these may play an overriding role in determining the injection characteristic, e.g., the "gate" electrode in field-effect devices. Thus the problem of SCL current injection into thin samples does not have a uniquely determined solution until the environment is completely specified.

On the basis of physical argument, as proposed first by Rose (1951), we may say the following about SCL current injection into a long needle or platelet in the case when there are no other close-by objects in the environment. In these extreme geometries, most of the lines of force will leave the sample before reaching the anode (or cathode) structure. Therefore it is reasonable to expect that the capacitance C which determines the total injected charge Q, via $Q = CV$, is roughly the same as the geometric capacitance C_0 of the sample to space (as if the entire sample were at one potential). If one knows further the fraction η of the injected charge which remains free (untrapped) to carry current—and this presumes knowledge of the trapping states in the sample—one can directly apply the phenomenological Eq. (2.5), corrected by the factor η:

$$J \simeq \eta v C_0 V/L \simeq \eta \mu C_0 V^2/L^2, \tag{8.64}$$

taking $v = \mu \mathscr{E} \simeq \mu V/L$. The factor η itself may or may not depend on the voltage depending on the distribution in energy of the traps. The characteristic described by (8.64), with constant η, conforms with the experimental observations referred to earlier as well as with the conclusions of a mathematical analysis of the sheet-flow problem carried out by Guerst (1966). The latter is complicated by the requirement of a nonlinear boundary condition on the tangential component of the electric field at the surface of the insulator in the interelectrode space. The first order result of Guerst's calculation, applied to the trap-free, perfect insulator in the shape of a thin, infinitely wide platelet is $J' = 2\mu\epsilon V^2/\pi L^2$ where J' is the current per unit width. It clearly exhibits the expected L-dependence. But there is a substantial, added complication in applying even the simple phenomenological result (8.64), as well as Guerst's analysis, and that is that the influence of surface trapping states as well as volume states must be taken into account in analyzing the factor η. Clearly, a great deal of more incisive work would have to be done before one could say more about these difficult problems.

9

THE EXACT THEORY

Our understanding of one-carrier injection currents in solids is based almost entirely on the simplified theory which neglects the diffusion current. This is certainly an adequate picture in almost all circumstances that are likely to be of practical interest. Nonetheless, before leaving the subject we want to give some consideration to the exact theory which takes into account the diffusion currents. Since the structure of the contact now comes into play it is necessary, once the formal preliminaries are out of the way, to carry out the discussion for particular contacts, in turn. This is done for the reservoir-type, tunneling-type, and n^+–n junction contacts. Then a situation is considered where the simplified theory is essentially exact, a universal geometric result is obtained, and limits of validity of the simplified theory are discussed. Finally, after some brief remarks on the perfect insulator problem in the context of the exact theory, some computer results are presented for the trap-filled limit problem in this context and the "validity" of the simplified theory for this problem thereby established.

9.1 Formulation

The equations governing the exact theory are (3.4), (3.8), and (3.11) which we write again here for convenience:

$$J = e\mu(n\mathscr{E} - V_{\mathrm{T}}\, dn/dx) = \text{constant}. \tag{9.1}$$

$$(\epsilon/e)\,(d\mathscr{E}/dx) = (n - n_0) + \sum_j (n_{tj} - n_{tj,0}). \tag{9.2}$$

$$n_{tj} = \frac{N_{tj}}{1 + (1/g_j)(N_j/n)}, \qquad N_j = N_c \exp[(E_{tj} - E_c)/kT]. \tag{9.3}$$

Using (9.1) to eliminate $\mathscr{E}$, and (9.3) to eliminate n_{tj}, Eq. (9.2) reduces to a nonlinear, second-order differential equation in n:

$$\frac{d^2n}{dx^2} - \left(\frac{J}{e\mu V_T} + \frac{dn}{dx}\right)\frac{1}{n}\frac{dn}{dx} - \frac{e}{\epsilon V_T}\,n(n - n_0)\left[1 + \frac{1}{n_0}\sum_j \frac{n_{tj,0}N_j}{N_j + g_j n}\right] = 0. \tag{9.4}$$

This is an extremely formidable equation. It is so formidable that, analytically speaking, nothing much can be done with it except in the simplest possible case, namely with $n_0 = 0$ and all $N_{tj}(n_{tj,0}) = 0$. Even for this perfect insulator problem, the analytic solution is so complicated that it is difficult to extract information from it. In general, the extreme complexity of Eq. (9.4) underscores the need for the simplified approach to the injection problem which has been used in the preceding chapters. Of course, numerical integration of Eq. (9.4) on high-speed computers is feasible once the values of the various parameters and the boundary conditions have been specified. This circumstance alone warrants a discussion of the exact theory.

Since Eq. (9.4) is a second-order differential equation, two boundary conditions are needed to determine a unique solution. In their broadest generality, these two conditions may be stated as:

the current entering at the cathode is J,

the current leaving at the anode is J. (9.5)

In order to express the boundary conditions in a mathematically useful form, it is necessary to take into account the actual structure of the contact. Therefore, there will be as many different cathode boundary conditions as there are varieties of injecting contacts, and likewise for the anode. We consider here three possibilities for the cathode, (i) the reservoir-type, metal-insulator contact, (ii) the tunneling-type, metal-insulator contact, and (iii) the n^+–n junction contact. For the anode, we consider only the nontunneling, metal-insulator contact.

9.2 The Reservoir-Type Metal-Insulator Contact

This contact is illustrated, for electrons, in the standard energy-band diagrams in Fig. 9.1a, corresponding to thermal equilibrium, in Fig. 9.1b

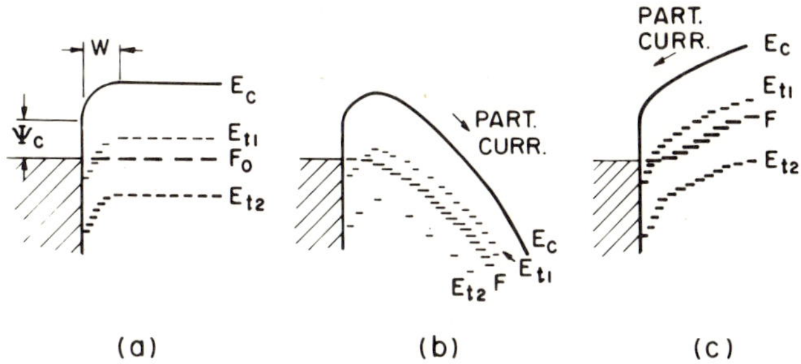

FIG. 9.1. Schematic energy-band diagrams for electron-injecting (reservoir) metal-insulator contacts: (a) in thermal equilibrium; (b) under applied voltage biased for injection; (c) under applied voltage biased for ejection.

corresponding to applied voltage biased for injection, and in Fig. 9.1c corresponding to applied voltage biased for ejection. If $n_{c,0}$ is the free-electron concentration at the contact interface in thermal equilibrium, then the random thermal current $J_{th,c}$ flowing from the insulator into the cathode, as derived in standard texts on gas-kinetic theory, is given by

$$J_{th,c} = en_{c,0}v_{th}/(6\pi)^{1/2}, \qquad v_{th} = (3kT/m)^{1/2}, \tag{9.6}$$

where m is the conduction-band effective mass (assumed scalar), and where reflection at the interface has been neglected. By detailed balance, $J_{th,c}$ is also the random thermal current, in thermal equilibrium, flowing from the cathode into the insulator. In fact, by replacing $n_{c,0}$ in (9.6) by $N_c \exp(-\Psi_c/kT)$, with N_c given in Chapter 3 [following Eq. (3.11)] and with Ψ_c the energy step at the contact interface (see Fig. 9.1a), Eq. (9.6) is directly transformed into Richardson's thermionic emission equation

$$J_{th,c} = J_{sat} = (4\pi emk^2/h^3)\, T^2 \exp - (\Psi_c/kT), \tag{9.7}$$

where h is Planck's constant. In the absence of field-enhanced emission $J_{th,c}$ is the temperature-limited or saturation current which can be drawn from the cathode, hence the subscript "sat" in (9.7).

Now, for $J \ll J_{th,c}$, the electron concentration n_c at the cathode interface can differ only negligibly from $n_{c,0}$; for, if $n_c \ll n_{c,0}$, the back-current flow from the insulator into the cathode, opposing the entering thermal current $J_{th,c}$, would be insufficient to give the small, net injection current J. Therefore, for this situation, the appropriate boundary condition at the cathode interface is:

$J \ll J_{th,c}$:

$$n_c = n_{c,0}. \tag{9.8}$$

As J approaches $J_{\text{th,c}} = J_{\text{sat}}$, say $J > J_{\text{th,c}}/3$, the boundary condition (9.8) is no longer adequate. In this case a formal statement of the appropriate boundary condition is:

$\frac{1}{3}J_{\text{th,c}} < J < J_{\text{th,c}}$:

$$n_{\text{c}} = (J_{\text{th,c}} - J)/e\tilde{v}, \tag{9.9}$$

with $\tilde{v} = \int_{-\infty}^{\infty} dv_z \int_{-\infty}^{\infty} dv_y \int_{-\infty}^{0} dv_x\, v_x f(v)$ and $f(v)$ normalized to unity, $\int d\mathbf{v}\, f(v) = 1$, where $f(v)$ is the velocity distribution function of the free electrons in the insulator at $x \simeq 0$. Note that (9.9) is simply a statement of the fact that the actual current J is the difference between the saturation current $J_{\text{th,c}}$ from the cathode and the random kinetic current $en_{\text{c}}\tilde{v}$ from the insulator back into the cathode. If $f(v)$ were Maxwellian, then $\tilde{v}$ would equal $v_{\text{th}}/(6\pi)^{1/2}$ with v_{th} given by (9.6). However, since the approach to saturation implies $v_{\text{dr}} \simeq v_{\text{th}}$, from (9.6), where v_{dr} is the electron drift velocity, substantial heating of the electrons in the applied field must be taking place. Therefore the actual velocity distribution of the electrons near $x = 0$ is a "hot carrier" distribution which may depart markedly from the Maxwellian. In any case, Eq. (9.9) can be made more precise only if $f(v)$ can be specified. Note that the current-flow equation (9.1) must also be modified because of the high fields, and Eq. (9.7) must apply near $x = 0$. Also note that in enforcing a hard saturation we are neglecting the Schottky image-force enhancement effect. This correction is incorporated in the following section on tunneling effects where the fields are even higher.

Similar reasoning applies if the anode is also an insulator-metal contact. If $n_{\text{a,0}}$ is the free-electron concentration at the contact interface at thermal equilibrium, then the appropriate boundary condition at the anode interface is:

$J \ll J_{\text{th,a}} = en_{\text{a,0}}v_{\text{th}}/(6\pi)^{1/2}$:

$$n_{\text{a}} = n_{\text{a,0}}\,. \tag{9.10}$$

On the other hand, for $J > J_{\text{th,a}}$, Eq. (9.10) must be replaced by

$J > J_{\text{th,a}}$:

$$n_{\text{a}} = (6\pi)^{1/2} J/ev_{\text{th}}\,. \tag{9.11}$$

Note that (9.10) or (9.11) is to be used whether the anode contact is of reservoir- (downward band-bending) or blocking- (upward band-bending) type. Again, the Schottky image-force enhancement effect has been neglected.

The applied voltage V is $(1/e)\,(F_{\text{C}} - F_{\text{A}})$ where F_{C}, F_{A} are the Fermi energies *inside* the cathode and anode, respectively. Note that $V \neq \int_0^L \mathscr{E}\, dx$ because of the presence of the built-in fields at the contacts. However, $e\int_0^L \mathscr{E}\, dx$ does give $E_{\text{c,c}} - E_{\text{c,a}}$ where $E_{\text{c,c}}$, $E_{\text{c,a}}$ are the energies of the conduction-band minimum at the cathode and anode interfaces respectively. $F_C - F_A$ is then obtained by adding to $E_{c,c} - E_{c,a}$ the quantity $\Psi_a - \Psi_c$ where Ψ_a, Ψ_c are the magnitudes of the energy steps at the cathode and anode interfaces respectively (see Fig. 9.1a) and it is assumed that these are not altered by the applied fields. Therefore we have, for the applied voltage:

$$V = \int_0^L \mathscr{E}\, dx + (1/e)(\Psi_a - \Psi_c) = \int_0^L \mathscr{E}\, dx + kT \ln(n_{c,0}/n_{a,0}). \qquad (9.12)$$

9.3 The Tunneling (Field Emission)-Type Metal-Insulator Contact

This contact, in thermal equilibrium, is simply a thin blocking contact, as illustrated in Fig. 9.2a. At low voltages virtually the entire applied

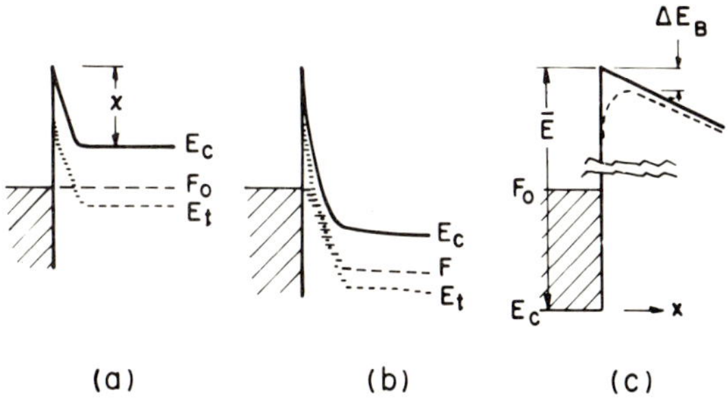

FIG. 9.2. Schematic energy-band diagrams for electron-tunneling (field-emission) metal-insulator contacts: (a) in thermal equilibrium; (b) under applied voltage, biased for injection; (c) effect of image force on shape of barrier ["blow-up" of (b) near the interface]. E_c is the energy at the bottom of the conduction band, ΔE_B is the amount of lowering of the barrier. The quantity $\bar{y}$ (see text) $= \Delta E_B/E$, with $E = \bar{E} - p_x^2/2m$, where p_x is the x-component of electron momentum and the Sommerfeld free-electron model is assumed.

voltage appears across the contact so that a very high field is built up in the neighborhood of the contact interface. When this field is sufficiently high—on the order of 10^6 to 10^7 V/cm—large electron currents can flow *through* the barrier by means of quantum-mechanical tunneling. Since barrier penetration depends on the area under the barrier (in the plane normal to the barrier) and this area, in turn, depends on the field $\mathscr{E}_c$ at the interface, as seen in Fig. 9.2b, the tunneling, or field-emission, current depends on $\mathscr{E}_c$:

$$J = \frac{e^3 \mathscr{E}_c^{\,2}}{16\pi^2 \hbar \Psi_c} \left[1 + \frac{4\pi^2 m \Psi_c}{3\hbar^2 e^2 \mathscr{E}_c^{\,2}} (kT)^2\right] \exp\left[-\frac{4(2m)^{1/2}}{3\hbar} \frac{\Psi_c^{3/2}}{e\mathscr{E}_c} v(\bar{y})\right], \qquad (9.13)$$

with $\bar{y} = (e^3\mathscr{E}_c/4\pi\epsilon')^{1/2}/\Psi_c$, where $\hbar = h/2\pi$ and ϵ' is the high-frequency (optical) dielectric constant of the insulator. The quantity $\bar{y}$ is the relative amount of lowering of the barrier Ψ_c by the image force, as shown in Fig. 9.2c. The function $v(\bar{y})$, appearing in the exponent in (9.13), accordingly is the correction factor due to the image force. As $\bar{y}$ varies from 0 to 1, $v(\bar{y})$ varies from 1 to 0. The function $v(\bar{y})$, which involves elliptic integrals, is to be found in the standard literature on field emission, and is tabulated by Good and Müller (1956, Table 1). The temperature correction factor, given by the quantity in brackets in (9.13), is an approximation which is valid so long as $J(T) < 2.7J\,(T = 0)$.

An idea of the magnitude and field dependence of tunneling injection currents is given by the J vs $\mathscr{E}_c$ plots in Fig. 9.3, calculated for $\Psi_c = 2.0$ eV and for several values of $\kappa = \epsilon'/\epsilon_0$. The maximum effect of the image force is seen in the $\kappa = 1$ curve, corresponding to field emission into vacuum, as compared to the $\kappa = \infty$ curve, corresponding to no image-force correction at all.

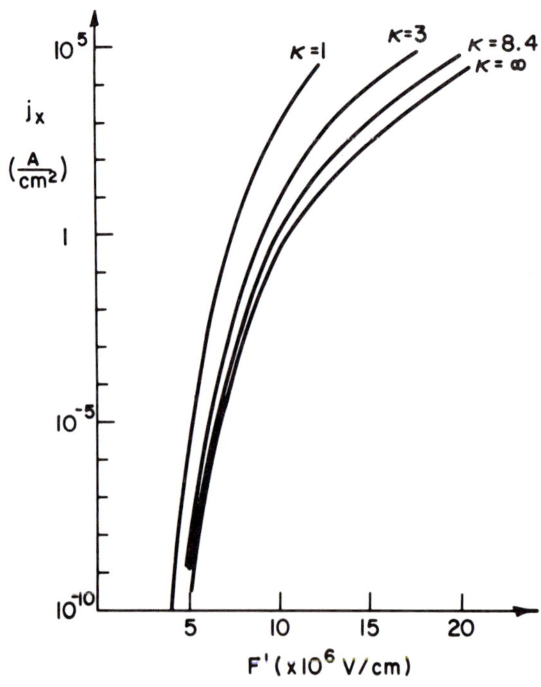

FIG. 9.3. Effect of the image force on tunneling injection currents. $\Psi_c = 2.0$ eV, $\kappa = \epsilon'/\epsilon_0$, F' is the same as $\mathscr{E}_c$ in the text. [From unpublished calculations of Meyerhofer, based on Eq. (9.13).]

SCL current injection into a perfect insulator via a tunnel-emission contact has been studied theoretically by Adirovich (1961).

Where the cathode–anode spacing L is very small, say $L \lesssim 50$ Å, tunneling may take place directly from the cathode into the anode, without the electrons having to enter the conduction band of the insulator. In this case, the distinction between "blocking" and "injecting" contacts loses significance. Such phenomena are beyond the scope of our text.

9.4 The n^+–n Junction Contact

This contact, under thermal equilibrium, is illustrated in Fig. 9.4a and under applied voltage in Fig. 9.4b. For convenience we have assumed an

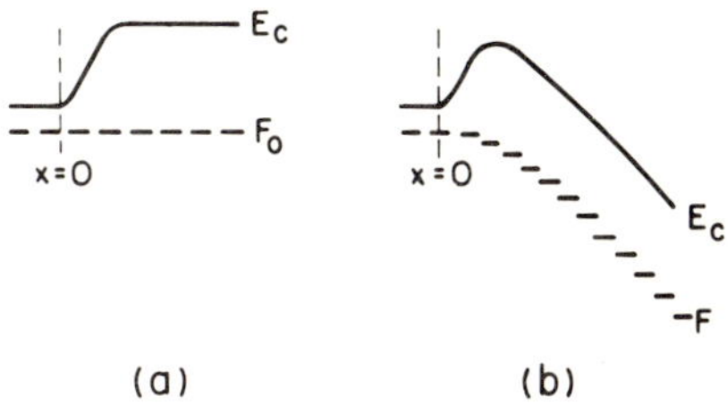

Fig. 9.4. Schematic energy-band diagrams for n^+–n junction contacts: (a), in thermal equilibrium; (b), under applied voltage, biased for injection.

abrupt junction, the material being heavily doped to the left of the plane at $x = 0$ and lightly doped to the right of this plane. (Note that the n^+ region, shown nondegenerate in Fig. 9.4a, could also be degenerate, as in Fig. 1.3b, in which case the contact remains injecting down to the lowest temperature.) Virtually all of the band-bending takes place in the "insulator," i.e., to the right of the plane at $x = 0$. So long as there is, under applied voltage, a potential minimum (energy maximum) in the insulator, as in Fig. 9.4b, the free electron concentration everywhere in the n^+ region will be essentially undisturbed. In this case the sensible boundary condition at the injecting contact is

$$n = n_0^+ \qquad \text{at} \quad x = 0, \tag{9.14}$$

where we have ignored the slight penetration of the junction region into the n^+ material. After the potential minimum disappears, with increasing voltage, the free-electron concentration "deep" in the n^+ material is still undisturbed but the junction region has begun to penetrate the n^+ material. Whether the amount of penetration is significant or insignificant depends on other details, such as geometry, n_0/n_0^+, etc.

Note that if the anode contact is likewise an n–n^+ abrupt junction with the dividing plane at $x = L$, then similar considerations apply and a useful boundary condition, at not too-high current levels, is: $n = n_0^+$ at $x = L$.

9.5 The Simplified Theory with a Finite Reservoir Contact

The mathematical complexities of the exact SCL current theory stem from the inclusion of the diffusion current in the analysis. If there is a potential minimum in the problem, as in Fig. 9.1b, then the diffusion current must obviously be sizable in the region of the contact, since the drift current vanishes precisely at the minimum and opposes the net current J to the left of the minimum. Shortly beyond the potential minimum, to the right, the diffusion current becomes negligible compared to the drift current.

As J increases, the potential minimum moves closer to the cathode interface, finally reaching it at some critical current J_{cr} and corresponding voltage V_{cr}. Generally, J_{cr} will lie well below $J_{th,c} = J_{sat}$. Above the voltage V_{cr}, there is no longer a potential minimum in the insulator and, accordingly, the diffusion current will everywhere be negligible compared to the drift current. Therefore, above V_{cr}, the simplified theory is valid and the appropriate boundary condition, replacing (9.8), is

$J_{cr} < J < \frac{1}{3} J_{th,c}$:

$$\mathscr{E}_c = \mathscr{E}(0) = J/en_{c,0}\mu. \tag{9.15}$$

Above J_{cr}, as J increases towards $J_{th,c}$, the J–V characteristic will approach the simple Ohm's law: $J = en_{c,0}\mu V/L$. The free-carrier concentration at the cathode has now spread completely throughout the insulator, and, concomitantly, the fields at the cathode and anode, $\mathscr{E}_c$ and $\mathscr{E}_a$, respectively, are of comparable magnitude. This ohmic behavior is illustrated in Fig. 9.7 (p. 203) where computer calculations of the J–V characteristic for the trap-filled-limit (TFL) problem are exhibited on a log–log plot.

If the simplified theory with a finite-reservoir contact, namely (9.15), is used at low currents (in other words, if the potential minimum and the diffusion current are simply ignored in the theory though present in fact) then $n_{c,0} \gg \bar{n}$, the average free, injected plus thermal, electron concentration in the bulk, and accordingly $\mathscr{E}_c \ll \mathscr{E}_\Omega = V/L$. In this range of currents, the boundary conditions (9.15) and (3.21), namely $\mathscr{E}_c = 0$, are equivalent for practical purposes; that is, the simplified theories using boundary conditions (9.15) and (3.21), respectively will give essentially the same J–V characteristics up to the current J_{cr} where the potential minimum leaves the insulator. This has been checked by a calculation on the aforementioned TFL problem, namely the dotted curve in Fig. 9.7.

9.6 A Geometric Result

In Section 3.6 important universal geometric results were derived in the framework of the simplified theory which neglects diffusion. We now show that these same results hold in the exact theory. As in Section 3.6, it is convenient to describe the universal properties in terms of current flow into a semi-infinite insulator, for which the boundary condition replacing (9.10) or (9.11) is

$$n \to n_0 \qquad \text{as} \quad x \to \infty. \tag{9.16}$$

Assuming that the cathode is a reservoir contact, i.e., that $n_{c,0} \gg n_0$, and, further, that $J \ll J_{th,c}$ (or, equivalently, that boundary condition (9.8) obtains) so that we are dealing with an injection problem, then the universal geometric properties of the solutions in the exact theory are precisely (i) and (ii) of Section 3.6, which we rewrite here for convenience:

(i) $n(x)$ is monotonic decreasing and approaches n_0 as $x \to \infty$.
(ii) $\mathscr{E}(x)$ is monotonic increasing and convex, that is, $d\mathscr{E}/dx$ is monotonic decreasing.

The strategy for proving these properties is similar to the strategy employed in Section 3.6, namely proof by contradiction, but the details are here a bit more involved because of the additional presence of diffusion currents. In establishing property (i) first assume, for example, that the behavior of n vs x is given by curve $\mathscr{C}_1$ in Fig. 9.5. Let x_1, x_2, with $x_1 < x_2$, be two positions for which $n(x_1) = n(x_2)$, as shown in Fig. 9.5. Since, in the interval $0 \leqslant x \leqslant x_2$, $\mathscr{C}_1$ lies everywhere above $n = n_0$, it follows from (9.2), and the fact that $n > n_0$ implies $n_{tj} > n_{tj,0}$ [from (9.3)], that $\mathscr{E}$ is monotonic increasing in the same interval, starting from $\mathscr{E}_c$ at $x = 0$ (where $\mathscr{E}_c < 0$, if there is a potential minimum in the insulator, as assumed in Fig. 9.5). Thus $\mathscr{E}(x_2) > \mathscr{E}(x_1)$, and therefore $J_{dr}(x_2) > J_{dr}(x_1)$, $J_{dr}(x_i) = en(x_i)\,\mu\mathscr{E}(x_i)$. But $J_{di}(x_2)$ is aiding $J_{dr}(x_2)$ whereas $J_{di}(x_1)$ is opposing $J_{dr}(x_1)$, where $J_{di}(x_i) = -e\mu V_T(dn/dx)_{x_i}$. Hence $J(x_2) > J(x_1)$ which violates the constancy of J, (9.1). In similar manner, it is shown that the behavior of n vs x cannot be given by a curve like $\mathscr{C}_2$ in Fig. 9.5, namely by considering two positions x_3 and x_4 for which $n(x_3) = n(x_4)$. Thus the correct behavior of n vs x is given by a curve like $\mathscr{C}$ in Fig. 9.5 which monotonically decreases from $n_c = n_{c,0}$ at $x = 0$ toward n_0 as $x \to \infty$. Having established the first geometric property, the second geometric property follows simply as a corollary. Since $n > n_0$ everywhere, it follows from (9.2) that $d\mathscr{E}/dx > 0$ everywhere, and that, since n is monotonic decreasing, $d\mathscr{E}/dx$ must be likewise.

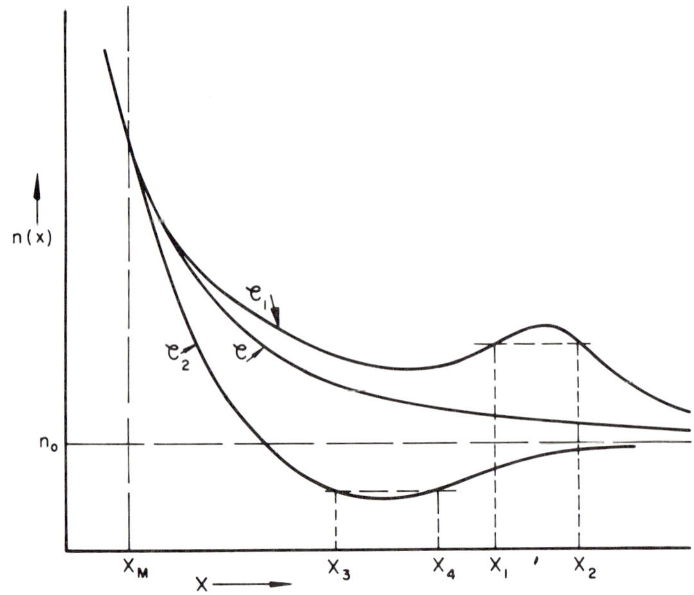

FIG. 9.5. Schematic plot of n vs. x for a semi-infinite crystal with an electron-injecting contact. $\mathscr{C}_1$ and $\mathscr{C}_2$ are impossible situations. An allowed situation is $\mathscr{C}$: $n(x)$ monotonically decreasing toward n_0 as x approaches ∞.

Finally, we note that, as in the simplified theory, the above arguments are valid under much more general conditions than (9.1) and (9.3), namely under conditions (a) and (b) of Section 3.6.

Since the entire discussion above pertains to the semi-infinite insulator, the conclusions reached are valid also for the finite insulator only if the actual structure of the anode contact is ignored, that is, if the solutions for the semi-infinite insulator are used, unperturbed, out to the anode interface at $x = L$. In this case, the applied voltage V is $(1/e)(F_C - F_A')$ where F_C is the Fermi level inside the cathode and F_A' is the quasi-Fermi level at the anode inside the insulator: $E_{c,a} - F_A' = kT \ln(N_c/n_a)$ where $n_a = n(L)$ as given by the solution for the semi-infinite insulator. Thus, in place of (9.12), we have

$$V = \int_0^L \mathscr{E}\, dx + kT \ln(n_{c,0}/n_a). \tag{9.17}$$

The geometric properties (i) and (ii) above obviously need not hold near the actual anode contact of a real, finite insulator. Thus, if the anode contact is of the injecting type, Fig. 9.1c, n increases rapidly near the anode at thermal equilibrium and therefore necessarily also over a finite range of currents. Likewise, if the anode contact is of the barrier type, Fig. 9.2a, then $\mathscr{E}$ decreases rapidly (algebraically) near the anode at thermal equilibrium and therefore necessarily also over a finite range of currents. Nonetheless, these geometric properties are useful because the solution for the finite insulator will behave like that for the semi-infinite insulator everywhere except in the immediate vicinity of the anode. In the simplified theory, which ignores all contact structure, the two solutions are, of course, very close to each other over their common range.

9.7 Limits of Validity of the Simplified Theory

Since a reasonably complete picture of one-carrier SCL current flow is available only via the simplified theory neglecting diffusion currents, it would be highly desirable to delineate precisely the limits of validity of the simplified theory. Although this cannot be done with completeness at the present time, it is not difficult to establish this validity for what is the most important case in practice, namely the case of well separated contacts.

Consider an insulator with reservoir contacts. We first define the width W of a reservoir contact: W is the distance over which the energy bands rise from their position at the interface to their bulk, neutral position, as shown in Fig. 9.1a. W is determined by the amount of band-bending η at the contact and by the space charge which "supports" the band-bending. In an insulator, this space charge will be primarily contained in traps. Assuming a constant density eN_t of trapped space charge in the neighborhood of the contact, the Poisson equation (9.2) can be written: $d\mathscr{E}/dx = eN_t/\epsilon$. Taking the interface at $x = 0$ and the "edge" of the contact at $x = W$, using the boundary condition: $\mathscr{E} = 0$ at $x = W$, and noting that $\int_0^W \mathscr{E}\, dx = \eta$, two successive integrations of the Poisson equation give

$$W = \left(\frac{2\epsilon\eta}{eN_t}\right)^{1/2} = 0.33 \times 10^{-4} \left[\left(\frac{\kappa}{10}\right)\left(\frac{\eta}{0.1}\right)\left(\frac{10^{15}}{N_t}\right)\right]^{1/2} \quad \text{(pr. units)}, \qquad (9.18)$$

where the first relation is in mks units and the second in practical units, κ being the relative dielectric constant, and W, η, N_t being measured in centimeters, electron volts, and centimeters^{-3}, respectively. Taking $\kappa = 11$, $\eta = 0.1$ eV, and $N_t = 10^{15}$ cm^{-3}, this last corresponding to a highly purified insulator, we see that $W = 0.33\ \mu$; further, the "dirtier" the insulator, i.e, the higher the N_t, the smaller is W.

Letting W_c, W_a denote the widths of the cathode and anode contacts respectively, L the separation of cathode and anode interfaces, these two contacts are said to be well separated, as in Fig. 9.6a, if $L \gg W_c + W_a$. We now show that for this case the simplified theory is a good approximation. First note that up to the voltage V_x, given by (2.32)–(2.35),

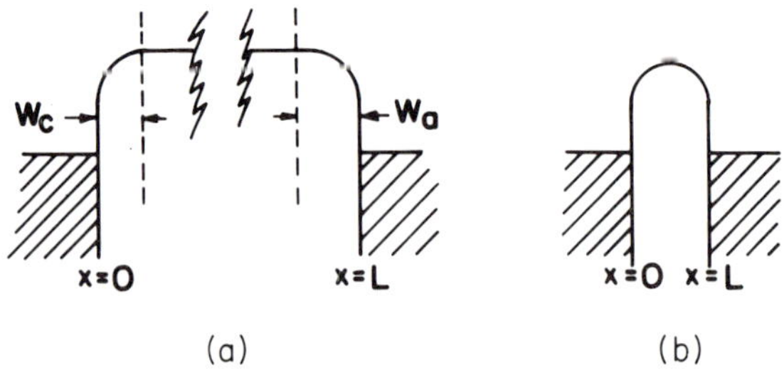

FIG. 9.6. Schematic energy-band diagrams: (a) well-separated cathode and anode contacts, $L \gg W_c + W_a$; (b) overlapping contacts, $L \lesssim W_c + W_a$.

insufficient excess carriers are injected to cause any significant deviation from Ohm's law. Ohm's law will be observed because the applied voltage is absorbed almost entirely across the bulk, the contacts being of lower resistivity and also much thinner than the bulk. Now assume that $V_x \gg kT/e$, the normal circumstance. We wish to show that for $V > V_x$, hence $V \gg kT/e$, the simplified theory is adequate.

Close to the cathode interface, at $x = x_m(V) < W_c$ is a maximum in the energy-band profile, or minimum in the potential, as shown in Fig. 9.1b. To the left of this minimum, i.e., for $0 < x < x_m$, the drift current opposes current flow and the net current J is the difference between the opposing diffusion and drift currents: $J = J_{di} - |J_{dr}|$. At the potential minimum J is a pure diffusion current: $J = J_{di}$. To the right of the potential minimum, i.e., for $x > x_m$, the drift current J_{dr} and diffusion current J_{di} are both contributing positively to J. We now show that the assumption $J \simeq J_{dr} \gg J_{di}$ in the bulk is self-consistent, whereas the opposite assumption $J \simeq J_{di} \gg J_{dr}$ is inconsistent; thus the use of the simplified theory is justified.

First we note that the assumption $J \simeq J_{dr} \gg J_{di}$ always leads to a free-carrier concentration distribution $n(x)$ which does not vary drastically over the bulk of the insulator, as has been shown in Chapter 4. (An exponential distribution is an example of a "drastic" variation of n with position.) Because of this relatively gradual variation of concentration with position an adequate, rough estimate of

$$J_{di} = -e\mu V_T \, dn/dx$$

in the bulk, to the right of x_m, is $J_{di} \simeq e\mu V_T \bar{n}/L$, with $\bar{n}$ the average value of n. But $J_{dr} \approx e\mu\bar{n}V/L$, and since $V \gg V_T$ by assumption, $J_{dr} \gg J_{di}$, self-consistently. Note that we have used the assumption $L \gg W_c + W_a$ in replacing dn/dx by $\bar{n}/L$ and $\mathscr{E}$ by V/L.

On the other hand, the opposite assumption $J \simeq J_{di} \gg J_{dr}$ quickly leads to a contradiction. For, integrating $J \simeq -e\mu V_T\, dn/dx$ from x_m to L, we see that n varies linearly with x, from n_m at $x = x_m$ to $n_a = n(L) < n_m$ at $x = L$, and that $J \simeq e\mu(n_m - n_a)\, V_T/L$. But, at the midpoint between x_m and L, i.e., at $x_m' = (x_m + L)/2$, $n = n_m' > n_m/2$ and $\mathscr{E} = \mathscr{E}_m' > \mathscr{E}_{max}/2 = \mathscr{E}(L)/2$, the latter inequality following from the convexity of $\mathscr{E}$ vs x, established in the previous section. Thus, without serious error, we may take $n_m' \simeq n_m$ and $\mathscr{E}_m' \simeq V/L$, so that $J_{dr} \simeq e\mu n_m V/L \gg J$, which is impossible. (In the above argument we have ignored the detailed structure of the anode contact, which is justified since $W_a \ll L$. In any case, any effects due to the anode-contact structure may be presumed to be, at most, comparable to those due to the cathode-contact structure, which we have just shown to be negligible. The fact that $x_m < W_c$, used in the above argument, derives from the circumstance that inside the reservoir contact, namely in the region $0 \leqslant x < x_m$, $n(x)$ does not differ appreciably from its thermal equilibrium value. This is true because the equation $J = J_{di} - |J_{dr}|$ with $J \ll J_{di} \approx |J_{dr}|$ does not differ significantly from the equation $0 = J_{di} - |J_{dr}|$ which characterizes the contact in thermal equilibrium. In the latter condition, the inequality $n_m \gg n_0$ can obviously be realized only inside the cathode reservoir, i.e., for $0 \leqslant x < W_c$. Finally, we note that if there is no potential minimum in the insulator, then the above argument-by-contradiction still applies, as presented. There is then simply no region $0 \leqslant x \leqslant x_m$ to be concerned about.)

If the anode contact is blocking, as in Fig. 9.2a, then a bulk-dominated current-flow is not applicable until sufficient voltage $V \approx \chi/e$ is applied to overcome the anode barrier of height χ. Beyond this point, the simplified theory is valid provided again that $L \gg W_c + W_a$.

If the two contacts overlap, as in Fig. 9.6b, i.e., if $L \lesssim W_c + W_a$, then further analysis is needed to clarify this situation. Such an analysis has not been provided up to the present time.

9.8 The Perfect Insulator Problem

The simplest problem that can be formulated in the framework of the exact theory (as well, of course, in the simplified theory) is that of injection into the perfect insulator, free of traps and of thermal free carriers. This is also the only problem which has been solved analytically. A first integral of the current equation (9.1) and Poisson equation (9.2), which is now simply $(\epsilon/e)(d\mathscr{E}/dx) = n$, is trivial to obtain: $Jx + \text{constant} = \epsilon\mu\mathscr{E}^2/2 - e\mu V_T n$. Further progress beyond this point involves considerably more mathematics than would be profitable to discuss here. The problem can be transformed, to the left of the potential minimum, into a Bessel equation of order $\frac{1}{3}$, and, to the right of the potential minimum, into a modified Bessel equation of order $\frac{1}{3}$. A complete description of the current flow then requires the use of the standard Bessel functions $J_{1/3}$, $J_{4/3}$, $Y_{1/3}$, $Y_{4/3}$ to the left of the potential minimum, and the standard modified Bessel functions $I_{1/3}$, $I_{4/3}$, $K_{1/3}$, and $K_{4/3}$ to the right of the potential minimum. A detailed discussion is given by Shockley and Prim (1953) for the particular problem of injection into a semi-infinite insulator. The earliest treatment of the problem in the literature appears to be that of Fan (1948) who transformed the problem into Airy's equation. Fan's treatment is incomplete and approximate. Landsberg (1951) pointed out the connection of Fan's solution with the Bessel functions. Skinner's (1955) solution is much along the lines of Shockley and Prim. However, his boundary conditions were such as to yield a negative resistance in the current-voltage characteristic. Other contributors to the problem are Suits (1957), Wright (1961), Adirovich (1961), and Lindmayer *et al.* (1963), the last reference giving a straightforward computer solution under rather unusual boundary conditions ($\mathscr{E}_c = 0$ and $n_a = 0$). The discussion of the boundary conditions in Adirovich's paper is particularly useful. In general, there is a great deal of ovelap in the above references. Related problems have been studied, by computer, by MacDonald (1962) and by Roberts and Tredgold (1962).

A relatively large amount of attention has been paid to the perfect insulator problem presumably because it is the analogue of the vacuum diode problem. However, an important difference exists between these two prototype problems; a close facsimile of the idealized vacuum diode can be built but only a crude approximation to the perfect insulator is possible in the current state of technology. The traps inevitably present in any insulator will make themselves felt both in the bulk (particularly at low voltages where deviations from the simplified theory are most marked) and at the contacts. The thermal equilibrium, and hence also the steady-state, behavior of the contacts in any real insulator will certainly be determined by traps.

9.9 The Trap-Filled-Limit (TFL) Problem

This problem has been solved analytically in Section 4.4 within the framework of the simplified theory neglecting diffusion. The corrections to the J–V characteristic produced by inclusion of the diffusion current in the problem have been studied by Lampert and Edelman (1964) with the aid of a high-speed, digital computer. These corrections are of practical interest in that they establish the magnitude of error in the determination of trap density, using (5.1), from measurement of the voltage V_{TFL}. The following is a summary of the above-cited work.

The equations defining the problem are (9.1) and, in place of (9.2), Eq. (4.46) which is rewritten here for convenience:

$$(\epsilon/e)\,(d\mathscr{E}/dx) = n + N_t\,. \tag{9.19}$$

Assuming that the current J is well below the cathode saturation current, the appropriate boundary condition at the cathode is (9.8).

Instead of taking a comparable boundary condition at the anode, namely (9.10), the above authors followed the approach of Shockley and Prim (1953) and considered the problem for a semi-infinite insulator. The appropriate boundary condition at infinity for the TFL problem is then

$$n(x) \to 0 \qquad \text{as} \quad x \to \infty. \tag{9.20}$$

For an insulator of finite thickness, the solution for the semi-infinite insulator is simply used out to the anode and the applied voltage is then given by (9.17). If the anode contact is of the reservoir type, as is often true in practice, the anode boundary condition should generally not be critical, judging from the studies of the perfect insulator problem mentioned in the previous section. At the worst, the corrections to the simplified theory imposed by the anode boundary condition may be presumed to be comparable, in magnitude, to those imposed by the cathode boundary condition.

The pair of equations (9.1) and (9.19) are not analytically tractable and therefore the use of a digital computer was necessary. It is convenient to rewrite the equations in terms of dimensionless variables, and the same units of voltage, distance, etc., that Shockley and Prim (1953) used for the perfect-insulator problem, are used here, namely

$$V^* = V_T = kT/e, \qquad L^* = (\epsilon\mu V_T^2/J)^{1/3}, \qquad \mathscr{E}^* = V_T/L^*, \qquad n^* = \epsilon V_T/eL^{*2}. \tag{9.21}$$

The dimensionless variables are

$$z = x/L^*, \qquad q = n/n^*, \qquad y = \mathscr{E}/\mathscr{E}^*, \qquad s = \Phi/V^*, \tag{9.22}$$

where Φ is the electrostatic potential: $\Phi_2 - \Phi_1 = \int_{x_1}^{x_2} \mathscr{E}\, dx$.

In terms of these variables, the Eqs. (9.1) and (9.19) become, respectively

$$1 = qy - (dq/dz), \tag{9.23}$$

$$dy/dz = q + B; \qquad B = N_t/n^*, \tag{9.24}$$

and the boundary conditions (9.8) and (9.20) become, respectively

$$q_c = q_{c,0}; \qquad q(z) \to 0 \quad \text{as} \quad z \to \infty. \tag{9.25}$$

The origin for position, $z = 0$, is arbitrarily taken at the location of the potential minimum: $y = y_0 = 0$ at $z = 0$.

The strategy for finding a computer solution to (9.23) and (9.24) was to convert the problem to an eigenvalue problem. From the discussion in Section 9.6 we know that the correct solution $q(z)$ must decrease monotonically from its value $q = q_0$ at $z = 0$ to $q = 0$ at $z = \infty$. Now q_0 is an eigenvalue for this problem in the sense that for $q(0) \neq q_0$, the solution to (9.23) and (9.24) will violate this prescribed behavior. For $q = q_{0,\mathrm{h}} > q_0$ at $z = 0$, the corresponding solution $q_{\mathrm{h}}(z)$ will first decrease, going away from $z = 0$, but will then bend upwards. For $q = q_{0,\mathrm{l}} < q_0$ at $z = 0$, the corresponding solution $q_{\mathrm{l}}(z)$ will properly decrease, going away from $z = 0$, but will sooner or later become negative. A trial-and-error procedure was followed in which a guess was made for q_0, the incorrect behavior for the corresponding computer solution ascertained, a second guess made, and so on. Finally the eigenvalue q_0 is found bracketed between two very close values: $q_{0,\bar{\mathrm{l}}} < q_0 < q_{0,\bar{\mathrm{h}}}$ and correspondingly the correct solution is bracketed between the two incorrect solutions $q_{\bar{\mathrm{l}}}(z) < q(z) < q_{\bar{\mathrm{h}}}(z)$. Since $q_{\bar{\mathrm{h}}}(z) - q_{\bar{\mathrm{l}}}(z)$ increases monotonically with z, the computer solutions so obtained are useful only out to some finite, and not very large z, say $z = z_2$.

At large z, the diffusion current is utterly negligible and the pure drift solution given in Section 4.4 is valid. The computer solution obtained for the interval $0 \leqslant z < z_2$ must be matched to this pure drift solution. Now, at z_2, the diffusion current makes a small, but certainly not negligible contribution to the total amount—in practice a contribution of between 5 and 10%, for the z_2's used. Therefore a set of auxiliary functions were introduced to obtain a smooth match at $z = z_2$. It was established that the choice of auxiliary functions is not critical; they need only have the property that their influence is confined to the matching region around $z = z_2$. In practice auxiliary functions closely related to the pure drift solution are chosen.

With the eigenvalue q_0 determined to high accuracy, straightforward computer integration of the differential equations (9.23) and (9.24) gives the solution to the left of the potential minimum out to some convenient distance $-z_1$, chosen so that beyond $-z_1$ a purely analytic, quasi-thermal solution is valid. Finally at some position $-z_\infty < -z_1$, which is determined to very high accuracy, q, $-y$, and $-p$ all become infinite.

Further details concerning the procedure for obtaining the solution are given in the published paper. Here we present, in Table 9.1, the computed results for two characteristic quantities, q_0 and $-z_\infty$. The value of q_0 for $B = 0$, which corresponds to the perfect insulator problem, is consistent with the Shockley–Prim (1953) solution.

TABLE 9.1

COMPUTER-DETERMINED QUANTITIES[a] FOR THE TFL PROBLEM ($B > 0$) AND THE TRAP-FREE INSULATOR WITH THERMAL FREE CARRIERS ($B < 0$)

M	$B = M^{-2/3}$	q_0	$-z_\infty$
∞	0	1.283598	
10^5	0.00046415880	1.283396	1.6621948
$10^{9/2}$	0.001	1.283160	1.6621508
10^4	0.0021544340	1.282654	1.6620552
$10^{7/2}$	0.0046415880	1.281564	1.6618495
10^3	0.01	1.279222	1.6614048
$10^{5/2}$	0.021544340	1.274205	1.6604404
10^2	0.046415880	1.263521	1.6583359
$10^{3/2}$	0.1	1.241075	1.6536804
10	0.21544340	1.195291	1.6431375
$10^{1/2}$	0.46415880	1.107629	1.6185484
1	1	0.9601078	1.5611445
$10^{-1/2}$	2.1544340	0.7613064	1.4409698
10^{-1}	4.6415880	0.5578439	1.2459873
$10^{-3/2}$	10	0.3908169	1.0104555
10^{-2}	21.544340	0.2688021	0.78206748
$10^{-5/2}$	46.415880	0.1836974	0.58744148
10^{-3}	100	0.1252744	0.43298785
	−0.01	1.288003	
	−1	1.868586	
	−20	20.22319	

[a] From Lampert and Edelman (1964). The maximum error in all quantities is one part in the last decimal place.

The monotonic decreasing behavior of the injected free carrier density is a universal property of one-carrier SCL current injection into a semi-infinite insulator, as established in Section 9.6. Therefore the technique used to obtain a computer solution is completely general and so presumably applicable to any problem in this class. Of course, where there are thermal free carriers n_0 present in the insulator, the correct boundary condition in place of (9.24), is: $n(x) \to n_0$ as $x \to \infty$. As an example, the same program used to obtain the eigenvalues q_0 for the TFL problem was used to obtain the eigenvalues q_0 for the problem of the trap-free insulator with thermal free carriers. For this latter problem, $B = -n_0/n^*$. Results for q_0 are given in Table 9.1 for three values of B.

Having obtained the computer solution for the semi-infinite insulator, the procedure for adapting this solution to an insulator of finite thickness L is quite straightforward. Recalling that $x = 0$ is, by definition, the position of the potential minimum, the position of the cathode x_c is located from the $n(x)$ vs x curve using the boundary condition (9.8). (There is, or is not, a potential minimum in the insulator depending on whether $x_c < 0$, or $x_c > 0$.) The location of x_a of the anode is given by: $x_a = x_c + L$. The entire solution is now known for the finite insulator, the voltage being given by (9.17). As already pointed out, this procedure does not yield the true behavior of the solution in the immediate vicinity of the anode, but at least the cathode is treated with great circumspection.

In order to see how the solution for a TFL problem neglecting diffusion compares to solutions including diffusion, detailed calculations were made with parameters chosen to correspond to a very pure, thin crystal of CdS, namely: $\kappa = 11$, $\mu_n = 200\ \text{cm}^2/\text{V sec}$, $N_t = 10^{12}/\text{cm}^3$, and $L = 5 \times 10^{-3}$ cm. The computed current-voltage characteristics are exhibited on a log–log plot in Fig. 9.7. The short-dashed curve is that computed from the simplified theory, neglecting diffusion and taking $\mathscr{E} = 0$ at the cathode. The solid curves are those calculated from the exact theory, as described above, for different values of $n_{c,0}$ (in cm^{-3}). The long-dashed straight lines are the Ohm's law characteristics: $J = en_{c,0}\mu_n V/L$, which are approached by the true characteristics below saturation but well beyond the point where the potential minimum has reached the cathode interface. The latter point, for each curve, is indicated by a cross. In each case this point coincides closely with the intersection of the exact solution with the simplified solution. The horizontal dashed lines are the saturation currents: $J_{sat} \simeq en_{c,0}v_{th}/4$ for different $n_{c,0}$'s. The dotted curve is that computed from the simplified theory, taking as the boundary condition at the cathode: $n_c = n_{c,0} = 10^{10}\ \text{cm}^{-3}$.

It is seen that the inclusion of diffusion in the problem "softens" somewhat the slope of the steep rise of current with voltage—an effect

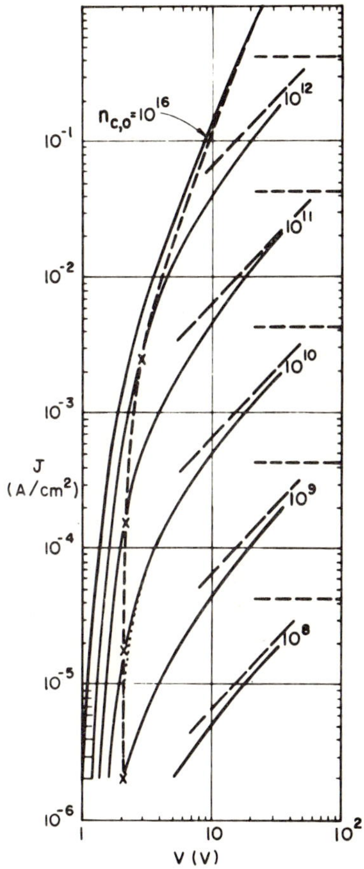

FIG. 9.7. Log–log plot of computed current-voltage characteristics. Short-dashed curve is from simplified theory; solid curves calculated from exact theory; and long-dashed straight lines are the Ohm's law characteristics.

associated with the shifting position of the potential minimum with voltage. Also the location of the steep rise is somewhat dependent on the "strength" of the cathode reservoir, i.e., on the boundary condition at the cathode—again, an effect associated with a (small) dependence of the position of the minimum on the boundary condition. However we note that the uncertainty in the location of an equivalent V_{TFL} is at most a factor of two over the large range of boundary conditions studied. This is a quite satisfactory situation in that the corresponding (identical) uncertainty in the determination of N_{t} is, in any case, commensurate with the contemporary state of insulator technology.

PART II

TWO-CARRIER CURRENTS

10

GENERALITIES AND PHENOMENOLOGY

10.1 General Remarks

A one-carrier injection current, consisting exclusively of injected electrons or exclusively of injected holes, is necessarily a space-charge current since each injected carrier contributes one excess electronic charge, negative for electrons and positive for holes, to the insulator. By making one contact to an insulator electron-injecting and the other hole-injecting and applying a voltage of proper polarity, it is possible to obtain double injection, that is, simultaneous injection of electrons and holes as illustrated in Fig. 10.1. Because the injected electrons and holes can largely neutralize each other, a two-carrier injection current will be

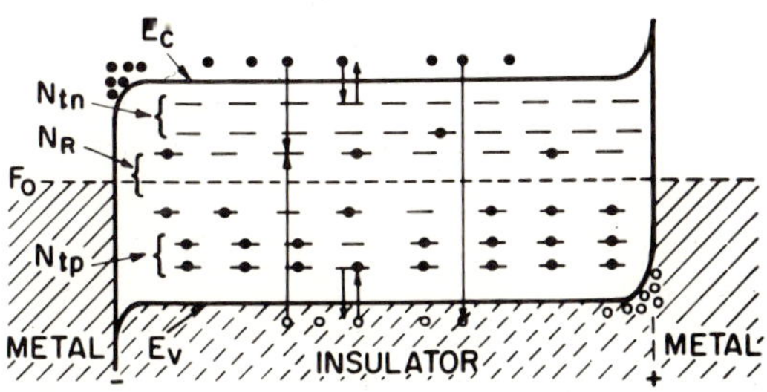

Fig. 10.1. Schematic energy-band diagram for double injection into an insulator with metal-insulator injecting contacts. N_{tn} denote electron traps, N_{tp} hole traps, N_R recombination centers.

larger than either one-carrier current in the same crystal. On the other hand, a new limitation on current flow makes its appearance—loss of current carriers through recombination. The injected electrons and holes can mutually recombine before they complete their respective transits between cathode and anode. Normally this recombination is a two-step process which takes place through localized recombination centers, labeled N_R in Fig. 10.1. First a center captures an electron, then a hole, or vice versa. In the steady state, the net rates of electron capture and hole capture by each set of recombination centers must, of course, be equal. Because of this requirement of kinetic equilibrium, the occupancies of recombination centers are usually not quasi-thermally tied to either the conduction band or the valence band, that is, they are usually not determined by a quasi-Fermi level. In addition to the normally dominant, two-step recombination process there can also occur a one-step recombination directly across the bandgap, as illustrated in Fig. 10.1. Also included on the diagram of Fig. 10.1 are localized electron traps, labeled N_{tn}, and localized hole traps, labeled N_{tp}.

The vacuum analog of insulator double injection, namely the simultaneous emission of electrons from a hot cathode and positive ions from an anode, in contrast to the insulator problem, cannot lead to a total current I substantially larger than the one-carrier, electron SCL current I_e as given by Child's law. Indeed, as shown by Langmuir (1929), the maximum possible value of I/I_e, for a planar diode, is 3.7. The reason why only limited space-charge neutralization, and therefore limited current enhancement, are achieved in this case is that in the vacuum diode the carrier drift velocity depends on the potential difference the carrier has traversed. Thus, where carriers of one sign are drifting slowly and therefore have high concentration, near their emitting electrode, the carriers of opposite sign are drifting at high velocity and have low concentration. In a solid, on the other hand, because of frequent scattering of the carriers, their drift velocity is proportional to the local electric field intensity. Thus, both electrons *and* holes drift slowly in regions of low field intensity and there is maximum opportunity for mutual neutralization. Correspondingly, there can be much greater current enhancement from the presence of carriers of both signs of charge. In this respect the difference between collision-dominated current flow and collision-free current flow has a more interesting consequence for double injection than for SCL single injection.

The preceding remarks can be paraphrased in another way. The two essential features which lend an inherent simplicity to the physics of steady-state, one-carrier injection currents are: (i) All defect states are "in communication" with only the one band into which carriers have been injected. In consequence the kinetic equilibrium between free and trapped carriers reduces to the simple quasithermal equilibrium. (ii) There is the simplest possible tie between injected carriers and injected space charge, namely all injected carriers contribute to space charge. Because of recombination and charge neutralization, both of these features are absent from the physics of two-carrier injection currents. In consequence, double injection provides a generally richer brew than one-carrier SCL current injection. A greater complexity and diversity of behavior is encountered whose limits have not yet been delineated even for one-dimensional, planar current flow. Further, the role of the contacts in double injection problems is a more sensitive one than in single injection problems. For we may generally expect that a contact which is injecting for carriers of one sign of charge, say the electrons, is blocking for carriers of the opposite sign, the holes. Therefore the simplified theory of current flow which neglects contact constraints and the diffusion currents associated with them gives a wholly inadequate description of the properties of the current flow within a distance from a contact of the order of an ambipolar diffusion length. For cathode–anode spacings which greatly exceed this diffusion length, the simplified theory, which we shall use extensively in the study of double injection, does give a useful description of the current-flow properties in the bulk of the insulator and of the current-voltage characteristic, just as in the case of one-carrier injection.

The subtlety of the physics underlying double injection is revealed if we try to answer the question; "What limits the double-injection current?" We confine our discussion to the simplified theory which automatically eliminates contact constraints, which are not of concern to us here, as a possibility. The simple, unqualified answer, "space charge," which suffices to answer the same question for single-injection currents, is inadequate and, in many cases, completely incorrect. Among other difficulties, it leaves unanswered the question as to why there must be *any* space charge, since carriers of both sign of charge are being injected. In fact, in the absence of any recombination, there is no mechanism to limit the current—it would be infinite. This provides a clue to the correct answer to our question, which is, "recombination, alone or in conjunction with space charge." A way of understanding this answer is provided by a closer look as to how space charge limits a one-carrier current. In Chapter 2, the $Q = CV$ relation provided a visible mechanism to limit the single-injection current. However, we saw in Chapter 3 that $Q = CV$ is simply an integral form of Poisson's equation:

$$\epsilon \, d\mathscr{E}/dx = \rho_{\text{total}} = \rho + \rho_t .$$

The important point is that the presence of injected carriers, hence of space charge, enforces a finite nonvanishing $d\mathscr{E}/dx$. This prevents $\mathscr{E}$ from being vanishingly small over the bulk of the insulator, hence ensures that a given current J be associated with finite voltage V. The point is that, with double injection, recombination plays a very similar role. For recombination constitutes a "disappearance" of injected carriers, and the "bookkeeping" of carriers, namely the continuity equations for electron and hole currents, J_n and J_p, respectively, therefore requires that these currents have nonvanishing divergences. But since, in the simplified theory $J_n \propto \mathscr{E}$, and $J_p \propto \mathscr{E}$, nonvanishing divergences of the currents imply a nonvanishing $d\mathscr{E}/dx$. It is in this sense that recombination can play, for double injection, a role very similar to that of space charge, for single injection, in limiting the current at fixed applied voltage.

Of course, the Poisson equation is still valid. In effect there are two equations involving $d\mathscr{E}/dx$ in double-injection theory. Now, there are some double-injection problems, or parts of problems, where the Poisson equation is ignored in obtaining the solution. (Indeed, in certain cases, electrical neutrality is assumed to hold throughout the solid in treating the other equations.) For these problems, $d\mathscr{E}/dx$ is determined solely by the recombination alone. The local space-charge density is still given by $\epsilon\,(d\mathscr{E}/dx)$, and Q still equals CV, but this space charge is accommodated by slight, inessential perturbations of the recombination-determined electron and hole distributions. An example of this type of situation is provided by double injection into a semiconductor, where there is a sufficient number of thermally generated free carriers to relax injected space charge. On the other hand, there are other problems where both the Poisson equation and the recombination equations are essential to the determination of $d\mathscr{E}/dx$. Here we say that the current is limited by recombination and space charge together. An example of this type of problem is the plasma injected into an insulator.

The double-injection problems that are simplest to analyze and are best understood at the present time are the injected plasma problems. These are problems in which the injected electrons and holes remain free and *approximately* neutralize each other—hence the description "injected plasma." It is assumed, in discussing these problems, that significant trapping does not occur and further that the *only* role of the recombination centers, if effective at all, is that of providing a funnel through which the recombination traffic may take place; that is, we need not be concerned about changes in occupancy of these centers. An injected plasma will always be the limiting regime reached in any double injection problem at applied voltages sufficiently high that the injected carriers outnumber the defect states. The injected plasma is a simple enough object that many of its salient properties can be exhibited via phenomenological analysis. This is done in the remaining sections of this chapter.

10.2 The Insulator Injected Plasma

A great deal of useful insight into the insulator injected-plasma problem can be obtained by the same kind of simple, phenomenological analysis which proved so powerful, in Chapter 2, in the study of one-carrier injection currents. Here, as in that chapter, we need deal only with the average values of the physical quantities of interest: injected free-carrier concentrations, space charge, drift velocities, and lifetime. Since the injected free electrons and free holes approximately neutralize each other, this being the injected-plasma condition, their respective average concentrations are equal and they have a common lifetime. The concentrations of localized defect states are assumed small enough that we can neglect any changes in their occupancies. Finally, the thermally generated electron and hole concentrations are assumed to be negligible.

The total current density J is the sum of the electron and hole drift currents:

$$J = e\bar{n}\bar{v}_{\mathrm{n}} + e\bar{p}\bar{v}_{\mathrm{p}} = e\bar{n}(\bar{v}_{\mathrm{n}} + \bar{v}_{\mathrm{p}}) \tag{10.1}$$

with $\bar{n}$, $\bar{p}$ the average injected free electron and hole concentrations respectively, $\bar{n} = \bar{p}$ (the injected plasma condition), and with $\bar{v}_{\mathrm{n}}$, $\bar{v}_{\mathrm{p}}$ the average electron and hole drift velocities, respectively.

Because there are no thermally generated free carriers available to relax injected space charge, the latter necessarily plays an important role in limiting the injected plasma. The relationship between injected charge and voltage is the same as for single injection, namely (2.4) which we write again here for convenience:

$$Q = \rho L \simeq C_0 V = (\epsilon/L)V. \tag{10.2}$$

In order to obtain the current-voltage characteristic, it is clearly necessary to establish the relationship between the injected charge Q and the injected plasma $e\bar{n}L$, that is, to calculate their ratio $\delta = Q/e\bar{n}L$.

The neutralization of the plasma will be incomplete because the electrons and holes, injected at opposite electrodes, must travel a finite distancc to reach each other. A reasonable estimate of δ is $\delta \approx (\bar{t}_{\mathrm{n}} + \bar{t}_{\mathrm{p}})/\bar{\tau}$ where $\bar{\tau}$ is the common average lifetime for the injected electrons and holes, and $\bar{t}_{\mathrm{n}}$, $\bar{t}_{\mathrm{p}}$ their respective average transit times. For, the smaller the lifetime, the greater is the attrition of carriers through mutual recombination, hence the greater will be the departure from neutrality. Conversely, the smaller the transit times, the more carriers can survive in the face of recombination to achieve neutralization. The simplest measure of these opposing, competitive tendencies is just the ratio: transit time/lifetime. Additivity of the ratios $\bar{t}_n/\bar{\tau}$ and $\bar{t}_p/\bar{\tau}$ derives from the fact that the electrons and holes are injected at opposite electrodes and drift in opposing senses. This entire picture of an approximately neutral, injected plasma is self consistent only if $\bar{t}_n/\bar{\tau} \ll 1$ and $\bar{t}_p/\bar{\tau} \ll 1$. Using this estimate for δ we can write

$$Q \approx e\bar{n}L[(\bar{t}_n + \bar{t}_p)/\bar{\tau}]. \tag{10.3}$$

Noting that $\bar{t}_n = L/\bar{v}_n$ and $\bar{t}_p = L/\bar{v}_p$, and combining (10.1)–(10.3), we obtain

$$J \approx \frac{C\bar{\tau}\bar{v}_n\bar{v}_p V}{L^2} \approx \frac{\epsilon\bar{\tau}\bar{v}_n\bar{v}_p V}{L^3} \approx \epsilon\bar{\tau}\mu_n\mu_p \frac{V^3}{L^5}. \tag{10.4}$$

This result was first derived by Rose (unpublished, 1955) for the case of equal mobilities and later rediscovered by one of the authors (Lampert, 1959).

The argument leading up to (10.4), in particular the relation (10.3), can be sharpened a bit by taking a closer look at the actual distribution of space charge. The hole-injecting contact, the anode, is taken at $x = 0$, the electron-injecting contact, the cathode at $x = L$. If $n(x)$, $p(x)$ are the free electron and free hole concentration distributions, respectively, $p(x) > n(x)$ on the anode side of the insulator, i.e., for $x < L/2$; $n(x) > p(x)$ on the cathode side of the insulator, i.e., for $L/2 < x$, and $p(x) = n(x)$ at $\bar{x}$ near the middle, $\bar{x} \simeq L/2$. In the spirit of the above argument, it is appropriate to estimate the space-charge distribution by the linear relation

$$e(p - n) \simeq e\bar{n}[(\bar{t}_n + \bar{t}_p)/\bar{\tau}][(\bar{x} - x)/L]. \tag{10.5}$$

The total uncompensated charge of one sign in the insulator, per unit area, is the integral

$$Q_+ = e\int_0^{L/2} (p - n)\, dx \simeq e\bar{n}[(\bar{t}_n + \bar{t}_p)/\bar{\tau}](L/8). \tag{10.6}$$

Note that $Q_+ = |Q_-| = e\int_{L/2}^{L} (n - p)\, dx$. This is necessarily true in a simplified theory which, in taking $\mathscr{E} = 0$ at both contacts, requires that, overall, the insulator be neutral under injection conditions. In the linear approximation (10.5), $\bar{x} = L/2$ *precisely* to assure this overall neutrality. Using (10.6) instead of (10.3), Eq. (10.4) is replaced by

$$J \simeq 8\epsilon\bar{\tau}\mu_n\mu_p(V^3/L^5). \tag{10.7}$$

The solution (11.14) obtained analytically from the simplified theory, for the case that $\bar{\tau}$ is a constant, differs only in that the 8 is replaced by $125/18 \simeq 7$.

In the derivation of the current-voltage relationship (10.7) [or (10.4)], no qualification has been entered concerning the possible dependence of the average lifetime $\bar{\tau}$ on the average injection level $\bar{n}$. If there is no such dependence, then (10.7) represents the final answer. Where there is such a dependence, $\bar{\tau} = \bar{\tau}(\bar{n})$, it is obviously necessary to eliminate $\bar{\tau}$ from (10.7) to obtain the current-voltage characteristic in terms of constants of the material. Now, if $\bar{\tau}$ is a function of $\bar{n}$, then the converse is true and we can equally well regard $\bar{n}$ as a function of $\bar{\tau}$, $\bar{n} = \bar{n}(\bar{\tau})$. Using (10.6), (10.2) with $Q = Q_+$, and $t = L^2/\mu V$, we obtain

$$\frac{\bar{\tau}}{\bar{n}(\bar{\tau})} \simeq \frac{e(\bar{t}_n + \bar{t}_p)L}{8Q_+} = \frac{eL^4}{8\epsilon V^2}\left(\frac{\mu_n + \mu_p}{\mu_n\mu_p}\right). \tag{10.8}$$

Once the function $\bar{n}(\bar{\tau})$ is specified, Eq. (10.8) is then solved for $\bar{\tau}$ and the result inserted into (10.7) to obtain the final current-voltage characteristic.

As an example of the use of (10.8), we consider the case of the direct recombination of free electrons and free holes. This is a bimolecular process in which the lifetime varies inversely with concentration: $\bar{\tau} = (\langle v\sigma_R\rangle \bar{n})^{-1}$, with v the magnitude of the microscopic relative velocity of electron and hole, $\sigma_R = \sigma_R(v)$ their recombination cross section, and $\langle\ \rangle$ denoting the average over the two velocity distributions. Thus $\bar{n}(\bar{\tau}) = (\langle v\sigma_R\rangle \bar{\tau})^{-1}$ and Eq. (10.8) yields

$\bar{\tau} = 1/\langle v\sigma_R\rangle \bar{n}$:

$$\bar{\tau} = (L^2/4V)[(\mu_n + \mu_p)/\mu_n\mu_p\mu_R]^{1/2}, \qquad \mu_R = (\epsilon\langle v\sigma_R\rangle/2e). \tag{10.9}$$

Substitution of (10.9) into (10.7) gives

$$J \simeq 2\epsilon\{\mu_p\mu_n(\mu_p + \mu_n)/\mu_R\}^{1/2}(V^2/L^3). \tag{10.10}$$

The approximate result (10.10) is very close to the exact result (11.35). It has the identical functional form and the numerical coefficient in (10.10) is larger than that in (11.35) by only the factor $(32/9\pi)^{1/2} = 1.06$.

10.3 Ambipolar Mobility and Another Point of View

It is illuminating to relate the above results to the ambipolar mobility μ_a. This quantity can be derived in a simple, physical manner as follows. Consider an n-type semiconductor as in Fig. 10.2, with cathode–anode spacing L. Imagine a plasma "injected," e.g., by optical illumination or

FIG. 10.2. A bounded volume of plasma injected into an n-type semiconductor (derivation of ambipolar mobility).

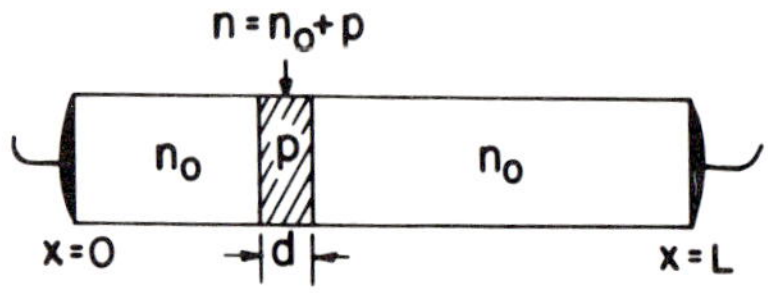

electron bombardment, over a thickness $d \ll L$, with the plasma otherwise filling the cross section. The free-carrier concentrations inside the "injected" plasma are p, for the holes, and $n = n_0 + p$ for the electrons, with n_0 being the (thermal) concentration elsewhere in the semiconductor. With a voltage V applied across the semiconductor, let the electric field intensity inside the plasma be $\mathscr{E}_p$; the field intensity elsewhere is approximately V/L (because of the assumption: $d \ll L$). The steady-state currents inside and outside the plasma must be identical:

$$J = e(n_0 + p)\,\mu_n \mathscr{E}_p + ep\mu_p\mathscr{E}_p = en_0\mu_n(V/L). \tag{10.11}$$

Solving (10.11) for $\mathscr{E}_p$, we obtain

$$\mathscr{E}_p = \{n_0\mu_n/[(n_0 + p)\,\mu_n + p\mu_p]\}(V/L). \tag{10.12}$$

Referring to Fig. 10.2, the holes are driven from left to right by the field $\mathscr{E}_p$. At the advancing front of the hole motion, the holes moving into a new region of semiconductor are space-charge neutralized by the thermal electrons in a time (the ohmic relaxation time $t_\Omega = \epsilon/en_0\mu_n$) assumed short compared to all other significant times in the problem. At the receding back of the plasma slab, the excess electrons are drained off in a comparably short time. Thus the motion of the holes is simultaneously the motion of the plasma slab, i.e., the plasma or ambipolar drift velocity v_a equals the hole drift velocity v_p. Then writing

$$v_a = \mu_p\mathscr{E}_p = \mu_a(V/L) \tag{10.13}$$

as the definition of the ambipolar drift mobility, and using (10.13), we obtain

$$\mu_a = \frac{n_0\mu_n\mu_p}{(n_0 + p)\,\mu_n + p\mu_p} = \frac{n - p}{(n/\mu_p) + (p/\mu_n)}. \tag{10.14}$$

It is the average magnitude (absolute value) of the ambipolar mobility that we need for the further discussion. (An algebraic average of μ_a would not be useful in the present problem since $\bar{\mu}_a \propto (\overline{n - p}) = 0$.) Taking $\bar{n} = \bar{p}$ in (10.14), we obtain

$$\overline{|\mu_a|} \simeq \frac{\overline{|n - p|}}{\bar{n}} \frac{1}{(1/\mu_n) + (1/\mu_p)} \simeq \frac{Q}{e\bar{n}L} \frac{1}{[(1/\mu_n) + (1/\mu_p)]}, \tag{10.15}$$

where we have replaced $\overline{|n - p|}$ by Q/eL. Finally, substituting for $Q/e\bar{n}L$ from (10.3) into (10.15), we obtain the important relation:

$$\bar{t}_a \simeq \bar{\tau} \qquad \text{with} \quad \bar{t}_a = L^2/\overline{|\mu_a|}\, V. \tag{10.16}$$

Since $\overline{|\mu_a|}$ is the average *pair* drift mobility, $\bar{t}_a$ is the average *pair* drift transit time or ambipolar transit time. In effect, Eq. (10.16) states that, at any given voltage, the injected plasma concentration builds up to the point where, approximately, the pair transit time equals the pair lifetime. We can understand this result by arguing that, on the one hand, there is nothing to prevent the injected plasma level from building up to this point, and on the other hand, that beyond this point, i.e., for $\bar{t}_a > \bar{\tau}$, the large (in fact, exponentially growing) $\mathscr{E}$ that would be required to keep the total current J constant with position, in the face of an exponential attrition of pairs, would be too great to be accommodated by the fixed voltage V. Depending on one's preferences, Eq. (10.16) might be considered as logical, and as physical, a starting point as (10.3).

10.4 The Semiconductor Injected Plasma

We now present, using Eq. (10.16) as the starting point, a simple phenomenological derivation of the current-voltage characteristic for a plasma injected into a semiconductor. Consider an n-type semiconductor with equilibrium electron and hole concentrations n_0 and p_0 respectively: $n_0 - p_0 = N_D - N_A$ with N_D, N_A the donor and (compensating) acceptor concentrations respectively. We assume that the excess thermal electron concentration $n_0 - p_0$ is large enough that excess injected charge is dissipated by ohmic relaxation in a time short compared to all other significant times in the problem. Then in place of (10.5) we have local neutrality everywhere; with n and p the total free electron and hole concentrations as previously,

$$n - p = n_0 - p_0 = N_D - N_A\,. \tag{10.17}$$

Using (10.17) in the right-hand side of (10.14) we obtain, in place of (10.15)

$$\bar{\mu}_a = \frac{(n_0 - p_0)}{\bar{n}} \frac{1}{(1/\mu_n) + (1/\mu_p)}\,, \tag{10.18}$$

where we are assuming sufficiently high injection level, $n \gg n_0$ and $p \gg n_0$, that $\bar{n} \approx \bar{p}$. Combining (10.16) with (10.18), $\bar{n}$ is obtained as

$$\bar{n} \simeq \frac{(n_0 - p_0)\,\mu_n\mu_p\bar{\tau}}{\mu_n + \mu_p} \frac{V}{L^2}\,. \tag{10.19}$$

Since, from (10.1), the current J is given by: $J \simeq e\bar{n}(\mu_n + \mu_p)(V/L)$, substitution for $\bar{n}$ by (10.19) gives for the current-voltage characteristic:

$$J \simeq e(n_0 - p_0)\, \mu_n \mu_p \bar{\tau}(V^2/L^3). \tag{10.20}$$

Of course, this result presupposes that $n_0 - p_0 \neq 0$ (although $n_0 - p_0$ may be much less than n_0, p_0). This is not a significant restriction in practice since compensation cannot be perfect. Indeed $n_0 - p_0 < 10^{10}\ \text{cm}^{-3}$ is a most unusual achievement (for the temperature range where all donors are ionized). The approximate result (10.20) differs from the exact result (12.7) only in that the latter is multiplied by the numerical coefficient 9/8.

The use of (10.16) as a starting point for the discussion of injected plasmas makes it clear at the outset that injected plasmas are recombination limited. For semiconductors the assumption of local neutrality (10.17) precludes any role for space charge:

The plasma injected into a semiconductor is purely recombination limited.

On the other hand, for insulators, the use of (10.15) incisively brings in the role of space charge:

The plasma injected into an insulator is simultaneously recombination and space-charge limited.

11

PLASMAS INJECTED INTO INSULATORS

The simplest double injection problems are the injected plasma problems. Indeed, we saw in the previous chapter that these problems yield to a relatively straightforward phenomenological treatment. In this chapter we obtain further information about plasmas injected into insulators, e.g., concentration and field distributions, through analytic studies. In the following chapter we do likewise for plasmas injected into semiconductors.

Although the simplest injected-plasma problem, in concept, is that of the plasma injected into a perfect insulator, this is a biomolecular-recombination problem which involves severe analytical difficulties. Therefore we take up first the constant-lifetime, injected plasma which is a simpler analytic problem.

11.1 The Constant-Lifetime Plasma

The plasma is formed from holes injected at the anode, taken at $x = 0$, and electrons injected at the cathode, taken at $x = L$. Since the holes move from left to right and electrons from right to left, the natural sign conventions are also the most convenient, namely

$$\mathbf{J}_\mathrm{p} = J_\mathrm{p}\mathbf{x}, \quad \mathbf{J}_\mathrm{n} = J_\mathrm{n}\mathbf{x}, \qquad \text{and} \qquad \boldsymbol{\mathscr{E}} = \mathscr{E}\mathbf{x}. \tag{11.1}$$

The hole and electron current-flow equations are, respectively, in the simplified theory which we study here,

$$J_\mathrm{p} = ep\mu_\mathrm{p}\mathscr{E} \qquad \text{and} \qquad J_\mathrm{n} = en\mu_\mathrm{n}\mathscr{E}. \tag{11.2}$$

The errors incurred through the neglect of diffusion currents are discussed in Chapter 13.

The Poisson equation is

$$(\epsilon/e)\,(d\mathscr{E}/dx) = p - n. \tag{11.3}$$

The particle-conservation equations appropriate to the steady state are

$$(1/e)\,(dJ_{\mathrm{n}}/dx) = \mu_{\mathrm{n}}(d/dx)(n\mathscr{E}) = r = n/\tau_{\mathrm{n}} = p/\tau_{\mathrm{p}}\,, \tag{11.4a}$$

and

$$-(1/e)\,(dJ_{\mathrm{p}}/dx) = \mu_{\mathrm{p}}(d/dx)(p\mathscr{E}) = r = n/\tau_{\mathrm{n}} = p/\tau_{\mathrm{p}}\,, \tag{11.4b}$$

where r is the recombination rate density and τ_{n}, τ_{p} the electron and hole lifetime respectively. Constant lifetime implies recombination through localized centers. From (11.3), it is clear that we are here neglecting any changes in occupancy of these centers due to the recombination.

The boundary conditions appropriate to the simplified theory are the usual ones:

$$\mathscr{E} = 0 \quad \text{at} \quad x = 0 \quad \text{and at} \quad x = L. \tag{11.5}$$

11.1.1 *The Injected-Plasma Regime*

In this regime $n \approx p$ everywhere, and accordingly $\tau_{\mathrm{n}} \approx \tau_{\mathrm{p}}$ in (11.4). We can therefore drop the subscripts on the lifetimes and denote the common lifetime simply by τ. Further, we assume that τ is a constant independent of injection level, hence of position.

Total current density J is a constant

$$J = J_{\mathrm{n}} + J_{\mathrm{p}} = e(b+1)\,\mu_{\mathrm{p}} n\mathscr{E} = \text{constant}; \qquad b = \mu_{\mathrm{n}}/\mu_{\mathrm{p}}\,, \tag{11.6}$$

which follows on subtracting (11.4b) from (11.4a) and using (11.2).

Multiplying (11.4b) by b and adding to (11.4a) gives, using (11.3) and (11.6),

$$\mathscr{E}\,(d/dx)\,[\mathscr{E}(d/dx)\,\mathscr{E}] = -J/\epsilon\mu_{\mathrm{p}}\mu_{\mathrm{n}}\tau. \tag{11.7}$$

Note that we have taken $p = n$ everywhere except where we encounter $p - n$, and then we have eliminated $p - n$ via the Poisson equation (11.3). It is useful to employ the following dimensionless variables:

$$w = \frac{\epsilon x}{Jb^{1/2}\mu_{\mathrm{p}}\tau^2}, \qquad u = \frac{dw}{dy} = \frac{\epsilon}{e(b+1)\,\mu_{\mathrm{p}}\tau n} = \frac{\epsilon\mathscr{E}}{J\tau}, \qquad v = \frac{\epsilon^2 V}{J^2 b^{1/2}\mu_{\mathrm{p}}\tau^3}, \tag{11.8}$$

where V is the applied potential: $V(x) = \int_0^x \mathscr{E}\, dx$. Substitution of (11.8) into (11.7) gives

$$u\,(d/dw)\,[u\,(d/dw)\,u] = d^2u/dy^2 = -1, \qquad u\,(d/dw) = d/dy, \tag{11.9}$$

with solution

$$u = \tfrac{1}{2}(y_c y - y^2). \tag{11.10}$$

Here we have arbitrarily taken $y = 0$ at $w = w_a = 0$ corresponding to $x = 0$; $y = y_c = y(w_c)$ is the value of y corresponding to $x = L$. Subscripts "a" and "c" refer to the anode and cathode, respectively. Equation (11.10) satisfies the boundary conditions (11.5).

Integration of (11.10) gives

$$w = \int_0^y u\, dy = \tfrac{1}{4}(y_c y^2 - \tfrac{2}{3}y^3), \qquad w_c = \tfrac{1}{12}y_c^3. \tag{11.11}$$

For the potential we have

$$v = \int_0^w u\, dw = \int_0^y (dw/dy)^2\, dy = \tfrac{1}{12}[y_c^2 y^3 - \tfrac{3}{2}y_c y^4 + \tfrac{3}{5}y^5], \qquad v_c = (1/120)\, y_c^5. \tag{11.12}$$

In terms of the dimensionless variables,

$$J = (\epsilon L/b^{1/2}\mu_p\tau^2)(1/w_c), \qquad V = V(L) = (L^2/b^{1/2}\mu_p\tau)(v_c/w_c^2), \tag{11.13}$$

where $V = V(L)$ is the applied voltage. Since $1/w_c$ is a function of v_c/w_c^2 we immediately have, from (11.13), the scaling relation: J/L is a function of V/L^2. Indeed, combining (11.11) and (11.12), we obtain $v_c/w_c^2 = 6/5y_c$, which gives finally

$$1/w_c = (125/18)(v_c^3/w_c^2) \qquad \text{or} \qquad J = (125/18)\,\epsilon\tau\mu_p\mu_n(V^3/L^5). \tag{11.14}$$

Some significant properties of the solution are as follows:

(i) The maximum value of the $\mathscr{E}$-field is $\mathscr{E}_m = 5\mathscr{E}_\Omega/4$ with $\mathscr{E}_\Omega = V_c/L$. This maximum occurs at $x = x_m = L/2$ corresponding to $w_m = w_c/2$, $y_m = y_c/2$ with $y_c = 6(\bar{t}_n\bar{t}_p)^{1/2}/5\tau$, $\bar{t}_n = L^2/\mu_n V$, $\bar{t}_p = L^2/\mu_p V$. Thus the midplane is the plane of neutrality.

(ii) The ratio of space-charge concentration to injected plasma concentration is $(p - n)/n = (b + 1)(y_c - 2y)/2b^{1/2}$ which has a maximum absolute value equal to $3(\bar{t}_n + \bar{t}_p)/5\tau$. Thus the above injected-plasma solution is self-consistent only at voltages high enough that $\bar{t}_p/\tau \ll 1$, $\bar{t}_n/\tau \ll 1$.

(iii) In the neighborhood of the neutrality plane, $x = x_{\mathrm{m}} = L/2$, the space-charge concentration is given by:

$$p - n = (4/5)n_{\mathrm{m}}\{(\bar{t}_{\mathrm{n}} + \bar{t}_{\mathrm{p}})/\tau\}(x_{\mathrm{m}} - x)/L.$$

This is very close to the "physical" estimate (10.5) used earlier, differing from it only (slightly) in the numerical coefficient. Here $n_{\mathrm{m}} = n(x_{\mathrm{m}})$ and n_{m} is the minimum value of n.

(iv) The average injected plasma concentration is

$$n = \int_0^L n\, dx/L = 25\epsilon\tau\mu_{\mathrm{n}}V^2/3e(b+1)L^4.$$

This can be rewritten as $e\bar{n}L = (25/3)\{\tau/(\bar{t}_{\mathrm{n}} + t_{\mathrm{p}})\}\, C_0V$ with $C_0 = \epsilon/L$. Furthermore, $\bar{n}/n_{\mathrm{m}} = \frac{3}{2}$.

(v) The true transit-time t_{n} is given by

$$t_{\mathrm{n}} = \int_0^L v_{\mathrm{n}}^{-1}\, dx = \mu_{\mathrm{n}}^{-1} \int_0^L \mathscr{E}^{-1}\, dx = 6\bar{t}_{\mathrm{n}}/5.$$

(vi) At a small Δx from either contact, the ratio of the neglected electron diffusion current to the electron drift current is

$$\{e\mu_{\mathrm{n}}V_{\mathrm{T}}\, dn/dx\}/\{bJ/(b+1)\} = (8\sqrt{3} \times 10^{-2})(V_{\mathrm{T}}/V)(L/\Delta x)^{1/2}(\bar{t}_{\mathrm{n}}\bar{t}_{\mathrm{p}}/\tau^2)^{5/2}.$$

Thus, the condition that there be an injected plasma, $(\bar{t}_{\mathrm{n}} + \bar{t}_{\mathrm{p}})/\tau \ll 1$, is essentially also the condition for self-consistency with respect to the neglect of diffusion except very close to the contacts.

The relative electric field and potential distributions are plotted (the solid curves) versus x/L in Fig. 11.1. In plotting these curves, use is made of (11.10), (11.11), (11.12), and the relations $x/L = w/w_{\mathrm{c}}$, $\mathscr{E}(x)/\mathscr{E}_{\mathrm{m}} = u/u_{\mathrm{m}}$, and $V(x)/V = v/v_{\mathrm{c}}$. Since $n/n_{\mathrm{m}} = \mathscr{E}_{\mathrm{m}}/\mathscr{E}$, the carrier concentration distribution need not be plotted. The most striking characteristic of the injected plasma is its relative uniformity. Over approximately 90% of the insulator, the field changes by only a factor of two: $\mathscr{E}/\mathscr{E}_{\mathrm{m}} = \frac{1}{2}$ at $x/L = 0.06$. Also plotted in Fig. 11.1 (solid line) is a quantity proportional to the space-charge distribution, namely the quantity $(2/3)\{\tau/(\bar{t}_{\mathrm{n}} + \bar{t}_{\mathrm{p}})\}\{(p - n)/n_{\mathrm{m}}\}$. Note that the linear approximation (short-dashed line) to this expression is good over most of the insulator; a factor-of-two deviation from linearity occurs at $x/L = 0.07$.

Experimental verification of the "insulator cubic law," Eq. (11.14), is taken up in Chapter 13.

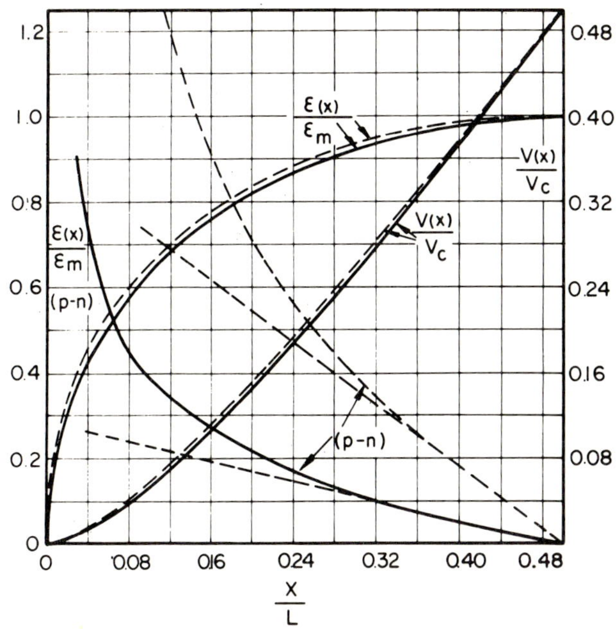

FIG. 11.1. Theoretical normalized electric field intensity, potential, and space-charge distributions for a plasma injected into an insulator. The solid curves pertain to the constant-lifetime case, the dashed curves to the bimolecular recombination case. The quantity labeled (p–n) is actually, for the constant-lifetime case, $(2/3)\{\tau/(\bar{t}_n + \bar{t}_p)\} \times \{(p - n)/n_m\}$, and, for the bimolecular recombination case, $\{(2\mu_R/\mu_p) + (2\mu_R/\mu_n)\}^{-1} \times \{(p - n)/n_m\}$. τ is the (constant) lifetime, $\bar{t}_n$, $\bar{t}_p$ the average free electron and free hole transit times, respectively, and n_m is the free electron density at the midplane, $x = L/2$. The dashed straight lines are linear approximations to the (p–n) curves.

11.1.2 *The Transition from a One-Carrier,* SCL *Regime to the Injected Plasma Regime*

In the above discussion of the injected-plasma problem for the constant lifetime case, we saw by examination of the accompanying space charge that the injected-plasma assumption necessarily breaks down at voltages sufficiently low that $(\bar{t}_n + \bar{t}_p)/\tau > 1$. What does happen at low voltages? This question does not have a unique answer; what happens depends on the detailed insulator model assumed. There is a particularly simple model for which a complete solution can be given analytically, and further the solution is of unusual interest because it traces the transition from a predominantly one-carrier SCL current at lower voltages to an injected plasma current at higher voltages. This is the problem in which *one* of the free-carrier lifetimes is a constant independent of injection level. Such a situation is readily realized by either of the simple models illustrated in Fig. 11.2. In the model of Fig. 11.2a, there is a single set of recombination centers, concentration N_R, lying well above the Fermi level and so essentially completely empty of electrons in thermal

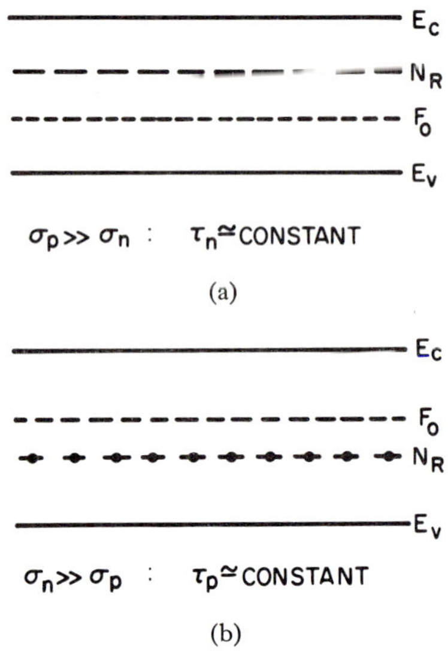

FIG. 11.2. Simple insulator models for which *one* of the free carrier lifetimes will be approximately constant at all levels of (double) injection; σ_n, σ_p are the electron and hole capture cross sections, respectively, of the recombination centers, τ_n and τ_p, the electron and hole lifetimes, respectively: (a) states N_R acceptor-like and well above the Fermi level; (b) states N_R donor-like and well below the Fermi level.

equilibrium. Further, the electron and hole capture cross sections, σ_n and σ_p, respectively, of these centers are assumed such that $\sigma_p \gg \sigma_n$. This would be true if the states N_R are acceptor-like; i.e., negatively charged when (electron) occupied and neutral when empty. Clearly under injected plasma conditions, $n \approx p$, the recombination centers will remain empty. For with $n \approx p$ and $\sigma_p \gg \sigma_n$, this is the only way in which the electron and hole recombination rates can be equal, as required by the steady state. Thus we conclude that the states N_R will remain empty at all voltages, whence the electron lifetime $\tau_n \simeq (\langle v\sigma_n \rangle N_R)^{-1}$ will be a constant independent of injection level. Figure 11.2b is the dual of Fig. 11.2a. Here the recombination centers N_R lie well below the Fermi level and are assumed donor-like, $\sigma_n \gg \sigma_p$, so that the hole lifetime $\tau_p \simeq (\langle v\sigma_p \rangle N_R)^{-1}$ will be a constant independent of injection level.

For the sake of definiteness, we take the electron lifetime constant, corresponding, e.g., to Fig. 11.2a. The equations characterizing the problem are the same equations, (11.1)–(11.5), used for the injected-plasma discussion except that we now use for r in (11.4) only the relation, $r = n/\tau_n$ with $\tau_n =$ constant. Furthermore, here we *nowhere* take $n = p$.

Thus, in place of (11.6), we write

$$J = J_{\text{n}} + J_{\text{p}} = e\mu_{\text{n}} n\mathscr{E} + e\mu_{\text{p}} p\mathscr{E} = \text{constant}. \tag{11.15}$$

Using (11.15) and (11.37), n and p are separately determined in terms of $\mathscr{E}$ and $d\mathscr{E}/dx$:

$$(b+1)n = \frac{J}{e\mu_{\text{p}}\mathscr{E}} - \frac{\epsilon}{e}\frac{d\mathscr{E}}{dx}; \qquad (b+1)p = \frac{J}{e\mu_{\text{p}}\mathscr{E}} + b\frac{\epsilon}{e}\frac{d\mathscr{E}}{dx}. \tag{11.16}$$

Now, multiplying (11.4b) by b and adding to (11.4a) gives, in place of (11.7), using (11.3) and (11.16):

$$\epsilon\mu_{\text{n}}\tau_{\text{n}}\mathscr{E}\frac{d}{dx}\left[\mathscr{E}\frac{d\mathscr{E}}{dx}\right] - \epsilon\mathscr{E}\frac{d\mathscr{E}}{dx} + \frac{J}{\mu_{\text{p}}} = 0. \tag{11.17}$$

It is now useful to employ the following dimensionless variables:

$$w = \frac{\epsilon x}{J\tau_{\text{n}}^2 b\mu_{\text{n}}}, \qquad u = \frac{dw}{dy} = \frac{\epsilon\mathscr{E}}{bJ\tau_{\text{n}}}, \qquad v = \frac{\epsilon^2 V}{J^2\tau_{\text{n}}^3 b^2\mu_{\text{n}}}. \tag{11.18}$$

Substitution of (11.18) into (11.19) gives

$$u\frac{d}{dw}\left[u\frac{d}{dw}u\right] - u\frac{d}{dw}u = \frac{d^2u}{dy^2} - \frac{du}{dy} = -1, \qquad u\frac{d}{dw} = \frac{d}{dy}, \tag{11.19}$$

with solution

$$u = y - y_{\text{c}}[(\exp y - 1)/(\exp y_{\text{c}} - 1)]. \tag{11.20}$$

Here, as in going from (11.9) to (11.10), we have arbitrarily taken $y = 0$ at $w = w_{\text{a}} = 0$ corresponding to $x = 0$; $y = y_{\text{c}} = y(w_{\text{c}})$ is the value of y corresponding to $x = L$. Equation (11.20) satisfies the boundary conditions $u = 0$ at $y = 0$ and at $y = y_{\text{c}}$, corresponding to the boundary conditions (11.5). Integration of (11.20) gives

$$w = \int_0^y u\,dy = \tfrac{1}{2}y^2 + M_{\text{c}}(y - \exp y + 1); \qquad M_{\text{c}} = y_{\text{c}}/(\exp y_{\text{c}} - 1). \tag{11.21}$$

$$w_{\text{c}} = w(y_{\text{c}}) = \tfrac{1}{2}y_{\text{c}}^2 - (1 - M_{\text{c}})y_{\text{c}}. \tag{11.22}$$

For the potential we have:

$$\begin{aligned} v &= \int_0^w u\,dw = \int_0^y (dw/dy)^2\,dy \\ &= M_{\text{c}}^2\{y - 2(\exp y - 1) + \tfrac{1}{2}(\exp 2y - 1)\} \\ &\quad + M_{\text{c}}\{y^2 + 2(\exp y - 1) - 2y\exp y\} + \tfrac{1}{3}y^3, \end{aligned} \tag{11.23}$$

$$v_{\text{c}} = v(y_{\text{c}}) = \tfrac{1}{3}y_{\text{c}}^3 - (\tfrac{3}{2} - M_{\text{c}})y_{\text{c}}^2 + (2 - M_{\text{c}})(1 - M_{\text{c}})y_{\text{c}}. \tag{11.24}$$

In terms of the dimensionless variables,

$$J = (\epsilon L/\tau_n^2 b\mu_n)(1/w_c), \qquad V = V(L) = (L^2/\mu_n\tau_n)(v_c/w_c^2). \tag{11.25}$$

As before, we have the scaling law: (J/L) is a function of (V/L^2).

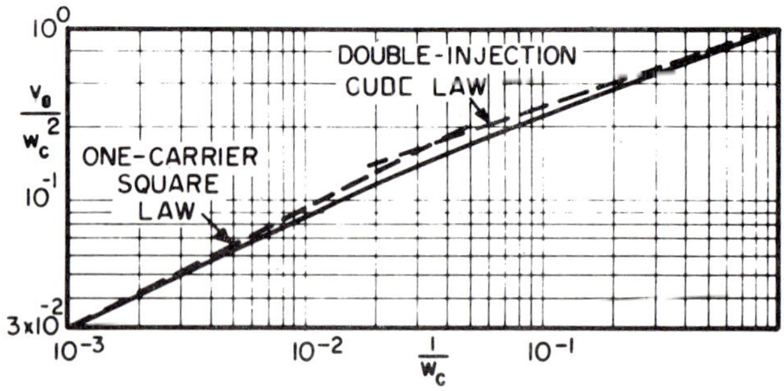

FIG. 11.3. Universal current-voltage characteristic (solid curve) for double injection into an insulator with the electron lifetime constant at all injection levels. The dashed lines are approximations characteristic of the two "pure" regimes, appropriately labeled. $J \propto 1/w_c$ and $V \propto v_c/w_c^2$.

A log–log plot of $1/w_c$ vs v_c/w_c^2 is shown in Fig. 11.3. The two regimes are as follows.

The One-Carrier, Square-Law Regime: $y_c \gg 1$

Then $w_c \simeq y_c^2/2$ and $v_c \simeq y_c^3/3$ so that $(v_c/w_c^2)^2 \simeq 8/9w_c$, shown plotted in Fig. 11.3 as a dashed line labeled "one-carrier square law." For this corresponds to the one-carrier, trap-free square law (4.9), with μ replaced by μ_p since the carriers are now holes. The physical origin of this result is clear. At low voltages, the electrons cannot significantly penetrate the insulator because of their finite lifetime against capture by the empty recombination centers. By the same token, except very close to the cathode, these centers are empty and so cannot capture the injected holes. Thus, the only limitation on the injection of holes is their own space charge. According to the model, the trap-free, hole square law holds down to the lowest current. The reason for this is the neglect of any charge whatever in the population of recombination centers. Actually this is an unrealistic approximation. Some of the recombination centers, even if only a small percentage, are filled with electrons at the outset, hence will function as hole traps at low current levels. Thus we would expect a trap-filled-limit threshold for SCL hole current flow. It is only at voltages exceeding this that the trap-free square law will be seen, up to the point of effective double injection.

The Double-Injection, Cube-Law Regime: $y_c \ll 1$

Then $w_c \simeq y_c^3/12$ and $v_c \simeq y_c^5/120$ so that $(v_c/w_c^2)^3 = 18/125w_c$, shown plotted in Fig. 11.3 as a dashed line labeled "double-injection cube-law." For this corresponds to the insulator double-injection current-voltage characteristic (11.14). The square law and cube law cross over at the voltage V_X corresponding to

$$t_{n,X} = L^2/\mu_n V_X = 500\tau_n/81 = 6.17\tau_n .$$

Several additional physical features of interest have been calculated for this problem and the results plotted in Fig. 11.4.

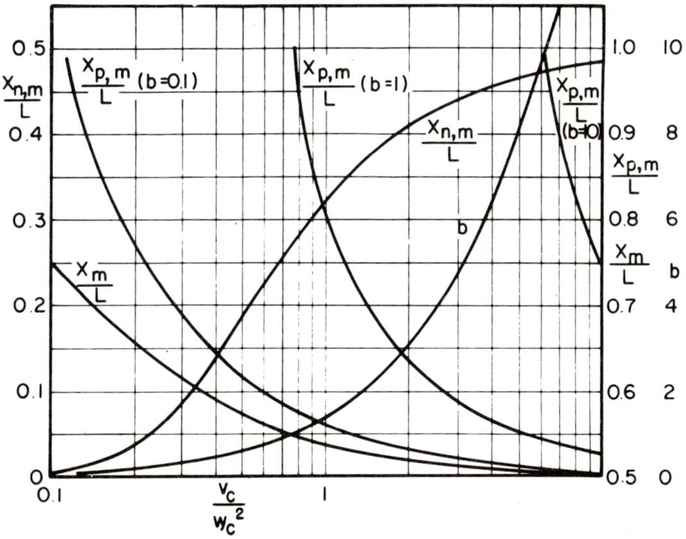

FIG. 11.4. Location of field and concentration extrema for double injection into an insulator with the electron lifetime constant at all injection levels; x_m is the location of the electric field intensity maximum, $x_{n,m}$ that of the injected electron concentration minimum, and $x_{p,m}$ that of the injected hole concentration minimum (dependent on the mobility ratio $b = \mu_n/\mu_p$). The curve labeled "b" is interpreted such that for any b the corresponding v_c/w_c^2 determines the minimum voltage ($V \propto v_c/w_c^2$) for which $x_{p,m} < L$; i.e., for smaller v_c/w_c^2 (voltage), $x_{p,m} = L$.

(i) The location x_m of the electric field maximum shifts with voltage from near the cathode, $x_m \simeq L$, at low voltages, to $x_m \simeq L/2$ at high voltages. A plot of $x_m/L = w_m/w_c$ vs v_c/w_c^2 is given. Here $w_m = w(y_m)$, given by (11.21) with $y_m = \ln M_c$. This, in turn, represents the solution to $d\mathscr{E}/dx = 0$ or $du/dy = 0$, using (11.20).

(ii) The location $x_{n,m}$ of the minimum in the injected electron density shifts with voltage from near the anode, $x_{n,m} \simeq 0$, at low voltages, to $x_{n,m} \simeq L/2$ at high voltages. A plot of $x_{n,m}/L = w_{n,m}/w_c$ vs v_c/w_c^2 is given. Here $w_{n,m} = w(y_{n,m})$, where $y_{n,m} = 1 - M_c$. This, in turn, represents the solution to $dn/dx = 0$ or $du/dy = u$.

(iii) The location $x_{p,m}$ of the minimum in the injected hole density shifts with voltage from *precisely* at the cathode, $x_{p,m} = L$, at low voltages to $x_{p,m} \simeq L/2$ at high voltages. Unlike the extrema discussed in (i) and (ii), $x_{p,m}/L = w_{p,m}/w_c$ is not a unique function of v_c/w_c^2, but depends also on the mobility ratio b. Plots of $x_{p,m}/L$ vs v_c/w_c^2 are given for $b = 0.1$, 1, and 10, respectively. For each value of b, there is some minimum value of v_c/w_c^2 below which $x_{p,m}/L = 1$. The corresponding curve of b vs v_c/w_c^2 is also given. Note that $w_{p,m} = w(y_{p,m})$ where $y_{p,m}$ is the solution to the transcendental equation:

$$(1 + 2b - y_{p,m}b - bM_c) \exp y_{p,m} = (1 + b).$$

This, in turn, represents the solution to $dp/dx = 0$.

11.2 The Bimolecular-Recombination Plasma

A plasma injected into a perfect insulator is a bimolecular-recombination plasma. Since there are no recombination centers in the insulator, the recombination of injected electrons and holes takes place as a one-step process and is therefore necessarily bimolecular. (At extremely high injection levels, Auger recombination, which is a trimolecular process, may dominate. We do not pursue this possibility further.) The equations characterizing the bimolecular-recombination plasma are (11.1)–(11.5) except that in Eqs. (11.4) the recombination-rate density r is now given by: $r = \langle v\sigma_R \rangle np$, where v is the magnitude of relative velocity between a free electron and a free hole, $\sigma_R = \sigma_R(v)$ is the recombination cross section at velocity v, and $\langle\ \rangle$ denotes an average over the velocity distributions of the electrons and holes. For convenience we rewrite Eqs. (11.4) here:

$$(1/e)\,(dJ_n/dx) = \mu_n\,(d/dx)(n\mathscr{E}) = r = \langle v\sigma_R \rangle np, \tag{11.26a}$$

$$-(1/e)\,(dJ_p/dx) = -\mu_p\,(d/dx)(p\mathscr{E}) = r = \langle v\sigma_R \rangle np. \tag{11.26b}$$

Equation (11.15) holds, as previously.

A fundamental property of the system of equations [(11.1)–(11.3), (11.26a, b), and (11.5)] *characterizing the bimolecular-recombination plasma is that the functional form of the solution* $[p(x), n(x), \mathscr{E}(x), V(x), J_n(x), J_p(x)]$ *is independent of the applied voltage.*

That is, if the applied voltage is doubled, p, n, $\mathscr{E}$, and V retain their functional form but are each doubled in magnitude; J_p and J_n retain their functional form but are each quadrupled in magnitude. From this, we conclude immediately that $J \propto V^2$. These observations are obvious by inspection.

At the outset we have already obtained the important result that if there is an injected plasma at any voltage, there is one at all voltages. This result contrasts sharply with the constant-lifetime problem where a threshold voltage must be reached before a plasma is injected, namely a voltage so as to satisfy the conditions: $t_n < \tau$, $t_p < \tau$. In the present case no threshold voltage is required, because at lower voltages, as the transit times get longer, the bimolecular lifetime gets correspondingly longer; that is, if the inequalities $t_n < \tau$, $t_p < \tau$ hold at one voltage, then they hold at all voltages.

Double injection into a perfect insulator does not necessarily lead to an injected plasma. It is intuitively obvious that if the recombination cross section σ_R is large enough, then the current will be primarily a space-charge-limited hole current near the anode and a space-charge-limited electron current near the cathode, there being a small region of overlap where these two space-charge currents "meet." We consider the more general problem of double injection into a perfect insulator later in this chapter. We first consider the limiting case of an injected plasma.

11.2.1 *The Injected-Plasma Limit*

Since the injected plasma is characterized by $p \approx n$ almost everywhere in the insulator, it is a legitimate approximation to replace p by n everywhere in the Eqs. (11.1)–(11.3) and (11.26) except where the difference $p - n$ occurs. This is the same approximation used in the analytic study of the injected plasma for the constant-lifetime case in the preceding chapter. The Poisson equation (11.3) is unchanged and Eqs. (11.26a), (11.26b), (12.4a, b), and (11.15) become, respectively

$$(1/e)(dJ_n/dx) = \mu_n(d/dx)(n\mathscr{E}) = r = \langle v\sigma_R\rangle n^2, \tag{11.27a}$$

$$-(1/e)(dJ_p/dx) = \mu_p(d/dx)(p\mathscr{E}) = r = \langle v\sigma_R\rangle n^2, \tag{11.27b}$$

$$J = J_n + J_p = e(b+1)\mu_p n\mathscr{E} = \text{constant}; \qquad b = \mu_n/\mu_p . \tag{11.28}$$

Multiplying (11.27b) by b and adding to (11.27a) gives, using (11.3) and (11.28),

$$\mathscr{E}^2\, d^2\mathscr{E}^2/dx^2 = -[4J^2\mu_R/\epsilon^2 b(b+1)\mu_p^3]; \qquad \mu_R = \epsilon\langle v\sigma_R\rangle/2e. \tag{11.29}$$

From this point on, it is convenient to employ dimensionless variables as follows:

$$w = \frac{8e^2\mu_R^{5/2}}{JL^6\epsilon\{b(b+1)\mu_p{}^3\}^{1/2}}\,x; \qquad u = \frac{2\mu_R}{(b+1)\mu_p L^3}\,\frac{1}{n} = \frac{2e\mu_R}{JL^3}\,\mathscr{E}. \tag{11.30}$$

Substitution of (11.30) into (11.29) gives the equation

$$u^2\, d^2u^2/dw^2 = -1. \tag{11.31}$$

The solution to (11.31) is most conveniently expressed in terms of a dummy variable s:

$$u = (w_c{}^2/2\pi)^{1/4}\exp(-\tfrac{1}{2}s^2); \qquad w = (w_c/\pi^{1/2})\int_{-\infty}^{s} ds\,\exp(-s^2), \tag{11.32}$$

where $s = -\infty$ corresponds to $w = 0$ and $s = \infty$ to $w = w_c$, and u (i.e., $\mathscr{E}$) vanishes at these limits, as it must to satisfy the boundary conditions (11.5).

In verifying that (11.32) is indeed a solution to (11.31), note that $du^2/ds = -2su$, $dw/ds = \sqrt{2}u^2$, $du^2/dw = (du^2/ds)/(dw/ds) = -\sqrt{2}s$ and $(d/dw)(du^2/dw) = -\sqrt{2}\,ds/dw$ which indeed equals $-1/u^2$.

It is now convenient to define a dimensionless potential variable v as follows:

$$v = \frac{(2\pi)^{3/4}\epsilon^{1/2}\{b(b+1)\mu_p{}^3\}^{1/4}}{2J^{1/2}L^{3/2}\mu_R^{1/4}}\,V. \tag{11.33}$$

Then the defining relation for potential,

$$V(x) = \int_0^x \mathscr{E}\,dx \propto \int_0^w u\,dw \propto \int_{-\infty}^{s} u(dw/ds)\,ds \propto \int_{-\infty}^{s} u^3\,ds,$$

gives for v

$$v = v(s) = \int_{-\infty}^{s} ds\,\exp(-\tfrac{3}{2}s^2), \qquad v_c = v(\infty) = (2\pi/3)^{1/2}. \tag{11.34}$$

Substitution for v_c from (11.34) into the defining relation for v, Eq. (11.33), gives finally

$$J = (9\pi/8)^{1/2}\epsilon\{\mu_p\mu_n(\mu_p+\mu_n)/\mu_R\}^{1/2}(V^2/L^3), \tag{11.35}$$

which is precisely the Parmenter–Ruppel (1959) result (11.52). The fact that we obtain *exactly* the same result is not surprising, since (11.52) is valid only in the limit $\mu_p/\mu_R \gg 1$, $\mu_n/\mu_R \gg 1$ which is also the criterion for an injected plasma, $n - p \ll n$. Correspondingly, we expect that the above solution also faithfully reproduces the details of the Parmenter–Ruppel solution, i.e., the as-yet-uncalculated distributions. Before going on to discuss detailed properties of the solution, we note that the double-injection plasma current (11.35) may be expected, generally, to be at least several orders of magnitude greater than the corresponding single-carrier, electron (or hole) space-charge-limited current, in an indirect bandgap, valence-bonded semiconductor at room temperature. At $T = 300°\text{K}$, in such a material, a reasonable value for $\langle v\sigma_R \rangle$ is 10^{-12} cm^3/sec, giving $\mu_R \approx 10^{-5}$ cm^2/V sec and a two-carrier to one-carrier current ratio of the order 10^3–10^4. For a direct bandgap semiconductor, the corresponding ratio will be at least an order of magnitude smaller since $\langle v\sigma_R \rangle$ is at least two orders of magnitude larger in such a material.

The injected plasma described by (11.30) and (11.32) has the following significant properties:

(i) The maximum value of the $\mathscr{E}$-field is $\mathscr{E}_m = (3/2)^{1/2}\,\mathscr{E}_\Omega$. This maximum occurs at $x = x_m = L/2$ corresponding to $w_m = w_c/2$, $s_m = 0$.

(ii) The ratio of space-charge concentration to injected plasma concentration is $(p - n)/n = -[(2\mu_R/\mu_p) + (2\mu_R/\mu_n)]^{1/2}\, s$. Thus for $|\, s \,|$ large enough, $|\, p - n \,| > n$, and the solution is inconsistent with the injected-plasma assumption over a very small distance adjacent to either contact.

(iii) The injected carrier concentration n_m at the midplane is given by $n_m = [3\pi\mu_n/4(b + 1)\,\mu_R]^{1/2}\,\epsilon V/eL^2$. Furthermore, $\bar{n} = 2^{1/2} n_m$ with $\bar{n} = \int_0^L n\, dx/L$.

(iv) Finally, $t_n = (4/3)^{1/2}\,\bar{t}_n$ where $t_n = \mu_n^{-1} \int_0^L \mathscr{E}^{-1}\, dx$, $\bar{t}_n = L^2/\mu_n V$.

The relative electric field and potential distributions are plotted (the dashed curves) versus x/L in Fig. 11.1. In plotting these curves, use is made of the relations

$$\frac{\mathscr{E}}{\mathscr{E}_m} = \frac{n_m}{n} = \frac{u}{u_m} = \exp(-\tfrac{1}{2}s^2); \qquad P(s) = \frac{2}{\pi^{1/2}} \int_0^s ds \exp(-s^2),$$
$$\frac{x}{L} = \frac{w}{w_c} = \tfrac{1}{2}[1 - P(s)]; \qquad \frac{V(x)}{V_c} = \frac{v}{v_c} = \tfrac{1}{2}\{1 - P[(\tfrac{3}{2})^{1/2}s]\}. \tag{11.36}$$

These forms are useful because $P(s)$ is, of course, the well-known probability integral which is tabulated in many books.

It is extraordinary how close these relative or normalized distributions are to those for the constant-lifetime injected-plasma problem. Also plotted in Fig. 11.1 (dashed line) is the quantity

$$[(2\mu_R/\mu_p) + (2\mu_R/\mu_n)]^{-1}(p - n)/n_m\,.$$

Note that the linear approximation (short-dashed line) is good over most of the insulator; a factor-of-two deviation from linearity occurs at $x/L = 0.08$. Note that, because of different constants of proportionality, there is no significance to be attached to the different slopes, near $x = L/2$, of the two $(p - n)$ curves, one for the constant lifetime case (solid curve) and the other for the bimolecular lifetime case (dashed curve).

11.2.2 *The One-Carrier, Space-Charge-Current Limit*

A second limiting case occurs for very large values of the recombination cross section σ_R. In fact, if we take σ_R infinite, then the current must be a pure electron SCL current J_n on the cathode side of the insulator and a pure hole SCL current J_p on the anode side; the two SCL currents meet and annihilate at a plane in the body of the insulator. There may be no volume (finite thickness) of overlap of the two currents since, if there were, then r would be infinite over a finite volume, which is impossible. Let x_m be the plane of annihilation. Defining $L_p = x_m$, $L_n = L - x_m$, $V_p = V_m = V(x_m)$, and $V_n = V - V_p$, we have $J_p = 9\epsilon\mu_p V_p^2/8L_p^3$, $J_n = 9\epsilon\mu_n V_n^2/8L_n^3$, $J = J_n = J_p$. Requiring the electric field to be continuous across x_m gives $V_p/L_p = V_n/L_n$. It is now easy to show that $V_n/V_p = L_n/L_p = (L - x_m)/x_m = \mu_n/\mu_p$ and that

$$(\mu_p + \mu_n)(V_p + V_n)^2/(L_p + L_n)^3 = \mu_p V_p^2/L_p^3 = \mu_n V_n^2/L_n^3.$$

These results, put together, yield the J–V characteristic:

$$J = (9/8)\,\epsilon(\mu_p + \mu_n)(V^2/L^3). \tag{11.37}$$

The total current is, in magnitude, simply the sum of the two separate SCL currents that would flow across the entire insulator, each in the absence of the other.

11.2.3 *The Parmenter–Ruppel Exact Solution*

The Eqs. (11.1)–(11.3), (11.27), and (11.5), characterizing double injection into a perfect insulator, have an exact, analytical solution as shown by Parmenter and Ruppel (1959). The work of these authors is of great historical importance in that their analytical solution is the first for *any* insulator double-injection problem. We have delayed discussion of this solution because its analytical complexities have not, as yet, yielded a detailed description of the double-injection current-flow properties. However, the solution does yield a precise current-voltage characteristic which is very useful indeed.

In the following, we reproduce the relevant portion of the Parmenter–Ruppel paper, with some minor differences of notation. They define the constants

$$\mu_R = \epsilon\langle v\sigma_R\rangle/2e, \qquad \nu_n = \mu_n/\mu_R, \qquad \nu_p = \mu_p/\mu_R,$$
$$\alpha = \epsilon\mu_n\mu_p/2\mu_R J, \qquad \beta_n = e\mu_n/J, \qquad \beta_p = e\mu_p/J, \tag{11.38}$$

and the dimensionless variables

$$A = \alpha\mathscr{E}^2, \qquad B = \beta_n n\mathscr{E}, \qquad C = \beta_p p\mathscr{E}. \tag{11.39}$$

Equations (11.15), (11.26), and (11.3) become, respectively

$$B + C = 1, \tag{11.40}$$

$$B' = -C' = BC/A, \tag{11.41}$$

$$A' = \nu_n C - \nu_p B, \tag{11.42}$$

where the prime superscript denotes differentiation with respect to x. Substitution of (11.41) into (11.42) yields

$$A'/A = \nu_n(B'/B) + \nu_p(C'/C). \tag{11.43}$$

This integrates directly to

$$A = KB^{\nu_n}(1 - B)^{\nu_p}, \tag{11.44}$$

where K is a constant of integration and we have used (11.40). Differentiation of (11.44) gives

$$A' = KB'(d/dB)[B^{\nu_n}(1 - B)^{\nu_p}] = -B'A[\nu_p B - \nu_n(1 - B)]/B(1 - B)$$

which, with (11.42) rewritten as $A' = \nu_n(1 - B) - \nu_p B$, gives

$$dx = [A/B(1 - B)]\, dB = KB^{(\nu_n-1)}(1 - B)^{(\nu_p-1)}\, dB, \tag{11.45}$$

where we have used (11.44).

Integrating from the anode ($x = 0$) to the cathode ($x = L$), we obtain

$$K = L\Big/\int_{B_a}^{B_c} B^{(\nu_n-1)}(1 - B)^{(\nu_p-1)}\, dB, \tag{11.46}$$

where B_c, B_a are the values of B at the cathode and anode, respectively. Combining Eqs. (11.39), (11.44), and (11.45), we obtain

$$\mathscr{E}\, dx = \alpha^{-1/2}K^{3/2}B^{[(3/2)\nu_n-1]}(1 - B)^{[(3/2)\nu_p-1]}\, dB.$$

Integrating this equation from the anode to the cathode, we obtain

$$V_c = \int_a^c \mathscr{E}\, dx = \frac{L^{3/2} \int_{B_a}^{B_c} B^{[(3/2)\nu_n - 1]}(1 - B)^{[(3/2)\nu_p - 1]}\, dB}{\alpha^{1/2}[\int_{B_a}^{B_c} B^{(\nu_n - 1)}(1 - B)^{(\nu_p - 1)}\, dB]^{3/2}}. \tag{11.47}$$

For injecting contacts and bimolecular recombination, the entire current at the anode is carried by holes, $B_a = 0$, and the entire current at the cathode is carried by electrons, $B_c = 1$. (Parmenter and Ruppel discuss, as well, other kinds of contacts in their paper, with correspondingly different values for B_c or B_a or both.) Taking $B_a = 0$, $B_c = 1$, and substituting for α from (11.38) into (11.47), and noting that

$$\int_0^1 B^m(1 - B)^n\, dB = m!n!/(m + n + 1)!; \qquad m \text{ and } n > -1, \tag{11.48}$$

Eq. (11.47) can finally be rewritten as

$$J = (9/8)\, \epsilon\mu_{\text{eff}}(V^2/L^3) \tag{11.49}$$

with

$$\mu_{\text{eff}} = \mu_R \nu_n \nu_p \frac{4}{9} \left[\frac{(\frac{3}{2}[\nu_n + \nu_p] - 1)!}{(\frac{3}{2}\nu_n - 1)!\,(\frac{3}{2}\nu_p - 1)!}\right]^2 \left[\frac{(\nu_n - 1)!\,(\nu_p - 1)!}{(\nu_n + \nu_p - 1)!}\right]^2. \tag{11.50}$$

Equations (11.49) and (11.50) are valid whatever the mobility ratios ν_n and ν_p. The expression for μ_{eff} simplifies substantially in three different limits.

Injected Plasma Limit: $\nu_p \gg 1$ and $\nu_n \gg 1$

In this case, it is valid to approximate the factorials in μ_{eff} using Stirling's formula:

$$n! \simeq (n/e)^n(2\pi n)^{1/2}. \tag{11.51}$$

Using (11.51) in (11.50), the final result is

$$\mu_{\text{eff}} = \tfrac{2}{3}\{2\pi[\mu_n\mu_p(\mu_n + \mu_p)/\mu_R]\}^{1/2}. \tag{11.52}$$

This result is identical to (11.35).

One-Carrier SCL current Limit: $\nu_p \ll 1$, but $\nu_n > 1$

Noting that for $0 < \delta \ll 1$, $(\delta - 1)! = \delta!/\delta \simeq 0!/\delta = 1/\delta$ and using this in (11.50), we obtain

$$\mu_{\text{eff}} \simeq \mu_n. \tag{11.53}$$

In this case, the current is a one-carrier, electron space-charge-limited current, as expected. The holes are so sluggish that they never get significantly away from the anode before they are annihilated by recombination.

Pair of One-Carrier-SCL current Limit: $\nu_p \ll 1$ and $\nu_n \ll 1$

Again we use $(\delta - 1)! \simeq 1/\delta$ for $0 < \delta \ll 1$ in (11.50), this time obtaining

$$\mu_{eff} = \mu_p + \mu_n . \tag{11.54}$$

This result, in conjunction with (11.49), is identical to (11.37). As already discussed in Section 11.2.2, it corresponds to the current being a pure electron SCL current on the cathode side of the insulator and a pure hole SCL current on the anode side of the insulator, the two SCL currents meeting and annihilating in the body of the insulator.

11.2.4 *The Regional-Approximation Solution*

For the problem of double injection into the perfect insulator we have, so far, a picture of details of the current flow only for limiting cases, namely those enumerated in the preceding sections. It can be useful to have a picture of the current flow for situations where these extreme limiting conditions are not realized. One successful approach to these situations is via the Regional Approximation Method. We here outline the application of this method to the perfect-insulator problem and present some results.

Our starting point is the observation that near the injecting anode, the total current is completely dominated by the contribution of the holes: $J \simeq J_p$ and, conversely, near the cathode, the total current is completely dominated by the electrons: $J \simeq J_n$. In each case, the domination is so complete that the current near an injecting contact is essentially a one carrier, space-charge-limited current corresponding to the carrier injected at that contact. More specifically, approaching the anode, $p/n \rightarrow \infty$, and approaching the cathode, $n/p \rightarrow \infty$. Building on this circumstance, it is convenient to single out three particular planes in the insulator, x_1, x_2, and x_3, defined as follows:

$$\begin{aligned} x_1: &\quad p_1 = p(x_1) = bn(x_1) = bn_1 , \\ x_2: &\quad p_2 = p(x_2) = 2n(x_2) = 2n_2 , \\ x_3: &\quad p_3 = p(x_3) = \tfrac{1}{2}n(x_3) = \tfrac{1}{2}n_3 , \end{aligned} \tag{11.55}$$

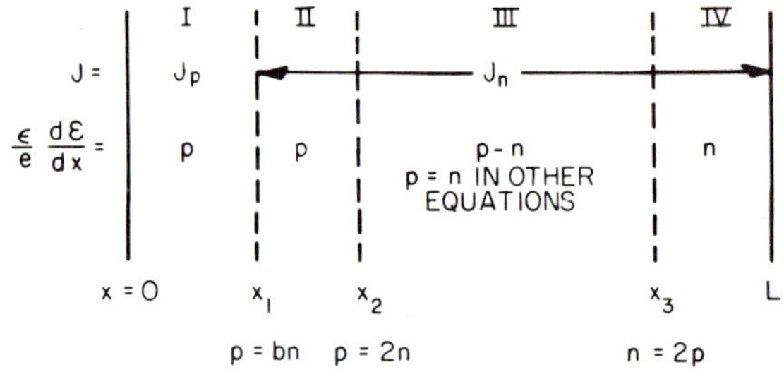

FIG. 11.5. Schematic Regional Approximation diagram for the problem of double injection into a perfect, trap-free insulator, for the case that $b = \mu_n/\mu_p > 2$.

with b defined by (11.6). Clearly we are here assuming that $b > 2$. The planes $x = 0$, x_1, x_2, x_3, and L, as illustrated in Fig. 11.5, delineate the boundaries of the separate regions in the insulator, in the sense of the Regional Approximation Method. These regions, and the governing equations in them, namely suitable approximations to (11.6), (11.3) and (11.26), are then:

Region I ($0 \leqslant x \leqslant x_1$, $p > bn$):

$$J = ep\mu_p\mathscr{E} = \text{constant}, \tag{11.56}$$

$$(\epsilon/e)\,(d\mathscr{E}/dx) = p. \tag{11.57}$$

These, of course, are the equations governing a trap-free, single-carrier (hole) SCL current, and they are solved subject to the usual boundary condition: $\mathscr{E} = 0$ at $x = 0$. Their solution is simple and is given in Section 4.1. Eq. (11.26b) is ignored, having been replaced by (11.56) with which it is, in a formal sense, inconsistent. The variation of n with x in Region I is obtained from (11.26a), regarding p and $\mathscr{E}$ as already determined from the solution to (11.56) and (11.57).

Region II ($x_1 \leqslant x \leqslant x_2$, $bn \geqslant p \geqslant 2n$):

$$J = en\mu_n\mathscr{E} = \text{constant}, \tag{11.58}$$

$$(\epsilon/e)\,d\mathscr{E}/dx = p, \tag{11.59}$$

$$-\mu_p(d/dx)(p\mathscr{E}) = \langle v\sigma_R\rangle np. \tag{11.60}$$

As with (11.26b) in Region I, we here ignore (11.26a), replacing it with (11.58), with which it is, in a formal sense, inconsistent. Region II is a rather unusual region in that the current is determined solely by the electrons and the space charge solely by the holes. Using (11.58) and

(11.59) to eliminate n and p from (11.50), we obtain the following equation in $\mathscr{E}$:

$$-\mathscr{E}^2\, d^2\mathscr{E}^2/dx^2 = A\, d\mathscr{E}^2/dx, \qquad A = \langle v\sigma_R\rangle J/e\mu_n\mu_p\,. \tag{11.61}$$

This differential equation is readily solved in terms of the well-known exponential integral function $E_1(y) = \int_y^\infty dt(\exp - t)/t$. The variation of p and n with x in Region II is then determined from (11.59) and (11.58), respectively.

Region III ($x_2 \leqslant x \leqslant x_3$, $2n > p > n/2$):

$$J = en\mu_n\mathscr{E} = \text{constant}, \tag{11.62}$$

$$(\epsilon/e)\,(d\mathscr{E}/dx) = p - n, \tag{11.63}$$

$$\mu_n(d/dx)(n\mathscr{E}) = \langle v\sigma_R\rangle n^2, \tag{11.64a}$$

$$-\mu_p(d/dx)(p\mathscr{E}) = \langle v\sigma_R\rangle n^2, \tag{11.64b}$$

Multiplying (11.64b) by b and adding to (11.64a) gives, after using (11.63) and (11.62) to eliminate $(p - n)$ and n, respectively,

$$-\mathscr{E}^2\, d^2\mathscr{E}^2/dx^2 = B, \qquad B = [4(b+1)\mu_R J^2]/[\epsilon^2 b^3\mu_p^{\,3}]. \tag{11.65}$$

Equation (11.65) has the same form as (11.29) and, in fact, is almost identical to it; if the factor b^3 in the denominator of (11.65) is replaced by $b(b+1)^2$, then the equations are indeed identical. The small difference between (11.65) and (11.29) arises because here Eq. (11.62) describes the total current, whereas in the derivation of (11.29), (11.15) was used. Since we have assumed $b > 2$, the difference is negligible. In both cases, an injected-plasma region is described, the other pertinent equations being identical.

The solution to (11.65) is given in terms of a Gaussian integral, namely (11.32). With $\mathscr{E}$ determined in Region III, the variation of n with x is obtained from (11.62), and then p with x for (11.63).

Region IV ($x_3 \leqslant x \leqslant L$, $\frac{1}{2}n > p$):

$$J = en\mu_n\mathscr{E}, \tag{11.66}$$

$$(\epsilon/e)\,(d\mathscr{E}/dx) = n. \tag{11.67}$$

These are the equations governing a trap-free, single-carrier (electron) SCL current; they are solved subject to the usual boundary condition:

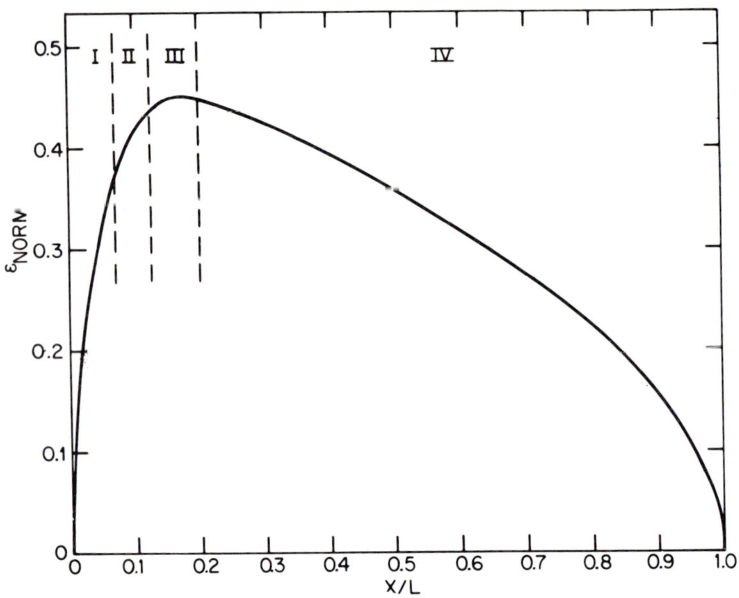

I
II
III
IV
0.5
0.4
0.3
0.2
0.1
0
ε NORM
0
0.1
0.2
0.3
0.4
0.5
0.6
0.7
0.8
0.9
1.0
X/L

(a)

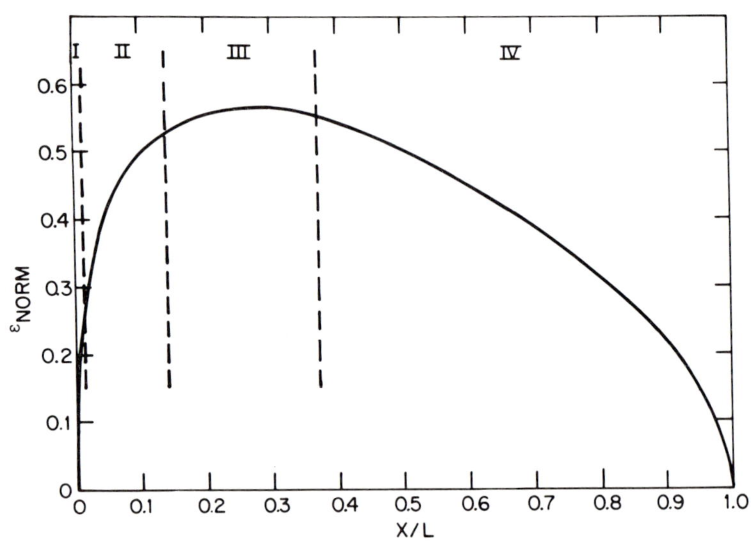

I
II
III
IV
0.6
0.5
0.4
0.3
0.2
0.1
0
ε NORM
0
0.1
0.2
0.3
0.4
0.5
0.6
0.7
0.8
0.9
1.0
X/L

(b)

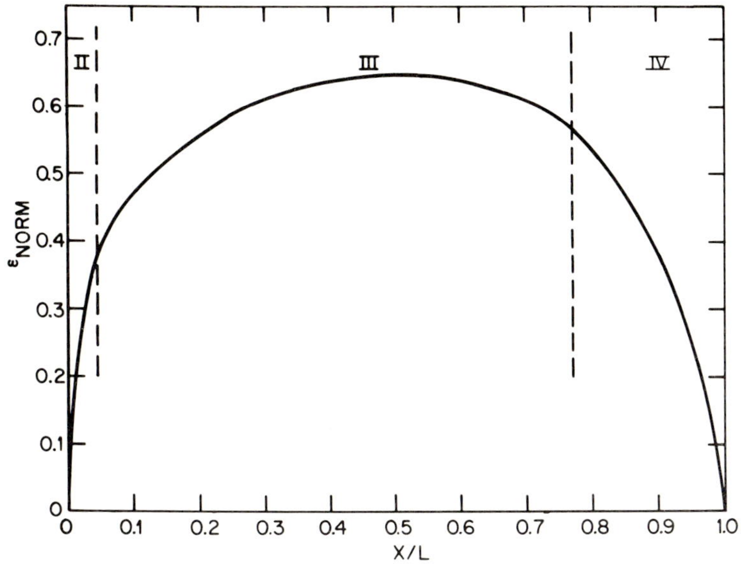

(c)

FIG. 11.6. Variation of the (normalized) field with position for the problem of double injection into a perfect insulator. $\nu_p = \mu_p/\mu_R$, $\nu_n = \mu_n/\mu_R$, $\mu_R = \epsilon\langle v\sigma_R\rangle/2e$, $b = 8$ for all cases: (a) (top left) $\nu_p = 1/16$, $\nu_n = \frac{1}{2}$; (b) (bottom left) $\nu_p = \frac{1}{4}$, $\nu_n = 2$; (c) (above) $\nu_p = 2$, $\nu_n = 16$.

$\mathscr{E} = 0$ at $x = L$, this being the location of the electron-injecting contact. The variation of p with x in this region is obtained using (11.26b), assuming n and $\mathscr{E}$ already determined from the solution to (11.66) and (11.67).

The solutions in the various regions are tied together in the usual manner, assuming continuity of the $\mathscr{E}$-field crossing a transition plane:

$$\mathscr{E}(x_1^-) = \mathscr{E}(x_1^+), \qquad \mathscr{E}(x_2^-) = \mathscr{E}(x_2^+), \qquad \mathscr{E}(x_3^-) = \mathscr{E}(x_3^+). \tag{11.68}$$

The above discussion obviously sketches only an outline of the solution using the Regional Approximation Method. Complete details are given in a paper by Rosenberg and Lampert (1969). Some concrete results for the relative electric field distribution, for three different cases, are plotted in Fig. 11.6. Case (a): $\nu_p = 1/16$, $\nu_n = \frac{1}{2}$, lies toward the one-carrier, electron SCL current limit (11.53); case (c): $\nu_p = 2$ and $\nu_n = 16$ toward the injected-plasma limit (11.52); and case (b): $\nu_p = \frac{1}{4}$ and $\nu_n = 2$, is somewhat transitional. Since the only quantity available from the exact, Parmenter Ruppel solution is μ_{eff}, given by (11.50), the Regional-Approximation solution can be compared to the exact solution only in terms of this quantity. Such a comparison is given in Fig. 11.7,

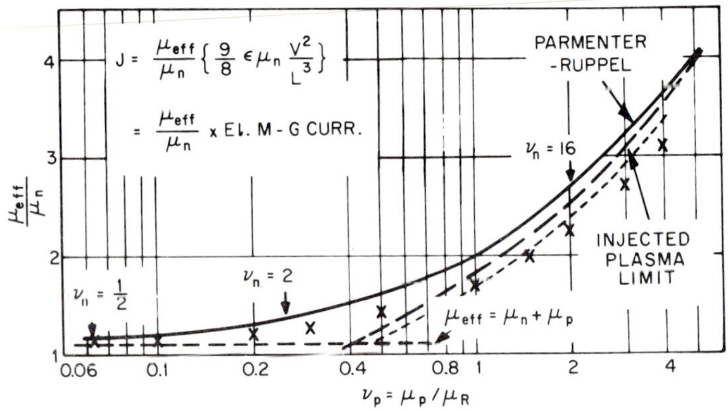

FIG. 11.7. The current "amplification" factor μ_{eff}/μ_n vs ν_p (μ_{eff} given by (11.50), ν_p and ν_n given by (11.38)) for the problem of double injection into a perfect insulator; $b = 8$.

for the case that $b = 8$. The smooth curve is, basically, a plot of the exact result (11.50), the heavy-dashed curve a corresponding plot of the injected-plasma limit (11.52), and the crosses are points obtained from the Regional Approximation Method, as outlined above. Because this last neglects the hole contribution to the injected-plasma current [see (11.62)], the crosses are asymptotes, at large ν_p, to the light-dashed curve, which represents the heavy-dashed curve multiplied by the constant $b/(b+1)$.

The entire discussion above is based on the assumption that $b > 2$. For $2 > b > \frac{1}{2}$, there is no need for a Region II, and the problem is solved much as before, with only the Regions I, III, and IV. With $\frac{1}{2} > b$, we are again back to a four-region problem, and the solution is identical to that sketched above for $b > 2$, only the roles of electrons and holes have been interchanged (so that Region I starts at the cathode, proceeding to Region IV, which ends at the anode).

12

PLASMAS INJECTED INTO SEMICONDUCTORS

The plasma injected into a semiconductor differs from that injected into an insulator mainly in that there are sufficient thermal free carriers in the semiconductor to prevent, by dielectric (ohmic) relaxation, the build-up of space charge. Correspondingly, in the analytical treatment of the semiconductor injected plasma, local neutrality is assumed. For the sake of definiteness, we consider a plasma injected into a n-type semiconductor, $n_0 > p_0$; the hole-injecting contact is taken at $x = 0$ and the electron-injecting contact at $x = L$.

12.1 The Constant-Lifetime Plasma: High Injection Levels

The equations characterizing current flow at high injection levels in the semiconductor ($n \approx p \gg n_0$, p_0) are the same as those, (11.1)–(11.4), characterizing current flow in the insulator except that the Poisson equation (11.3) is replaced by the condition of local neutrality (12.2). For the sake of convenience, we write again here the full set of equations corresponding to the simplified theory neglecting diffusion. The sign conventions specified by (11.1) hold here also:

$$J_p = ep\mu_p\mathscr{E}, \qquad J_n = en\mu_n\mathscr{E}, \tag{12.1}$$

$$n - p = n_0 - p_0 \,. \tag{12.2}$$

Note that (12.2) states that the *injected* electron and hole concentrations, $n - n_0$ and $p - p_0$, respectively, are everywhere equal.

$$(1/e)\,(dJ_{\mathrm{n}}/dx) = \mu_{\mathrm{n}}(d/dx)(n\mathscr{E}) = r = n/\tau, \tag{12.3a}$$

$$-(1/e)\,(dJ_{\mathrm{p}}/dx) = \mu_{\mathrm{p}}(d/dx)(p\mathscr{E}) = r = n/\tau, \tag{12.3b}$$

where the assumption of high injection levels has been used in the expression for recombination. From (12.3a) and (12.3b), once again (11.6) follows:

$$J = J_{\mathrm{n}} + J_{\mathrm{p}} = e(b+1)\mu_{\mathrm{p}}n\mathscr{E} = \text{constant.} \tag{12.4}$$

Note that in writing (12.4), we have neglected the contribution of the thermal carriers to the total current; we take $n \simeq p$ except where the difference $n - p$ appears.

Because of the replacement of the Poisson equation (11.3) by the condition of local neutrality (12.2), the characteristic differential equation for the problem, Eq. (12.6), is only of first order. Therefore, only one boundary condition may be imposed, which we take to be the vanishing of the electric field intensity at the minority-carrier (hole) injecting contact:

$$\mathscr{E} = 0 \quad \text{at} \quad x = 0. \tag{12.5}$$

This replaces the boundary conditions (11.51) for the insulator problem.

Multiplying (12.3b) by b and adding to (12.3a) gives, using (12.2) and (12.4),

$$\mathscr{E}\,(d\mathscr{E}/dx) = J/[e(n_0 - p_0)\,\mu_{\mathrm{n}}\mu_{\mathrm{p}}\tau]. \tag{12.6}$$

Now we need only note that (12.6) is identical to (4.6), which determines the electric field distribution ($\mathscr{E} \propto x^{1/2}$) in the trap-free insulator, if ϵ in the latter is replaced by $e(n_0 - p_0)\,\mu_{\mathrm{p}}\tau$. Further, the boundary conditions for the two problems are identical. Thus the current-voltage characteristics for the two problems are identical once ϵ in (4.9) is replaced as indicated above:

$$J = (9/8)\,e(n_0 - p_0)\,\mu_{\mathrm{n}}\mu_{\mathrm{p}}\tau(V^2/L^3). \tag{12.7}$$

This result was first obtained by Lampert and Rose (1961).

The relative distribution relations (4.10) are all valid, except that here $\bar{n} = \int_0^L n\,dx/L = 3(n_0 - p_0)\,\mu_{\mathrm{n}}\mu_{\mathrm{p}}\tau(V/2L^2)/(\mu_{\mathrm{n}} + \mu_{\mathrm{p}})$, in place of (4.11). This last differs from the phenomenological estimate (10.19) only in the numerical factor.

12.2 The Constant-Lifetime Plasma: Lower Injection Levels

At lower voltages, Ohm's law must obtain:

$$J = e(n_0\mu_n + p_0\mu_p)(V/L). \tag{12.8}$$

In order to study the details of the transition from Ohm's law to the semiconductor square law (12.7), the particle-conservation equations (12.3) must be modified to include the thermally generated free carriers n_0 and p_0. Since n and p in (12.3) are *total* free carrier concentrations, i.e., the sum of injected and thermally generated free carriers, the net recombination rate density r in (12.3a) and (12.3b) must be written as $(n - n_0)/\tau$:

$$\mu_n(d/dx)(n\mathscr{E}) = (n - n_0)/\tau, \tag{12.9a}$$

$$-\mu_p(d/dx)(p\mathscr{E}) = (n - n_0)/\tau. \tag{12.9b}$$

The equations (12.9), the current-flow equation (12.1), the neutrality condition (12.2), and the single boundary condition (12.5) characterize the problem. Subtracting (12.9b) from (12.9a) gives

$$J = J_n + J_p = e\mu_p\mathscr{E}(bn + p) = \text{constant}. \tag{12.10}$$

Adding (12.9b), multiplied by b, to (12.9a) gives

$$\frac{d\mathscr{E}}{dx} = \frac{(b+1)(n - n_0)}{(n_0 - p_0)\mu_n\tau} = \frac{J - e(bn_0 + p_0)\mu_p\mathscr{E}}{e(n_0 - p_0)\mu_p\mu_n\tau\mathscr{E}}, \tag{12.11}$$

where (12.2) and (12.10) have been used.

It is now useful to employ the following dimensionless variables:

$$u = \frac{bn_0 + p_0}{bn + p} = \frac{e(bn_0 + p_0)\mu_p\mathscr{E}}{J},$$
$$w = \frac{e(bn_0 + p_0)^2 x}{b\tau(n_0 - p_0)J}, \qquad v = \frac{e^2(bn_0 + p_0)^3\mu_p V}{b\tau(n_0 - p_0)J^2}. \tag{12.12}$$

With this choice of variables, (12.11) becomes

$$u\,du/(1 - u) = dw, \tag{12.13}$$

with solution

$$w = -u - \ln(1 - u), \tag{12.14}$$

corresponding to $u = 0$ at $w = 0$, i.e., $\mathscr{E} = 0$ at $x = 0$. The voltage relation $V = \int_0^x \mathscr{E}\,dx$ becomes

$$v = \int_0^w u\,dw = \int_0^u u\,(dw/du)\,du = -\tfrac{1}{2}u^2 - u - \ln(1 - u). \tag{12.15}$$

We note that (12.13)–(12.15) are identical to (4.16)–(4.18), respectively, and therefore the formal calculations for this problem are identical to that for the one-carrier "Trap-Free Insulator with Thermal Free Carriers" problem discussed in Section 4.2. The subscript "a" of that problem is here changed to "c" because the focus has switched from electron injection to hole injection.

Here current J and applied voltage V are given by

$$J = \frac{e(bn_0 + p_0)^2 L}{b\tau(n_0 - p_0)} \frac{1}{w_c}, \qquad V = \frac{(bn_0 + p_0)L^2}{(n_0 - p_0)\mu_n \tau} \frac{v_c}{w_c^2} = \frac{L^2}{\mu_{a,\text{low}}\tau} \frac{v_c}{w_c^2}, \tag{12.16}$$

where subscript "c" corresponds to taking $x = L$ in the defining relations and where $\mu_{a,\text{low}}$ denotes the ambipolar mobility at very low injection levels, $n \simeq n_0$, and $p \simeq p_0$. The plot of v/w^2 vs $1/w$ in Fig. 4.1 (p. 48) is equally useful for the present problem. The regime $u_c \simeq 1$, $v_c/w_c^2 \simeq 1/w_c$, corresponds to Ohm's law (12.8). The regime

$$u_c \ll 1, \qquad v_c/w_c^2 \simeq (2^{3/2}/3)(1/w_c)^{1/2},$$

corresponds to the semiconductor square-law (12.7). The quantity $(1/u_c) - 1$ is a measure of the injected free carrier concentration at the cathode: $(1/u_c) - 1 = (b + 1)(n_c - n_0)/(bn_0 + p_0)$. The quantity $w_a + u_a$ plotted in Fig. 4.1, which for the one-carrier problem was just the transit-time to ohmic-relaxation-time ratio t/t_Ω, is here a measure of the transit-time to lifetime ratio: $w_c + u_c = (bn_0 + p_0)\, t_p/b(n_0 - p_0)\,\tau$, with $t_p = (1/\mu_p) \int_0^L dx/\mathscr{E}$.

The quantities $u_a w_a/v_a$, $w_a(1 - u_a)/u_a^2$, $v_a(w_a + u_a)/w_a^2$ tabulated in Table 4.2 have the same significance here:

$$u_c w_c/v_c = \mathscr{E}_c/\mathscr{E}_\Omega, \qquad w_c(1 - u_c)/u_c^2 = (n_c - n_0)/(\overline{n - n_0}),$$
$$v_c(w_c + u_c)/w_c^2 = t_p/\bar{t}_p$$

where $\bar{t}_p = L^2/\mu_p V$. Likewise the normalized, injected, free-carrier concentration distributions in Fig. 4.2 are equally applicable to the present problem.

It is instructive to consider in detail ambipolar drift effects. The ambipolar mobility, given by (10.14), can be written as $\mu_a = \mu_{a,\text{low}}(bn_0 + p_0)/(bn + p) = \mu_{a,\text{low}} u$ from (12.12) with $\mu_{a,\text{low}} = (n_0 - p_0)\,\mu_n/(bn_0 + p_0)$. Because of the neglect of diffusion currents, u, and therefore μ_a, goes to 0 at $x = 0$. Since the electric field also goes to 0 at $x = 0$, the ambipolar drift time t_a goes to infinity:

$$t_a/\tau = \int_0^L dx/\mu_a \tau \mathscr{E} \propto \int_0^{x_a} \{du/u(1 - u)\} \to \infty.$$

This appears to indicate that the phenomenological relation (10.16) is not valid for this problem. However we must realize that the result $t_a/\tau \to \infty$ is a spurious result of the simplified theory, arising from the wholly incorrect description, by this theory, of the contact regions. Since the simplified theory does give a good description of quantities away from the contacts, we write

$$\bar{\mu}_a \simeq \mu_{a,\mathrm{low}}(bn_0 + p_0)/(b\bar{n} + \bar{p}) = \mu_{a,\mathrm{low}} w_c/(w_c + u_c),$$

this last following from a bit of algebra. Finally

$$\bar{t}_a/\tau = L^2/\bar{\mu}_a V\tau = w_c(w_c + u_c)/v_c\,.$$

As $w_c \to 0$, corresponding to high injection levels, $\bar{t}_a/\tau \to \frac{3}{2}$, and the use of (10.16) is thereby justified. At $w_c = 0.1$, 1, and 10, $\bar{t}_a/\tau = 1.8$, 2.8, and 11.6, respectively. As $w_c \to \infty$, $\bar{t}_a/\tau \to w_c \to \infty$.

A final point of interest is a comparison of the neglected diffusion term, evaluated in the framework of the simplified theory, with the field-driven term. As is made clear in the following chapter, which discusses the "unified theory," the corresponding terms are $2V_T\, d^2n/dx^2$ (diffusion) and $(n_0 - p_0)\,(d\mathscr{E}/dx)$ (drift). We may write symbolically,

$$\begin{aligned}|\,\text{diff./drift}\,| &= [2V_T/(n_0 - p_0)](d^2n/dx^2)/(d\mathscr{E}/dx) = (L_D/L)^2 w_c^2(3 - 2u)/u^4 \\ &\simeq (L_D/L)^2 3w_c^2/4w^2 = (3/4)(L_D/x)^2,\end{aligned}$$

where L_D is the ambipolar diffusion length $L_D^2 = 2V_T\mu_n\tau/(b+1)$. Thus, beyond a diffusion length, the simplified theory is at least self-consistent. We shall see that a more careful treatment of diffusion, given in the following chapter, shows that the diffusion term can dominate beyond one diffusion length, indicating that self-consistency is, in itself, an inadequate criterion for validity of a calculation.

12.3 The Bimolecular Lifetime Plasma: High Injection Levels

The equations characterizing the bimolecular-lifetime problem at high injection levels are identical to those used for the analysis of the constant-lifetime problem at high injection levels, namely (12.1)–(12.5), except for the particle-conservation equations. In these two equations, the recombination rate density r is now the bimolecular quantity: $r = \langle v\sigma_R\rangle\, np \simeq \langle v\sigma_R\rangle\, n^2$, precisely as in Section 11.2.1. Thus (12.3a) and (12.3b) are replaced, respectively, by (11.27a) and (11.27b), rewritten here for convenience:

$$(1/e)(dJ_n/dx) = \mu_n\,(d/dx)(n\mathscr{E}) = r = \langle v\sigma_R\rangle n^2, \qquad (12.17a)$$

$$-(1/e)(dJ_p/dx) = \mu_p\,(d/dx)(p\mathscr{E}) = r = \langle v\sigma_R\rangle n^2. \qquad (12.17b)$$

Multiplying (12.17b) by b and adding to (12.17a) gives, using (12.2) and (12.4),

$$\mathscr{E}^2\, d\mathscr{E}/dx = \langle v\sigma_R\rangle J^2/[e^2b(b+1)\mu_p{}^3(n_0-p_0)]. \tag{12.18}$$

This differential equation, subject to the boundary condition (13.6), is readily integrated to give $\mathscr{E} = \alpha x^{1/3}$, and from this, $V(x) = (3\alpha/4)\, x^{4/3}$ with $\alpha = \{3\langle v\sigma_R\rangle\, J^2/e^2b(b+1)\, \mu_p{}^3(n_0-p_0)\}^{1/3}$. Taking $x = L$ and $V = V(L)$ in the latter relation, we obtain

$$J = \frac{8}{9}\, e\mu_p \left\{\frac{(b+1)\,\mu_n(n_0-p_0)}{\langle v\sigma_R\rangle}\right\}^{1/2} \frac{V^{3/2}}{L^2}, \tag{12.19}$$

which is identical to the result obtained earlier by Rashba and Tolpygo (1955).

Further relations of interest are

$$\frac{\mathscr{E}}{\mathscr{E}_\Omega} = \frac{4}{3}\left(\frac{x}{L}\right)^{1/3}; \qquad \frac{V(x)}{V} = \left(\frac{x}{L}\right)^{4/3}; \qquad \frac{n}{\bar{n}} = \frac{2}{3}\left(\frac{L}{x}\right)^{1/3},$$
$$\bar{n} = \left\{\frac{\mu_n(n_0-p_0)V}{(b+1)\langle v\sigma_R\rangle L^2}\right\}^{1/2} = \left\{\frac{\mu_n}{2(b+1)\mu_R}\, \frac{\epsilon V}{eL^2}\,(n_0-p_0)\right\}^{1/2}. \tag{12.20}$$

Note that this value for $\bar{n}$ can be obtained directly from (10.19) simply by replacing $\bar{\tau}$ by $[\bar{n}\langle v\sigma_R\rangle]^{-1}$. Thus the current-voltage characteristic (12.19), except for the precise numerical factor 8/9, comes directly out of the earlier phenomenological argument of Chapter 10. This problem has been generalized, and the solution obtained, by including both constant-lifetime and bimolecular recombination processes (Hirota *et al.*, 1963, Appendix).

12.4 Surface Recombination Corrections: Two-Dimensional Theory

The injected-plasma theory presented up to this point has been entirely one-dimensional, the presence of bounding surfaces being ignored. These theories might be expected to give a good description for real semiconductors whose minimum, transverse, cross-sectional dimension d was large compared to the cathode-anode spacing L, and for which, therefore, surface recombination would not, ordinarily, be important. On the other hand, in order that diffusion corrections be negligible, it is necessary that $L \gg L_D$, as discussed above. Experimentally the injected plasma in semiconductors has usually been studied under conditions $d \ll L$, $L_D \ll L$, for which case, in the studies with germanium, surface recombination sometimes strongly dominates volume recombination. An outstanding result of these studies is that the one-dimensional characteristic (12.7) fits the observations if τ is replaced by an effective lifetime τ_{eff}. Further, in the germanium experiments, τ_{eff} is frequently considerably less than τ. In this section, we shall outline a two-dimensional theory for the injected plasma in semiconductors which derives this result and yields an explicit formula for τ_{eff} in terms of τ and s, the surface recombination velocity.

In place of the one-dimensional current-flow equations (12.1), we now have the vector equations

$$
\begin{aligned}
(1/e)\mathbf{J}_{\text{n}} &= \mu_{\text{n}} n \mathscr{E} + D_{\text{n}} \operatorname{grad} n, \\
(1/e)\mathbf{J}_{\text{p}} &= \mu_{\text{p}} p \mathscr{E} - D_{\text{p}} \operatorname{grad} p,
\end{aligned}
\tag{12.21}
$$

where we have assumed isotropic transport properties for the semiconductor. Including diffusion currents at the outset, we are anticipating the crucial importance of these currents in the transverse direction.

In place of the one-dimensional particle-conversation equation (12.9), we now have the vector equations

$$(1/e) \operatorname{div} \mathbf{J}_{\text{n}} = r = [(n - n_0)/\tau], \tag{12.22a}$$

$$-(1/e) \operatorname{div} \mathbf{J}_{\text{p}} = r = [(n - n_0)/\tau]. \tag{12.22b}$$

We retain the assumption of local neutrality (12.2), rewritten here for convenience:

$$n - n_0 = p - p_0 . \tag{12.23}$$

It is a direct consequence of Eqs. (12.22) that

$$\operatorname{div} \mathbf{J} = 0, \qquad \mathbf{J} = \mathbf{J}_{\text{n}} + \mathbf{J}_{\text{p}} . \tag{12.24}$$

We consider explicitly the problem of a semiconductor slab of n-type material, $n_0 > p_0$, with a hole-injecting contact at $x = 0$ and an electron-injecting contact at $x = L$, and with plane bounding surfaces at $y = +d$ and $y = -d$, as illustrated in Fig. 12.1. The slab is infinite in the z-direction, and no quantities are z-dependent. Finally it is assumed that the slab is symmetric, i.e., that s is the same for both surfaces. (The unsymmetric case in which s is not the same for the two surfaces merely increases the length of the calculation without introducing essentially new difficulties.)

The basic strategy for solving the problem is the effective decoupling, with the aid of certain key approximations, of the *average*, longitudinal,

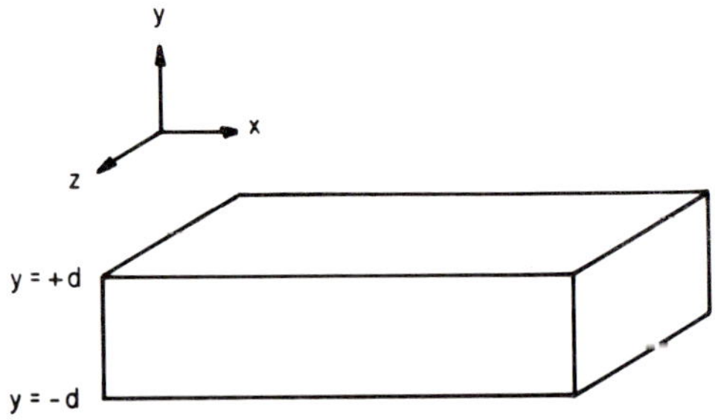

FIG. 12.1. Schematic configuration for double injection into a semiconductor slab immersed in a uniform magnetic field. $\mathbf{B}_0 = B_0\hat{z}$.

x-motion and the transverse, y-motion of injected carriers. The longitudinal motion is drift-dominated, for $L \gg L_D$, and the transverse motion is diffusion-dominated. The interaction between the two motions manifests itself through τ_{eff}.

Simple algebraic manipulation and an integration over the cross section yields the following x-dependent equation:

$$(n_0 - p_0)\frac{d\tilde{\mathscr{E}}_x}{dx} + 2V_T\frac{d^2\tilde{n}}{dx^2} = \frac{(b+1)(\tilde{n} - n_0)}{\mu_n \tau_{eff}}; \qquad \frac{1}{\tau_{eff}} = \frac{1}{\tau} + \frac{s n_d}{d\tilde{n}}, \tag{12.25}$$

where $n_d = n(y = d) = n(y = -d)$, s is defined in the usual way by $J_{py}(d) = en_d s = -J_{ny}(d)$ [similarly $J_{py}(-d) = -en_d s = J_{ny}(d)$], and a tilda over the quantity denotes the average of that quantity over the cross section: Thus $\tilde{n} = n(x) = (1/2d)\int_{-d}^{d} n(x, y)\,dy$.

Equation (12.25) is precisely Eq. (12.11) with the addition of a diffusion term (the second term on the left-hand side) and with τ replaced by τ_{eff}. For $L \gg L_D$, we neglect the diffusion term in (12.25), and use a simplified theory as we have done in the previous sections of this chapter. Now, the previous results for the semiconductor plasma problem will be valid here if two conditions are realized: (i) τ_{eff} is a constant, independent of x, and (ii) $\widetilde{n\mathscr{E}_x} \simeq \tilde{n}\tilde{\mathscr{E}}_x$. It turns out that these two conditions are closely related. As regards (i), n_d and $\tilde{n}$ are separately both functions of x, but we require that $n_d/\tilde{n}$ be independent of x. It turns out that this is indeed the case if the ambipolar diffusion constant D_a, given in (13.27), is a constant, independent of injection level, and if we further assume that J_y is separately conserved, $\partial J_y/\partial y = 0$, and that $\mathscr{E}_y$ is independent of x, $\partial\mathscr{E}_y/\partial x = 0$. Since there are no time-varying fields in the problem, curl $\mathscr{E} = 0$ so that $\partial\mathscr{E}_y/\partial x = 0$ implies $\partial\mathscr{E}_x/\partial y = 0$, in which case condition (ii) is automatically realized since $\mathscr{E}_x$ acts like a constant in averaging over the cross section at fixed x.

These assumptions lead to the y-dependent diffusion equation

$$-\frac{n-n_0}{\tau}+D_a\frac{\partial^2}{\partial y^2}(n-n_0)=-\frac{(\tilde{n}-n_0)}{\tau_{\text{eff}}}=\text{constant};$$
$$D_a=\frac{p+n}{(p/D_n)+(n/D_p)}, \tag{12.26}$$

where $D_n=V_T\mu_n$, $D_p=V_T\mu_p$, and $V_T=kT/e$. The boundary conditions appropriate to the symmetric slab are

$$-D_a\frac{\partial(n-n_0)}{\partial y}=(n-n_0)s \quad \text{at} \quad y=d$$
$$-D_a\frac{\partial(n-n_0)}{\partial y}=0 \quad \text{at} \quad y=0. \tag{12.27}$$

Equation (12.26) is readily solved and n_d/n calculated, giving finally the result

$$J\simeq(9/8)\,e(n_0-p_0)\,\mu_n\mu_p\tau_{\text{eff}}(V^2/L^3), \tag{12.28}$$

with

$$\frac{1}{\tau_{\text{eff}}}=\frac{1}{\tau}+\frac{1}{(d/s)+\{(d/L_a)\coth(d/L_a)-1\}}, \qquad L_a=(D_a\tau)^{1/2}. \tag{12.29}$$

Note that at high injection levels

$$n\gg n_0, \qquad L_a=L_D=\{2V_T\mu_n\tau/(b+1)\}^{1/2}.$$

At low injection levels ($n-n_0\ll n_0$) if $p_0\ll n_0$, then $L_a=(V_T\mu_p\tau)^{1/2}$. Thus, for $b>1$, $1<D_a$ (high level) / D_a (low level) <2, and it is a reasonable approximation to regard D_a as constant and the above theory applicable. On the other hand, for $b\ll 1$, D_a changes drastically between the two regimes of low and high injection level and the transition between the two regimes is very poorly described by the above theory.

For the case of a semiconducting cylinder of radius r_0, the theory goes through much as above, the final result being

$$\frac{1}{\tau_{\text{eff}}}=\frac{1}{\tau}+\left[\frac{r_0}{2s}+\tau\left\{\frac{r_0}{2L_a}\frac{I_0(r_0/L_a)}{I_1(r_0/L_a)}-1\right\}\right]^{-1}, \tag{12.30}$$

where I_0 and I_1 are the modified Bessel functions.

A more detailed derivation of (12.29) and (12.30) is given in the paper by Hirota *et al.* (1963). A similar, but not identical, result as (12.29) obtains for a semiconductor slab under purely diffusive current flow—the normal injection domain of transistor physics. Here the mathematical problem is rigorously separable and solvable as an eigenvalue problem, as shown by Shockley (1950, Section 12.6).

The two-dimensional injection theory for the insulator problem, taking into account the presence of the bounding surfaces, is much more complicated because of the important role of space charge. As in the comparable single injection SCL current problem, discussed in Section 8.8, one has the difficult problem of dealing with lines of force that leave the sample.

The two-dimensional theory for the plasma injected into a semiconductor slab immersed in a uniform magnetic field has been studied by Robinson (1969). The interesting configuration here is that in which the uniform field is along the z-direction: $\mathbf{B}_0 = B_0\mathbf{z}$, referring to Fig. 12.1, that is, transverse to the direction of current flow between electrodes, but parallel to the bounding surfaces of the slab. This is the configuration which produces the Suhl effect. The excess, injected electrons and holes, since they move in opposite directions, are driven by the Lorentz force toward the same surface of the slab. This can lead to a strong confinement of the injected plasma near that surface of the semiconductors. If B_0 is positive (the magnetic field pointing along the positive z-direction), the injected plasma is driven toward the lower slab surface, at $y = -d$ in Fig. 12.6. Assuming local neutrality and, further, that the transverse hole concentration gradient is everywhere much larger than the longitudinal hole concentration gradient at the same point in the sample, Robinson finds the exponential transverse density distribution

$2d \gg l$:

$$p \simeq \frac{2d}{l}\,\tilde{p}\exp -\frac{(y+d)}{l}, \qquad l = \frac{(p+n)}{n}\,\frac{kT}{(\mu_{\mathrm{n}}+\mu_{\mathrm{p}})\,eB_0E_x} \tag{12.31}$$

where $1 < (p+n)/n < 2$.

The longitudinal equation for the cross-sectional average density $\tilde{p}$ is now effectively decoupled from the transverse flow, and essentially Eq. (12.25) is obtained, neglecting the longitudinal diffusion term. The effective lifetime τ_{eff} for the present problem is given by

$$1/\tau_{\mathrm{eff}} = (1/\tau) + (s/l). \tag{12.32}$$

An unusual aspect of this problem is the appearance of E_x in the effective lifetime. Robinson, in applying his theory to electron injection into

p–InSb [interchange p and n in (12.31)] had also to be concerned with hot-electron effects; in this case, he took $T = T(E_x) \propto E_x$ so that the E_x-dependence of τ_{eff} canceled out and the final current-voltage characteristic was (12.28), with $n_0 - p_0$ replaced by $p_0 - n_0$, for the case that $\tau_{\text{eff}} \gg L^2/\mu_{\text{n}} V$.

13

UNIFIED THEORY OF THE INJECTED PLASMA AND EXPERIMENTAL RESULTS

In the previous two chapters dealing with the analytic study of injected plasmas, the insulator and semiconductor problems have been treated entirely separately. Yet these problems must obviously be related. For example, at some sufficiently high level of injection, the thermally generated free carriers in a semiconductor should become irrelevant and the semiconductor square law (12.7) should metamorphose into the insulator cube law (11.14). In order to ascertain the interrelationship between the two problems, we here formulate the injected plasma problem in a unified manner, including also diffusion. For simplicity, we consider only the constant lifetime problem. Correspondingly, as previously, we assume that changes in population of the localized defect states are small enough that they may safely be neglected.

13.1 Unified Theory

Following the same sign conventions as before, namely with the holes flowing from left to right and writing

$$\mathbf{J}_{\mathrm{p}} = J_{\mathrm{p}}\mathbf{x}, \qquad \mathbf{J}_{\mathrm{n}} = J_{\mathrm{n}}\mathbf{x}, \qquad \boldsymbol{\mathscr{E}} = \mathscr{E}\mathbf{x}, \tag{13.1}$$

the current flow equations are

$$J_{\mathrm{p}} = ep\mu_{\mathrm{p}}\mathscr{E} - eD_{\mathrm{p}}\,(dp/dx), \qquad D_{\mathrm{p}} = V_{\mathrm{T}}\mu_{\mathrm{p}}\,, \qquad V_{\mathrm{T}} = (kT/e), \tag{13.2a}$$

and

$$J_n = en\mu_n\mathscr{E} + eD_n\,(dn/dx), \qquad D_n = V_T\mu_n\,. \tag{13.2b}$$

The Poisson equation is

$$(\epsilon/e)(d\mathscr{E}/dx) = (p - p_0) - (n - n_0). \tag{13.3}$$

The particle-conservation equations are

$$\frac{1}{e}\frac{dJ_n}{dx} = \mu_n \frac{d}{dx}(n\mathscr{E}) + V_T\mu_n \frac{d^2n}{dx^2} = r = \frac{n - n_0}{\tau} = \frac{p - p_0}{\tau}, \tag{13.4a}$$

and

$$-\frac{1}{e}\frac{dJ_p}{dx} = -\mu_p \frac{d}{dx}(p\mathscr{E}) + V_T\mu_p \frac{d^2p}{dx^2} = r = \frac{n - n_0}{\tau} = \frac{p - p_0}{\tau}. \tag{13.4b}$$

If local neutrality holds exactly: $n - n_0 = p - p_0$, then the two forms of r written in (13.4) are precisely equivalent, and it is immaterial which one is used in the further development. If local neutrality is not a good approximation, then the choice for r corresponds to that carrier whose lifetime is most nearly constant with injection level, as discussed previously in Section 11.1.2.

Adding (13.4a) and (13.4b) gives

$$J = J_n + J_p = e\mu_p(p + bn)\,\mathscr{E} + eV_T\mu_p\,(d/dx)(bn - p) = \text{constant}. \tag{13.5}$$

Multiplying (13.4b) by $b = \mu_n/\mu_p$ and adding to (13.4a) gives

$$(d/dx)[(n - p)\mathscr{E}] + V_T[d^2(n + p)/dx^2] = (b + 1)(n - n_0)/\mu_n\tau, \tag{13.6}$$

where we have used the choice $r = (n - n_0)/\tau$ in (13.4). From (13.3), $(n - p) = (n_0 - p_0) - (\epsilon/e)(d\mathscr{E}/dx)$, so that (13.6) can be written:

$$-\frac{\epsilon}{e}\frac{d}{dx}\left(\mathscr{E}\frac{d\mathscr{E}}{dx}\right) + (n_0 - p_0)\frac{d\mathscr{E}}{dx} + 2V_T\frac{d^2n}{dx^2} = \frac{(b + 1)(n - n_0)}{\mu_n\tau}, \tag{13.7}$$

where we have dropped the term $V_T(d^2/dx^2)(p - n) = (\epsilon V_T/e)(d^3\mathscr{E}/dx^3)$ from the left-hand side, since it is generally small compared to the diffusion term retained, namely, $2V_T(d^2n/dx^2)$, where the latter is itself significant.

If the choice $r = (p - p_0)/\tau$ had been made in (13.4), then in place of (13.7), we would have

$$-\frac{\epsilon}{e}\frac{d}{dx}\left(\mathscr{E}\frac{d\mathscr{E}}{dx}\right) + (n_0 - p_0)\frac{d\mathscr{E}}{dx} + 2V_T\frac{d^2p}{dx^2} = \frac{(b + 1)(p - p_0)}{\mu_n\tau}, \tag{13.7a}$$

where the term $V_T(d^2/dx^2)(n - p)$ has been dropped from the left-hand side.

Equation (13.7) [or (13.7a)] is the basis for a unified consideration of the injected-plasma problem. First note that if the third term on the left-hand side (LHS) of (13.7) is dominant throughout the semiconductor, then a pure diffusion solution for the injected carriers $(n - n_0)$ is obtained with (ambipolar) diffusion length $L_a = \{2V_T\mu_n\tau/(b + 1)\}^{1/2}$. We are here concerned with problems where $L/L_a \gg 1$ so that the diffusion term is dominant only near the contacts, giving rise to "contact corrections" to the pure drift solution to (13.7). Such corrections are considered later in this chapter. For the remainder of this section, we neglect the diffusion term in (13.7), as well as in (13.2a) and (13.2b).

If the second term on the LHS of (13.7) is dominant throughout most of the crystal, then retention only of this term leads precisely to the semiconductor injected plasma discussed in Chapter 12. If the first term on the LHS of (13.7) is dominant throughout most of the crystal, then retention only of this term leads precisely to the insulator injected plasma discussed in Chapter 11.

If the first two terms on the LHS of (13.7) are retained, the problem can still be handled analytically and the solution was presented by Lampert and Rose (1961, Appendix A). Here we discuss only the different possible regimes in the current-voltage characteristic, which are four in number. At low voltages, Ohm's law, I (below) must obtain. At high voltages, the insulator cube law, III (below), must obtain. These *might* be the only two regimes in the characteristic. On the other hand, at intermediate voltages there are two other possibilities: for semiconductors, n_0 "large," the square law, II (below), or, for insulators, n_0 "small," the trap-free square law, II′ (below), for holes—holes because the electron lifetime has been taken constant, as in (13.7); if the hole lifetime were constant, as in (13.7a), we would use for II′, the trap-free electron square law.

$$\text{I}: \quad J = e(\mu_n n_0 + \mu_p p_0)(V/L); \tag{13.8I}$$

$$\text{II}: \quad J = (9/8)\, e(n_0 - p_0)\, \mu_n\mu_p\tau(V^2/L^3); \tag{13.8II}$$

$$\text{II}': \quad J = (9/8)\, \epsilon\mu_p(V^2/L^3); \tag{13.8II′}$$

$$\text{III}: \quad J = (125/18)\, \epsilon\tau\mu_n\mu_p(V^3/L^5). \tag{13.8III}$$

Because II and II′ have the same V^2/L^3 dependence, they are parallel, on a log J versus log V plot, and their separation depends only on parameters of the material. Thus II is higher than II′ if

$$\tau > t_{n,\Omega}/[1 - (p_0/n_0)] \qquad \text{where} \quad t_{n,\Omega} = \epsilon/en_0\mu_n\,.$$

Let (I, II) denote the intersection of curves I and II. Then this intersection corresponds to $t_p \approx \tau[1 - (p_0/n_0)]/[1 + (p_0/n_0 b)]$ with $t_p = L^2/\mu_p V$. In a similar vein, the various curve intersections have the following correspondences:

$$
\begin{aligned}
&\text{(I, II)}: \quad t_p \simeq \frac{1-\alpha}{1+(\alpha/b)}\,\tau; \qquad \text{(II, III)}: \quad t_p \simeq \frac{6bt_{n,\Omega}}{1-\alpha}\,;\\
&\text{(I, III)}: \quad t_p \simeq \left[\frac{6bt_{n,\Omega}\tau}{1+(\alpha/b)}\right]^{1/2};\\
&\text{(I, II')}: \quad t_p \simeq \frac{t_{n,\Omega}}{1+(\alpha/b)}\,; \qquad \text{(II', III)}: \quad t_p \simeq 6b\tau; \qquad \alpha = (p_0/n_0).
\end{aligned}
\tag{13.9}
$$

Three different situations may prevail:

(i) If $6(b+\alpha)\,\tau > t_{n,\Omega} > (1-\alpha)^2\,\tau/6(b+\alpha)$, then only branches I and III appear in the characteristic: I for $t_p > [6b^2 t_{n,\Omega}\tau/(b+\alpha)]^{1/2}$ and III when the inequality is reversed. The above sequence of inequalities is possible only if $6b > 1 - 7\alpha$. The sequence can also be written as $6(b+\alpha)\,t_{n,\Omega}/(1-\alpha)^2 > \tau > t_{n,\Omega}/6(b+\alpha)$.

(ii) If $\tau > t_{n,\Omega}/(1-\alpha)$ and $\tau > 6(b+\alpha)\,t_{n,\Omega}/(1-\alpha)^2$, then branches I, II, and III appear in the characteristic: I for $t_p > b(1-\alpha)\,\tau/(b+\alpha)$, II for

$$b(1-\alpha)\,\tau/(b+\alpha) > t_p > 6bt_{n,\Omega}/(1-\alpha),$$

and III for $6bt_{n,\Omega}/(1-\alpha) > t_p$.

(iii) If $\tau < t_{n,\Omega}/(1-\alpha)$ and $\tau < t_{n,\Omega}/6(b+\alpha)$, then branches I, II′, and III appear in the characteristic: I for $t_p > bt_{n,\Omega}/(b+\alpha)$, II′ for $bt_{n,\Omega}/(b+\alpha) > t_p > 6b\tau$, and III for $6b\tau > t_p$.

Before concluding this discussion of the unified theory, there is an interesting point about the self-consistency of approximations that can be made. If one neglects the first and third terms on the LHS of Eq. (13.7), then the simplified differential equation yields the semiconductor square-law regime (13.8II) in the limit of high injection levels. We have seen in Chapter 12 that the positional dependence of $\mathscr{E}$ in this regime is given by $\mathscr{E} \propto x^{1/2}$. Therefore $\mathscr{E}\,d\mathscr{E}/dx = \frac{1}{2}\,d\mathscr{E}^2/dx =$ constant, independent of x. Based on this solution, the first term of the LHS of (13.7) vanishes identically, independent of applied voltage. Thus the neglect of this term would appear to be valid up to arbitrarily high injection levels. Now we know that this conclusion is false, that, in fact, at sufficiently high voltage, the first term dominates and yields the insulator cube law. The moral evidently is that the comparison of neglected to retained terms in the "characteristic" differential equation (based on the solution obtained with the retained terms) yields a necessary, but not sufficient, condition for self-consistency. In the particular example cited, the loss of a boundary condition, in dropping the first term and thereby lowering the order of the differential equation, proved incisive in the determination of the limits of validity of the approximate solution.

13.2 Conventional Treatment

The conventional manipulations of the particle-conservation equations (13.4) are different from those employed to obtain (13.7); namely (13.4a) is multiplied by $\mu_p p$, (13.4b) by $\mu_n n$, and the two then added together to obtain

$$-\mu_n\mu_p\mathscr{E}\left[n\frac{dp}{dx}-p\frac{dn}{dx}\right]+V_T\mu_n\mu_p\left[p\frac{d^2n}{dx^2}+n\frac{d^2p}{dx^2}\right]=(\mu_p p+\mu_n n)\frac{(p-p_0)}{\tau}, \tag{13.10}$$

where we have used $r=(p-p_0)/\tau$ in (13.4), following convention. It is then noted that

$$n\,(dp/dx)-p\,(dn/dx)=(n-p)(dp/dx)-p\,(d/dx)(n-p), \tag{13.11a}$$

$$p\,(d^2n/dx^2)+n\,(d^2p/dx^2)=(p+n)(d^2p/dx^2)+p\,(d^2/dx^2)(n-p). \tag{13.11b}$$

If, now, both sides of (13.10) are divided by $(\mu_p p+\mu_n n)$ and (13.11a) and (13.11b) used, we obtain

$$-\mu_a\mathscr{E}\,(dp/dx)+D_a\,(d^2p/dx^2)=(p-p_0)/\tau, \tag{13.12}$$

with

$$\mu_a=\frac{n-p}{(p/\mu_n)+(n/\mu_p)}=\frac{n_0-p_0}{(p/\mu_n)+(n/\mu_p)}$$
$$D_a=\frac{n+p}{(p/\mu_n)+(n/\mu_p)}\,V_T=\frac{n+p}{(p/D_n)+(n/D_p)}. \tag{13.13}$$

Two terms have been neglected on the LHS of (13.12), namely

$$\{p\mathscr{E}/[(p/\mu_n)+(n/\mu_p)]\}\,d(n-p)/dx=-\{\ \}(\epsilon/e)(d^2\mathscr{E}/dx^2)$$

and

$$\{V_{\mathrm{T}}p/[(p/\mu_{\mathrm{n}}) + (n/\mu_{\mathrm{p}})]\}\, d^2(n - p)/dx^2 = \{\ \}(\epsilon/e)(d^3\mathscr{E}/dx^3),$$

where (13.3) has been used.

From the form of (13.12), it is clear that its main usefulness will come in problems where $\mathscr{E}$ is a constant, independent of position, e.g., in problems of low-level, minority-carrier injection into extrinsic semiconductors (i.e., the classical transistor problems). In such problems, the two neglected terms, listed above, will obviously be very small, and their neglect is justifiable.

Equation (13.12) bears a strong resemblance to (13.4) if the latter is rewritten as

$$-\mu_{\mathrm{p}}\mathscr{E}\,(dp/dx) + D_{\mathrm{p}}\,(d^2p/dx^2) = (p - p_0)/\tau, \tag{13.14}$$

where the term $-\mu_{\mathrm{p}}p\,(d\mathscr{E}/dx)$ has been dropped from the LHS.

For an extrinsic semiconductor, $n_0 \gg p_0$, and small-signal injection levels, $p \ll n \simeq n_0$, Eqs. (13.12) and (13.14) are identical, since then $\mu_{\mathrm{a}} = \mu_{\mathrm{p}}$ and $D_{\mathrm{a}} = D_{\mathrm{p}}$.

13.3 Diffusion Corrections

In the discussion of injected plasmas up to this point, diffusion has been neglected in the derivation of current-voltage characteristics. Yet, because a contact which is injecting for one sign of carrier is generally blocking for the other sign of carrier, we expect that diffusive current flow will actually dominate in the neighborhood of the contact. For, the blocking of egress of carriers at a contact, e.g., electrons at a hole-injecting junction, leads to a pileup of the electrons in the vicinity of the junction. The pileup of electrons, and holes by space-charge neutralization, leads to a suppression of the electric field intensity and a dominance of the diffusion term in (13.7). It is convenient to refer to the region adjacent to an injecting contact over which the diffusion solution dominates as an "accommodation region." This is appropriate terminology because such a region exists in order to accommodate the drift-type solution which is dominant over the bulk, for $L \gg L_{\mathrm{a}}$, to the boundary conditions imposed by the injecting contact. In terms of the Regional Approximation, then, we expect injected plasma problems to be three-region problems, the regions being (i) an anode accommodation region, (ii) a central, drift-dominated region, and (iii) a cathode accommodation region. The semiconductor and insulator injected-plasma problems have indeed been studied, in certain limits, by this method, as outlined in the following subsection, taken from Schilling and Lampert (1970).

13.3.1 *The Semiconductor Injected Plasma*

At high injection levels, $n \simeq p \gg n_0$, and at constant lifetime, the full working equations for this problem are

$$J = e\mu_p(b + 1)n\mathscr{E} = \text{constant}, \tag{13.15}$$

which is just (13.5) with $n = p$ and with the diffusion term dropped, and

$$\underbrace{(n_0 - p_0)\frac{d\mathscr{E}}{dx}}_{\text{SC term}} + \underbrace{2V_T\frac{d^2n}{dx^2}}_{\text{D term}} = \frac{(b + 1)n}{\mu_n\tau}, \tag{13.16}$$

which is just (13.7) with the insulator term (the first term on the LHS dropped.

The boundary conditions are

$$n_a = n(0) = \frac{JL_a}{2e\mu_p V_T}\theta_p, \qquad n_c = n(L) = \frac{JL_a}{2e\mu_n V_T}\theta_n,$$
$$L_a = \left\{\frac{2V_T\mu_n\tau}{b + 1}\right\}^{1/2}. \tag{13.17}$$

Taking $\theta_n = \theta_p = 1$, these boundary conditions correspond to taking $J_n(0) = 0$ and $J_p(L) = 0$ with J_n, J_p given by (13.2b), (13.2a), respectively. The quantities θ_n and θ_p are simply adjustable parameters, introduced into the theory by Baron (1965) in order to improve the agreement of theory with experiment.

The three regions of the problem are indicated in Fig. 13.1.

Region I: The D term dominates (13.16) and, dropping the SC term, the corresponding solution is a decaying exponential for $n(x)$ and a growing exponential for $\mathscr{E}(x)$. Using this solution, it is evident that the (neglected) SC term is positive and growing exponentially with increasing x, whereas the D term decays exponentially. Thus, the SC term must

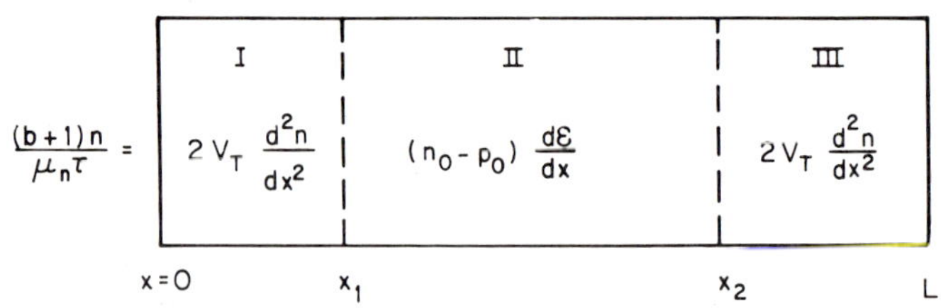

FIG. 13.1. Schematic Regional Approximation diagram for the problem of a constant-lifetime plasma injected into a semiconductor.

eventually overtake the D term and the end of Region I is defined as that plane x_1 where the two terms are equal.

Region II: The SC term dominates (13.16) and, dropping the D term, the corresponding solution gives $\mathscr{E}(x)$ increasing monotonically with increasing x. The left-hand boundary of Region II is plane x_1, the right-hand boundary plane x_2.

Region III: The D term dominates (13.16) and therefore, as in Region I, the SC term is dropped. Since $\mathscr{E}$ is large at x_2, and must become small again well into Region III, there must be a plane x_m near x_2 where $\mathscr{E}$ reaches a maximum: $(d\mathscr{E}/dx)_{x_m} = 0$. To the right of x_m, the (neglected) SC term is negative; to the left of x_m this term is positive. The plane x_2 is where the SC term in Region III equals the D term. In order that $\mathscr{E}$ have a maximum in Region III, it is necessary to use both solutions to the diffusion equation, a growing and a decaying exponential.

It finally remains to match the solutions in the several regions. Here it is required that the SC term be continuous crossing x_1 and x_2. It is readily seen that this is equivalent to requiring that $\mathscr{E}$, hence also n, be continuous crossing x_1 and x_2.

For the mathematical discussion of the problem, it is convenient to use dimensionless variables as follows:

$$\mathscr{E}(x) = \mathscr{E}^* u, \qquad n(x) = n^* s, \qquad x = x^* w, \qquad V(x) = V^* v,$$

$$\mathscr{E}^* = \left\{ \frac{2V_T J^2}{e^2 \mu_p{}^3 b(b+1)\,\tau(n_0 - p_0)^2} \right\}^{1/4}, \qquad x^* = L_a = \left\{ \frac{2V_T \mu_n \tau}{b+1} \right\}^{1/2},$$

$$n^* = \frac{J}{e\mu_p(b+1)\mathscr{E}^*} = \left\{ \frac{b\tau(n_0 - p_0)^2 J^2}{2e^2 \mu_p(b+1)^3 V_T} \right\}^{1/4}, \tag{13.18}$$

$$V^* = \mathscr{E}^* x^* = \left\{ \frac{8V_T{}^3 J^2 b\tau}{e^2 \mu_p (b+1)^3 (n_0 - p_0)^2} \right\}^{1/4}.$$

Equations (13.15) and (13.16) then become, respectively

$$1 = su, \tag{13.19}$$

$$\underbrace{du/dw}_{\text{SC term}} + \underbrace{d^2s/dw^2}_{\text{D term}} = s = 1/u. \tag{13.20}$$

Note that since x^* is the ambipolar diffusion length, $w_c = w(L) = L/x^* = L/L_a$ is a fixed quantity, independent of J and V. The current-voltage characteristic is given by a plot of

$$s_c{}^2 = 1/u_c{}^2 \qquad \text{vs} \qquad v_c s_c, \tag{13.21}$$

where subscript "c" denotes that the quantity is to be evaluated at the cathode, $x = L$.

$$s_c^2 = J\theta_n^2 \frac{\{(b+1)\tau\}^{1/2}}{e(n_0 - p_0)\{2\mu_p V_T b^3\}^{1/2}}; \qquad v_c s_c = V\theta_n \frac{b+1}{2bV_t}. \tag{13.22}$$

Note also that

$$s_c = \frac{n_c}{n^*} = \theta_n \left\{\frac{J^2\tau(b+1)}{2e^2 V_T \mu_p b^3 (n_0 - p_0)^2}\right\}^{1/4}, \qquad s_a = s_c b \frac{\theta_p}{\theta_n}. \tag{13.23}$$

The semiconductor square law (12.7) is now simply

$$s_c^2 = (9/8)(1/w_c^3)(v_c s_c)^2 \qquad \text{or} \qquad v_c^2 = (8/9)w_c^3. \tag{13.24}$$

The separate regions now are as follows.

Region I ($0 \leqslant w \leqslant w_1$)

The SC term is dropped from (13.20), giving

$$d^2s/dw^2 = s. \tag{13.25}$$

Only the exponentially decaying solution to (13.25) is used:

$$s = s_a \exp -w \qquad \text{or} \qquad u = u_a \exp w \qquad \text{with} \quad u_a = 1/s_a. \tag{13.26}$$

The D term is $d^2s/dw^2 = s = s_a \exp -w$. The SC term is $du/dw = u_a \exp w = (\exp w)/s_a$. The end w_1 of Region I is taken where the SC term equals the D term:

$$\frac{1}{s_a} \exp w_1 = s_a \exp -w_1 \rightarrow \exp w_1 = s_a \rightarrow s_1 = s(w_1) = u_1 = u(w_1) = 1. \tag{13.27}$$

For the voltage across Region I we have

$$v_I = \int_0^{w_1} u\, dw = 1 - (1/s_a). \tag{13.28}$$

Region II ($w_1 \leqslant w \leqslant w_2$)

The D term is dropped from (13.20), giving

$$du/dw = (1/u) \rightarrow \tfrac{1}{2}\, du^2/dw = 1, \tag{13.29}$$

with solution

$$u^2 - 1 = 2(w - w_1), \tag{13.30}$$

where we have used (13.27), matching u across w_1. For the voltage across Region II we have

$$v_{\text{II}} = \int_{w_1}^{w_2} u\,dw = \tfrac{1}{3}[\{2(w_2 - w_1) + 1\}^{3/2} - 1]. \tag{13.31}$$

Note that if we take Region II to fill the whole semiconductor, that is, take $w_1 \ll w_2$ and $w_2 \simeq w_c \gg 1$, then (13.31) reduces to the semiconductor square law (13.24), as expected.

Region III ($w_2 \leqslant w \leqslant w_c$)

As in Region I, the governing equation is the pure diffusion equation:

$$d^2s/dw^2 = s. \tag{13.32}$$

Here, unlike Region I, we use both the growing and decaying exponential solutions to (13.32):

$$s = A \exp - (w_c - w) + B \exp(w_c - w), \tag{13.33}$$

from which

$$s_c = s(w_c) = A + B. \tag{13.34}$$

The left-hand end w_2 of Region III is taken where the (neglected) SC term equals the D term:

$$(du/dw)_2 = (d^2s/dw^2)_2 = s_2 \to -(ds/dw)_2 = s_2{}^3. \tag{13.35}$$

Using (13.33) this can be written

$$B \exp w_{c2} - A \exp - w_{c2} = (A \exp - w_{c2} + B \exp w_{c2})^3, \qquad w_{c2} = w_c - w_2. \tag{13.36}$$

Continuity of s across w_2 gives, using (13.30) and (13.33),

$$(2w_{21} + 1)^{-1/2} = A \exp - w_{c2} + B \exp w_{c2}, \qquad w_{21} = w_2 - w_1. \tag{13.37}$$

Letting $\alpha = A \exp - w_{c2}$, $\beta = B \exp w_{c2}$, Eqs. (13.36) and (13.37) can be written, respectively, as

$$-\alpha + \beta = 1/(2w_{21} + 1)^{3/2}, \tag{13.38}$$

$$\alpha + \beta = 1/(2w_{21} + 1)^{1/2}, \tag{13.39}$$

which are easily solved to give

$$\alpha = w_{21}(2w_{21} + 1)^{-3/2} \qquad \text{and} \qquad \beta = (w_{21} + 1)(2w_{21} + 1)^{-3/2},$$

that is,

$$A = \frac{w_{21}}{(2w_{21}+1)^{3/2}} \exp w_{c2}\,, \qquad B = \frac{w_{21}+1}{(2w_{21}+1)^{3/2}} \exp - w_{c2}\,. \tag{13.40}$$

Equation (13.34) can now be written as

$$s_0 = \frac{w_{21}(\exp w_{c2} + \exp - w_{c2}) + \exp - w_{c2}}{(2w_{21}+1)^{3/2}}\,. \tag{13.41}$$

Since

$$w_c = w_{c2} + w_{21} + w_1 \to w_{c2} = (w_c - w_1) - w_{21}\,, \tag{13.42}$$

and since $w_1 = \ln(s_c b\theta_p/\theta_n)$, Eq. (13.41) uniquely relates w_{21} to s_c in terms of the fixed parameters w_c, b, θ_p, and θ_n.

The voltage v_{III} across Region III is $\int_{w_2}^{w_c} u\,dw = \int_{w_2}^{w_c} dw/s$. Using (13.33), this becomes

$$v_{III} = [1/(AB)^{1/2}]\{\tan^{-1}[(B/A)^{1/2} \exp w_{c2}] - \tan^{-1}(B/A)^{1/2}\}. \tag{13.43}$$

The total voltage across the semiconductor is $v_c = v_I + v_{II} + v_{III}$, where from (13.23) and (13.31)

$$v_I = 1 - (\theta_n/\theta_p b s_c) \tag{13.44}$$

and

$$v_{II} = \tfrac{1}{3}[(2w_{21}+1)^{3/2} - 1]. \tag{13.45}$$

With w_c, b, θ_p, and θ_n given, the dimensionless current-voltage characteristic (13.21) is computed by choosing a value for s_c, solving (13.42) and (13.41) for w_{21}, w_{c2}, obtaining A and B from (13.40), and finally v_c from (11.43) to (11.45). Specific calculations, taking $\theta_p = \theta_n$ and $b = 2$, for the two cases $L/L_a = 12$ and $L/L_a = 25$, namely the open circles, are compared to the corresponding computer calculations of Baron (1968), the solid lines, in Fig. 13.2.

As the voltage continues to increase, a critical voltage V_{cr} is reached at which Region II shrinks to zero width: $w_1 = w_2$:

$$v_{c,cr}s_{c,cr} \simeq \frac{\pi}{2(b\theta_p/\theta_n)^{1/2}} \exp w_c/2 \to V_{cr} \simeq V_T \frac{\pi}{b+1}\left(\frac{b}{\theta_p\theta_n}\right)^{1/2} \exp \frac{L}{2L_a}\,. \tag{13.46}$$

Beyond this point, current increases without further increase in voltage, as in the upper part of the curve labeled: $L/L_a = 12$ in Fig. 13.2.

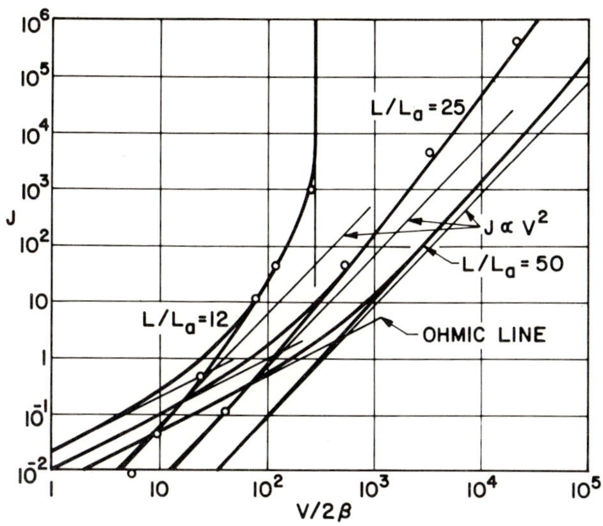

FIG. 13.2. Calculated current-voltage curves for the constant-lifetime plasma injected into a semiconductor. The heavy solid lines are computer characteristics obtained by Baron (1968, Fig. 2). The open circles are points obtained by the Regional Approximation Method for the two cases $L/L_a = 12$ and $L/L_a = 25$.

A simplified, approximate way to look at the above results is to regard the semiconductor square law (12.7) as valid, but with the cathode–anode spacing L replaced by an effective spacing L_{eff}:

$$J \simeq (9/8)\, e(n_0 - p_0)\, \mu_n \mu_p \tau (V^2/L_{\text{eff}}^3), \qquad L_{\text{eff}} \simeq L - W_a - W_c\,, \tag{13.47}$$

where $W_a = x_1$ and $W_c = L - x_2$ in Fig. 13.1. Since W_a and W_c both increase with voltage, Eq. (13.47) is equivalent to a power-law dependence of J on V with the exponent exceeding 2:

$$J \propto V^{\{2/[1-m(L_a/L)]\}} \qquad \text{with } m \text{ on the order of 2 or 3.} \tag{13.48}$$

13.3.2 *The Insulator Injected Plasma*

At high injection levels, $n \simeq p$, and at constant lifetime the full working equations for this problem are (13.15) and

$$\underbrace{-\frac{\epsilon}{e}\frac{d}{dx}\left(\mathscr{E}\,\frac{d\mathscr{E}}{dx}\right)}_{\text{INS term}} + \underbrace{2V_T\,\frac{d^2n}{dx^2}}_{\text{D term}} = \frac{(b+1)n}{\mu_n \tau}\,, \tag{13.49}$$

which is just (13.7) with the semiconductor term (the middle term on the LHS) dropped. The boundary conditions are the same as for the semiconductor problem, namely (13.17). Once again, as in that problem, we are dealing with a three-region problem. The three regions are indicated in Fig. 13.3.

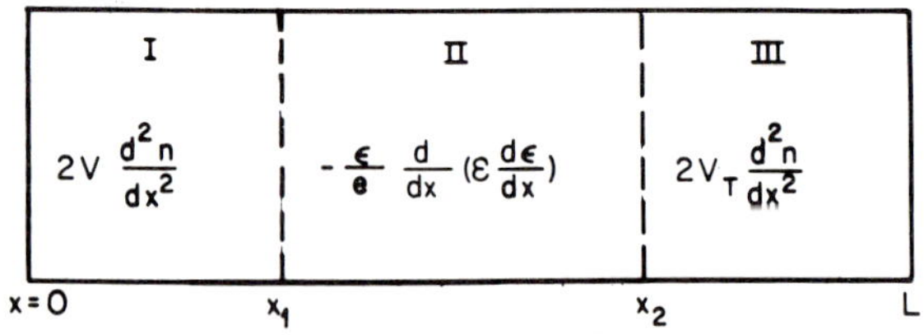

FIG. 13.3. Schematic Regional Approximation diagram for the problem of a constant-lifetime plasma injected into an insulator.

Region I: The D term dominates (13.49) and, dropping the INS term, pure diffusive behavior governs the current flow. However, it is not adequate to use just the decaying exponential solution since the (neglected) INS term corresponding to this solution is always negative. Consequently, the solution in Region I is an admixture of the decaying and growing exponential solutions. Over most of Region I, the INS term is negative. It goes through 0 at x_l, and for $x > x_l$ it is positive, finally overtaking the D term. The end of Region I is defined as that plane x_1 where the two terms are equal.

Region II: The INS term dominates (13.38) and, dropping the D term, the corresponding drift solution is qualitatively like that given in Section 11.1.1, namely with the $\mathscr{E}$ field rising to a maximum in the middle of the sample. The left-hand boundary of Region II is plane x_1, the right-hand boundary plane x_2.

Region III: Here, as in Region I, the D term dominates (13.38), and so the INS term is dropped. Region III is qualitatively, in every way, similar to Region I. Again, both the growing and decaying exponential solutions must be used in order that the (neglected) INS term not be negative everywhere in Region III. Proceeding from the cathode into the sample, the INS term is initially negative and remains negative over most of the region out to $x = x_r$, where it goes through zero. Beyond x_r, i.e., for $x < x_r$, the INS term is positive and it overtakes the D term at the plane x_2.

The solutions in adjacent regions are matched as in the previous problem; namely, it is required that the INS term be continuous crossing x_1 and x_2. This is equivalent to requiring that $\mathscr{E}$, hence n, be continuous crossing x_1 and x_2. A final point about matching conditions is to be made.

As defined above, x_1 is the plane where the neglected INS term in Region I equals the D term, and correspondingly for x_2 and Region III. These are not enough specifications to define a unique solution. Two further conditions are required, and these are: At x_1 the neglected D term in Region II equals the INS term, and likewise at x_2. (Neither of these conditions was needed in the semiconductor problem because (i) in Region II, for that problem, the SC term is a first-order differential as compared to the second-order differential INS term in the present problem, and (ii) in Region I for that problem, only the decaying exponential was retained.)

For the mathematical discussion of the problem, it is convenient to use the following dimensionless variables:

$$\begin{gathered}
\mathscr{E}(x) = \mathscr{E}^{\dagger} u, \qquad n(x) = n^{\dagger} s, \qquad x = x^{\dagger} w, \qquad V(x) = V^{\dagger} v, \\
\mathscr{E}^{\dagger} = \{2V_{\mathrm{T}} J/\epsilon(b+1)\mu_{\mathrm{p}}\}^{1/3}, \qquad x^{\dagger} = L_{\mathrm{a}} = \{2V_{\mathrm{T}}\mu_{\mathrm{n}}\tau/(b+1)\}^{1/2}, \\
n^{\dagger} = \frac{J}{e\mu_{\mathrm{p}}(b+1)\mathscr{E}^{\dagger}} = \left\{\frac{\epsilon J^2}{2V_{\mathrm{T}} e^3 (b+1)^2 \mu_{\mathrm{p}}^2}\right\}^{1/3}, \\
V^{\dagger} = \mathscr{E}^{\dagger} x^{\dagger} = \left\{\frac{32 b^3 \mu_{\mathrm{p}} V_{\mathrm{T}}^5 J^2 \tau^3}{\epsilon^2 (b+1)^5}\right\}^{1/6}.
\end{gathered} \tag{13.50}$$

Equations (13.15) and (13.48) become, respectively

$$1 = su, \tag{13.51}$$

$$\underbrace{-\frac{d}{dw}\left(u\frac{du}{dw}\right)}_{\text{INS term}} + \underbrace{\frac{d^2 s}{dw^2}}_{\text{D term}} = s = \frac{1}{u}. \tag{13.52}$$

Again, $w_{\mathrm{c}} = w(L) = L/x^{\dagger} = L/L_{\mathrm{a}}$ is a fixed quantity, independent of J and V. The current-voltage characteristic is given by a plot of

$$s_{\mathrm{c}}^3 = 1/u_{\mathrm{c}}^3 \qquad \text{vs} \qquad v_{\mathrm{c}} s_{\mathrm{c}}. \tag{13.53}$$

$$s_{\mathrm{c}}^3 = J\theta_{\mathrm{n}}^3\{(b+1)\mu_{\mathrm{p}}\tau^3/2b^3\epsilon^2 V_{\mathrm{T}}\}^{1/2}, \qquad v_{\mathrm{c}} s_{\mathrm{c}} = v\theta_{\mathrm{n}}[(b+1)/2bV_{\mathrm{T}}]. \tag{13.54}$$

The insulator cube-law (11.14) is now simply

$$s_{\mathrm{c}}^3 = (125/18)[(v_{\mathrm{c}} s_{\mathrm{c}})^3/w_{\mathrm{c}}^5] \qquad \text{or} \qquad v_{\mathrm{c}}^3 = (18/125) w_{\mathrm{c}}^5. \tag{13.55}$$

The separate regions now are as follows.

Region I ($0 \leqslant w \leqslant w_1$)

The INS term is dropped from (13.41) giving

$$d^2 s/dw^2 = s, \tag{13.56}$$

which has the solution

$$s = C \exp - w + D \exp w. \tag{13.57}$$

At $w = 0$,

$$s_a = C + D \simeq C. \tag{13.58}$$

The (neglected) INS term is

$$-\frac{d}{dw}\left(u\frac{du}{dw}\right) = -\frac{1}{2}\frac{d^2}{dw^2}\left(\frac{1}{s^2}\right) = \frac{1}{s^2}\left[1 - \frac{3}{s^2}\left(\frac{ds}{dw}\right)^2\right]. \tag{13.59}$$

This term is negative at small w, goes through zero inside Region I, and is thereafter positive. (The purpose of retaining the increasing exponential in (13.57) is precisely to enable the INS term to become positive in Region I.) The right-hand end w_1 of Region I is taken where the (neglected) INS term equals the D term:

$$-\{(d/dw)\,u(du/dw\}_1 = (d^2s/dw^2)_1 = s_1\,. \tag{13.60}$$

From (13.57),

$$s_1 = \gamma + \delta; \qquad \gamma = C \exp -w_1\,, \quad \delta = D \exp w_1\,. \tag{13.61}$$

Since $(ds/dw)_1 = \delta - \gamma$, Eqs. (13.59) and (13.60) give

$$(\delta + \gamma)^3 = 1 - 3[(\delta - \gamma)^2/(\delta + \gamma)^2]. \tag{13.62}$$

For the voltage v_I across Region I, $v_\mathrm{I} = \int_0^{w_1} u\, dw$, we have, in analogy with Region III of the preceding problem, and with (13.43) in particular,

$$v_\mathrm{I} = [1/(CD)^{1/2}][\tan^{-1}\{(D/C)^{1/2} \exp w_1\} - \tan^{-1}(D/C)^{1/2}]. \tag{13.63}$$

Region II ($w_1 \leqslant w \leqslant w_2$)

The D term is dropped from (13.52) giving

$$-(d/dw)(u\, du/dw) = (1/u) \qquad \text{or} \qquad u(d/dw)(u\, du/dw) = -1. \tag{13.64}$$

Making the substitution $u = dw/dy$, that is, $ud/dw = d/dy$, Eq. (13.64) becomes

$$d^2u/dy^2 = -1, \tag{13.65}$$

with solution

$$u - u_1 = -\tfrac{1}{2}y^2 + (y/y_2)\{(u_2 - u_1) + \tfrac{1}{2}y_2{}^2\}, \tag{13.66}$$

satisfying the boundary conditions

$$y = 0 \quad \text{at} \quad w = w_1, \qquad y = y_2 \quad \text{at} \quad w = w_2 \qquad (u = u_2), \tag{13.67}$$

where w_2 is the right-hand end of Region II, w_1 being the left-hand end. From $dw = u\,dy$, we get

$$w - w_1 = \int_0^y u\,dy = -\tfrac{1}{6}y^3 + (y^2/2y_2)\{(u_2 - u_1) + \tfrac{1}{2}y_2^2\} + u_1 y, \tag{13.68}$$

giving

$$w_2 - w_1 = \tfrac{1}{12}y_2^3 + \tfrac{1}{2}(u_2 + u_1)y_2 . \tag{13.69}$$

At this point, it is convenient to rewrite (13.66) as

$$u = ey^2 + fy + u_1 ; \qquad e = -\tfrac{1}{2}, \qquad f = (1/y_2)\{(u_2 - u_1) + \tfrac{1}{2}y_2^2\}. \tag{13.70}$$

The voltage v_{II} across Region II is

$$v_{II} = \int_{w_1}^{w_2} u\,dw = \int_{w_1}^{w_2} u\,(dw/dy)\,dy = \int_{w_1}^{w_2} u^2\,dy,$$

or, from (13.70),

$$v_{II} = \tfrac{1}{5}e^2y_2^5 + \tfrac{1}{2}efy_2^4 + \tfrac{1}{3}(f^2 + 2eu_1)\,y_2^3 + fu_1y_2^2 + u_1^2y_2 . \tag{13.71}$$

Note that if Region II is assumed to fill the entire insulator ($w_1 = 0$, $w_2 = w_c$, $y_2 = y_c$, and $u_1 = u_2 = 0$ are taken as boundary conditions), then, from (13.69), $w_c = y_c^3/12$, and from (13.71), $v_c = y_c^5/120$ so that $v_c^3 = 18w_c^5/125$, which is the insulator cube law (13.55). At both ends of Region II, we must satisfy the condition: the (neglected) D term equals the INS term, that is,

$$\frac{d^2s}{dw^2} = -\frac{d}{dw}\left(u\frac{du}{dw}\right) = \frac{1}{u} \qquad \text{or} \qquad u\frac{d^2}{dw^2}\left(\frac{1}{u}\right) = 1. \tag{13.72}$$

But $u(d^2/dw^2)(1/u) = (d/dy)[(1/u)(d/dy)(1/u)]$, so that (13.72) becomes

$$u^4 = -u(d^2u/dy^2) + 3(dw/dy)^2 \qquad \text{or} \qquad u^4 = u + 3(f - y)^2, \tag{13.73}$$

where wc have used (13.70). At w_1 and w_2, Eq. (13.73) becomes, respectively,

$$u_1^4 = u_1 + 3f^2 \qquad \text{and} \qquad u_2^4 = u_2 + 3(f - y_2)^2. \tag{13.74}$$

A solution to the two equations in (13.74) which is compatible with the defining relation for f in (13.70) is

$$u_1 = u_2 \qquad (s_1 = s_2) \qquad \text{and} \qquad y_2 = 2f. \tag{13.75}$$

Using (13.75) in (13.69) and (13.71), respectively, we obtain

$$w_2 - w_1 = (1/12)\, y_2^3 + u_1 y_2 \tag{13.76}$$

and

$$v_{\text{II}} = (1/120)\, y_2^5 + (1/6)\, u_1 y_2^3 + u_1^2 y_2\,. \tag{13.77}$$

Region III ($w_2 < w < w_c$)

This region is symmetric with Region I. It is governed by the diffusion equation

$$d^2s/dw^2 = s, \tag{13.78}$$

with solution

$$s = M \exp -(w_c - w) + N \exp(w_c - w), \tag{13.79}$$

from which

$$s_c = M + N \simeq M. \tag{13.80}$$

The (neglected) INS term is given by (13.59). The left-hand end of Region III corresponds to that position w_2 at which this term equals the D term:

$$-\{(d/dw)\, u(du/dw)\}_2 = (d^2s/dw^2)_2 = s_2\,, \tag{13.81}$$

which is, of course, the same as (13.60), only taken at w_2.

From (13.79),

$$s_2 = \mu + \nu; \qquad \mu = M \exp(-w_{c,2}), \qquad \nu = N \exp w_{c,2}\,, \qquad w_{c,2} = w_c - w_2\,. \tag{13.82}$$

In analogy with (13.62), we have

$$(\nu + \mu)^3 = 1 - 3[(\nu - \mu)^2/(\nu + \mu)^2]. \tag{13.83}$$

In analogy with (13.63),

$$\begin{aligned} v_{\text{III}} &= [1/(MN)^{1/2}][\tan^{-1}\{(N/M)^{1/2} \exp w_{c,2}\} - \tan^{-1}(N/M)^{1/2}] \\ &= [1/(MN)^{1/2}][\tan^{-1}(M/N)^{1/2} - \tan^{-1}\{(M/N)^{1/2} \exp - w_{c,2}\}]. \end{aligned} \tag{13.84}$$

From (13.61), (13.75), and (13.82), we have

$$\gamma + \delta = \mu + \nu. \tag{13.85}$$

The equation (13.83), relating $\nu - \mu$ to $\nu + \mu$, is identical to the one, Eq. (13.62), relating $\delta - \gamma$ to $\delta + \gamma$. Using this fact and (13.85), we conclude that

$$\delta - \gamma = \nu - \mu. \tag{13.86}$$

From (13.85) and (13.86), it is obvious that

$$\gamma = \mu \qquad \text{and} \qquad \delta = \nu, \tag{13.87}$$

whence, from (13.61) and (13.82),

$$\gamma\delta = \mu\nu = CD = MN. \tag{13.88}$$

The voltage expressions (13.63) and (13.84) now simplify, respectively, to

$$v_{\mathrm{I}} = [1/(\gamma\delta)^{1/2}][\tan^{-1}(\delta/\gamma)^{1/2} - \tan^{-1}(\gamma\delta)^{1/2}/C], \tag{13.89}$$

$$v_{\mathrm{III}} = [1/(\gamma\delta)^{1/2}][\tan^{-1}[M/(\gamma\delta)^{1/2}] - \tan^{-1}(\gamma/\delta)^{1/2}]. \tag{13.90}$$

Since $s_{\mathrm{a}} \simeq C$ and $s_{\mathrm{c}} \simeq M$, and $s_{\mathrm{a}} = s_{\mathrm{c}} b\theta_{\mathrm{p}}/\theta_{\mathrm{n}} = s_{\mathrm{c}} b$ for $\theta_{\mathrm{p}} = \theta_{\mathrm{n}}$, it follows that $C = bM$. Furthermore,

$$\gamma\mu = \gamma^2 = (C \exp - w_1)[M \exp(w_2 - w_{\mathrm{c}})] = bM^2 \exp(w_2 - w_1) \exp(-w_{\mathrm{c}}),$$

or

$$M = [\gamma/(b)^{1/2}] \exp(w_{\mathrm{c}}/2) \exp - [(w_2 - w_1)/2], \qquad C = bM. \tag{13.91}$$

With w_{c}, b, θ_{p}, and θ_{n} given (here it is assumed that $\theta_{\mathrm{p}} = \theta_{\mathrm{n}}$), the dimensionless current-voltage characteristic (13.54) is computed by choosing a value for $s_1 = s_2 = \gamma + \delta = \mu + \nu = 1/u_1 = 1/u_2$ and using (13.62) to solve for $\delta - \gamma$. Now γ and δ are known separately. From (13.74) we find f, hence $\gamma_2 = 2f$. Then (13.76) gives $w_2 - w_1$, and (13.91) gives M and C. Finally $s_{\mathrm{a}} \simeq C$, $s_{\mathrm{c}} \simeq M$, and $v_{\mathrm{c}} = v_{\mathrm{I}} + v_{\mathrm{II}} + v_{\mathrm{III}}$ is determined using (13.77), (13.89), and (13.90).

Specific calculations taking $\theta_{\mathrm{p}} = \theta_{\mathrm{n}}$ and $b = 2$ for the case $L/L_{\mathrm{a}} = 16$, namely the crosses are compared to the corresponding computer calculations of Baron (1965), the solid lines in Fig. 13.4.

As in the previous problem, a current and voltage are finally reached at which Region II shrinks to zero width: $w_1 = w_2$. Beyond this point, the current is independent of voltage. As in that problem, Eq. (13.46), this critical voltage V_{cr} is on the order of $V_{\mathrm{T}} \exp(L/2L_{\mathrm{a}})$.

Also, as with the previous problem, a simplified, approximate way to look at the above results is to regard the insulator cube law (11.14) as

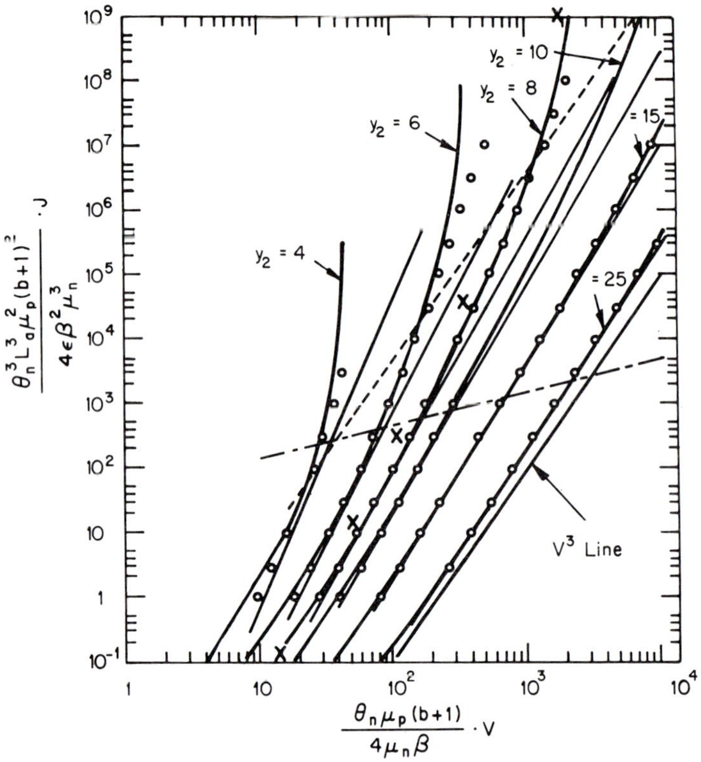

FIG. 13.4. Calculated current-voltage curves for the constant-lifetime plasma injected into an insulator. The heavy solid lines are computer characteristics (and the open circles are computed points) obtained by Baron (1965, Fig. 9). The crosses are points obtained by the Regional Approximation Method for the case $L/L_a = 2y_2 = 16$.

valid, but with the cathode–anode spacing L replaced by an effective spacing L_{eff}:

$$J \simeq (125/18)\,\epsilon\mu_n\mu_p\tau(V^3/L_{eff}^5), \qquad L_{eff} = L - W_a - W_c, \tag{13.92}$$

where $W_a = x_1$ and $W_c = L - x_2$ in Fig. 13.3. Since W_a and W_c both increase with voltage, Eq. (13.92) is equivalent to a power-law dependence of J on V with the exponent exceeding 3:

$$J \propto V^{\{3/[1-n(L_a/L)]\}} \qquad \text{with } n \text{ on the order of 2 or 3.} \tag{13.93}$$

13.4 Experimental Results

The main results of the injected plasma theory are, for sufficiently long samples, the square law (12.7) for the semiconductor and the cube law (11.14) for the insulator. Both have been confirmed experimentally. The semiconductor square law was verified immediately after its prediction by Larrabee's (1961) measurements on germanium. Figure 13.5, reproduced from this paper, shows the I–V characteristics for a long, n-type specimen ($L = 0.97$ cm) at three different temperatures.

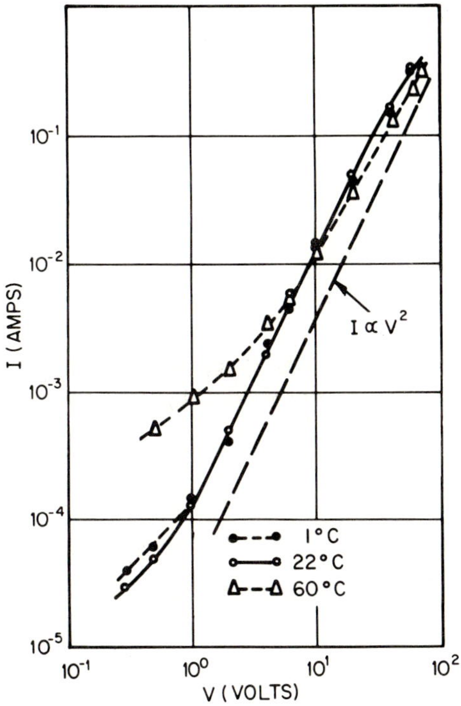

FIG. 13.5. Temperature-dependent current-voltage characteristics for the injected plasma in a single crystal of germanium ($L = 0.97$ cm, cross section 0.063×0.055 cm). The specimen was n-type and nearly intrinsic at room temperature ($n_0 \simeq 2 \times 10^{13}$ cm^{-3}). (From Larrabee, 1961, Fig. 3.)

Each of these characteristics consists of an Ohm's law going over into a square law. The important point to observe is that the high-temperature square law ($T = 60°$C) is actually a bit lower, by a factor of about 1.6, than the lower temperature square laws ($T = 0°$, $22°$C), whereas the $T = 60°$C Ohm's law is about an order of magnitude higher than the others due to the higher $n_0 \approx p_0$ at $T = 60°$C. This is direct evidence for the dependence of the square law current (12.7) or (12.28) on $n_0 - p_0$, which is independent of temperature at temperatures above carrier freeze-out. The slightly lower position of the higher-temperature square law is in accord with the temperature dependence of the mobilities: $\mu_p \propto T^{-2.3}$, $\mu_n \propto T^{-1.66}$, so that $\mu_n\mu_p(1°C)/\mu_n\mu_p(60°C) \simeq 2$, in good agreement with the observed displacement, considering an unknown, slight temperature dependence of the lifetime τ. Note that the displacement of the 1°C-Ohm's law above the 22°C-Ohm's law is in accord with the temperature dependence of the mobilities; here the lifetime is in no way involved. A quantitative fit of the square-law expression (12.7), rewritten in practical units as

$$J = (1.8 \times 10^{-19})\tau\mu_n\mu_p(n_0 - p_0)\, V^2/L^3,$$

gives $\tau = 1 \times 10^{-3}$ sec which is in excellent agreement with the observed, low-level bulk lifetime, 600 μsec; i.e., $\tau_{eff} \approx \tau$ in Eq. (12.28). (Note that here we have used the full specimen length $L = 0.97$ cm, without any diffusion correction.) The surface of the Ge was treated so as to provide a strong electron accumulation layer which acts to keep the injected holes away from surface recombination states. This same value of τ is consistent with the observed voltage of transition, $\simeq$1 V, from Ohm's law to the square law at 22°C, recalling that $L^2/\mu_p V_X \simeq \tau$. At the higher injection levels, the I–V curve is seen to be curving towards lower slope, a result consistent with an expected density-dependent lifetime at high injection levels, leading to lower lifetimes. The bulk ambipolar diffusion length is somewhat long, $L_a \simeq 2$ mm, which makes it somewhat surprising that the observed slope is so close to two. For reasons that are not clear, the diffusion corrections discussed in Section 13.3.1 for the case of one-dimensional, planar current flow played a minimal role in Larrabee's experiments. Perhaps the (equivalent) ambipolar diffusion length was very much smaller (due to very much shorter τ_{eff}) in the immediate vicinity of the contacts.

Marsh *et al.* (1965), working at the Hughes Research Laboratories, have also verified the semiconductor injected-plasma square law in room-temperature experiments with lightly doped p-type silicon, namely with forward-biased p-π-n structures. Their sample lengths ranged from 0.8 to 8.1 mm and the cross-sectional areas from 0.02 to 9.10 cm^2. Current-voltage characteristics obtained by these authors are presented in Fig. 13.6. The solid lines are square-law slope lines. It is seen that the higher the L/L_a ratio, the better the fit of the data to a square law, as expected from the discussion in Section 13.3.1. The solid point at 10 V in Fig. 13.6 were calculated from Eq. (12.28) using measured values of $p_0 - n_0 = p_0$, τ_{eff}, and L. For samples 4 and 5, agreement between experiment and calculation is excellent. The authors note that for

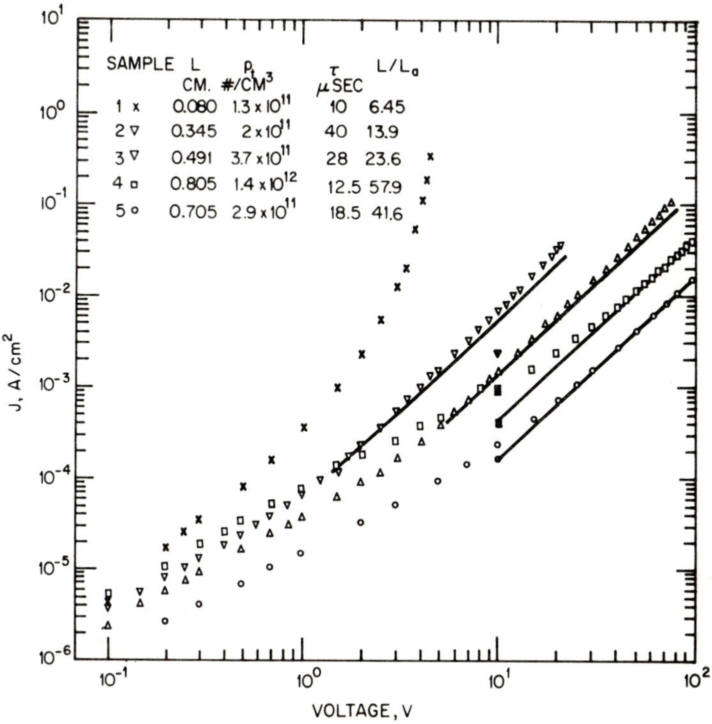

FIG. 13.6. Current-voltage characteristics for double injection into lightly doped p-type silicon. The solid-lines are square-law slope lines. (From Marsh *et al.*, 1965, Fig. 1.)

Samples 2 and 3, the discrepancy between the experimental and calculated current values is largely eliminated if L is replaced, in (12.7) or (12.28), by the effective length $L_{\text{eff}} = L - 2.2 \times (2L_a)$. [Compare with (13.47).] Experimental verification of the L^3-dependence of J for the semiconductor injected plasma is shown by Fig. 13.7, taken from the same authors. Here plotted against L^3 is the quantity $J/p_0\tau$ (where J is the experimental value of the V^2-dependent current density extrapolated to 10 V) for some 19 samples with large L/L_a ratios. The values of p_0 ranged from 10^{11} to 2×10^{12} cm^{-3} and τ_{eff} from 4 to 30 μsec.

Arizumi *et al.* (1965), working with a long thin rod of p-Ge, measured the injected plasma concentration as a function of distance along the rod, with results shown in Fig. 13.8. The theory, given by the dashed curve, is fitted to experiment at one point, namely the open circle. It is seen that a rather good fit elsewhere is obtained, with an effective lifetime $\tau_{\text{eff}} = 72$ μsec, as compared to the measured bulk lifetime $\tau = 840$ μsec.

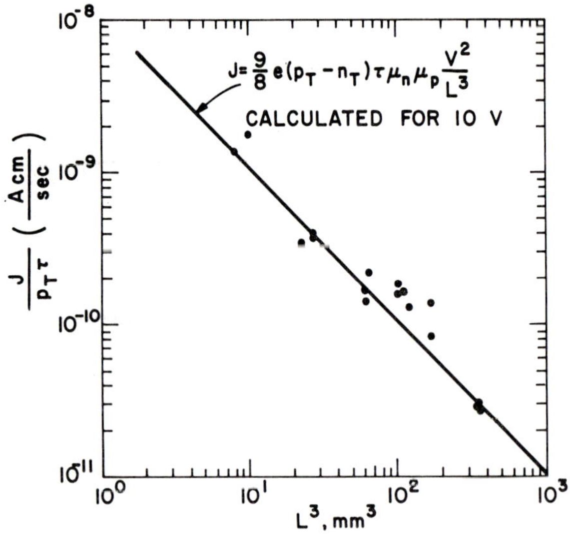

FIG. 13.7. Experimental verification of the L^{-3}-dependence of the current for the semiconductor injected plasma. (Log–log plot of $J/p_0\tau$ vs. L^3. Points obtained from 19 p-type silicon samples with large L/L_a ratios. Values of p_0 range from 10^{11} to 2×10^{12} cm^{-3}, τ_{eff} from 4 to 30 μsec.) (From Marsh *et al.*, 1965, Fig. 2.)

The corresponding value of the surface recombination velocity s is 400 cm/sec, as determined from a formula similar to (12.29) and (12.30).

It is of interest to note that data obtained early by Stafeev (1958, Fig. 2), and interpreted by him via a quite different model can be fitted

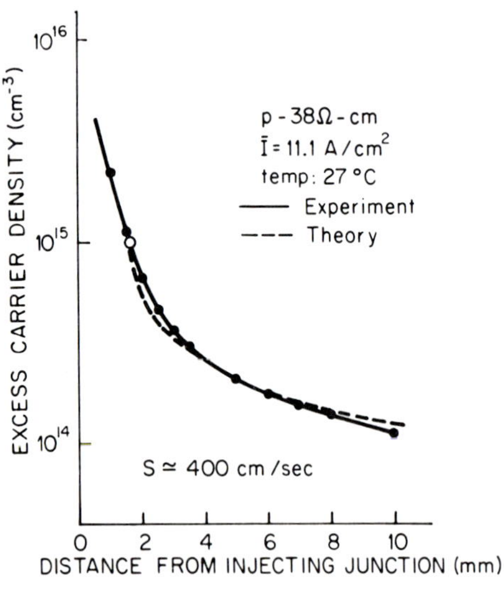

FIG. 13.8. Spatial distribution for the injected plasma in a long thin rod of p-Ge (L = 1.0 cm, a = 0.55 mm). The theoretical distribution is fitted to the experimental one at a single point (the open circle). The best fit is obtained for τ_{eff} = 72 μsec, as compared to the bulk τ = 840 μsec. (From Arizumi *et al.*, 1965, Fig. 3.)

very well with the semiconductor, injected-plasma square law (12.7), taking $n_0 - p_0 = 2 \times 10^{14}$ cm^{-3} (not given by Stafeev) in order to fit one point (Rose, unpublished).

The plasma injected into a semiconductor is a convenient vehicle for the study of plasma instabilities. Since this subject is beyond the scope of this book, we refer the reader to reviews published elsewhere for further information (Ioffe, 1960; Ancker-Johnson, 1966).

The insulator injected-plasma cube law was verified by the Hughes group (Baron *et al.*, 1964; Mayer *et al.*, 1965) working with specially prepared, high-resistivity silicon (compensated over a large volume by the lithium ion-drift technique). In Fig. 13.9, taken from this work, are shown current-voltage characteristics, exhibited in a log–log plot, taken at room temperature with five different PIN samples of Si with varying lengths of i-(intrinsic) region. The straight lines represent cube-law fits

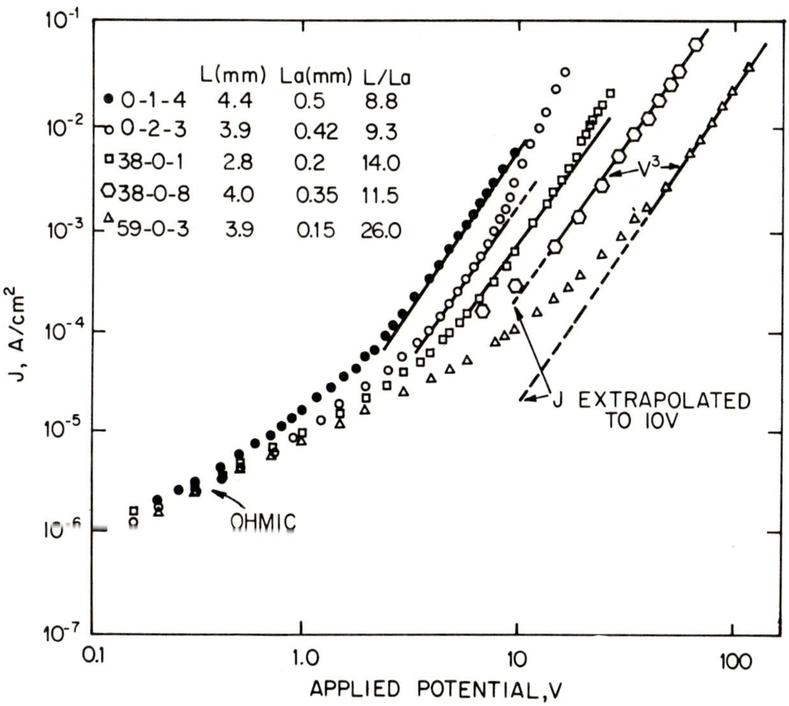

FIG. 13.9. The log current-density versus log applied voltage characteristics, taken at 300°K, for given PIN junctions whose intrinsic lengths L and ambipolar diffusion lengths L_a are shown in the enclosed table. The solid lines represent a V^3-current-dependence. The samples were Si PIN structures prepared by Li ion drift. (From Mayer *et al.*, 1965a, Fig. 4.)

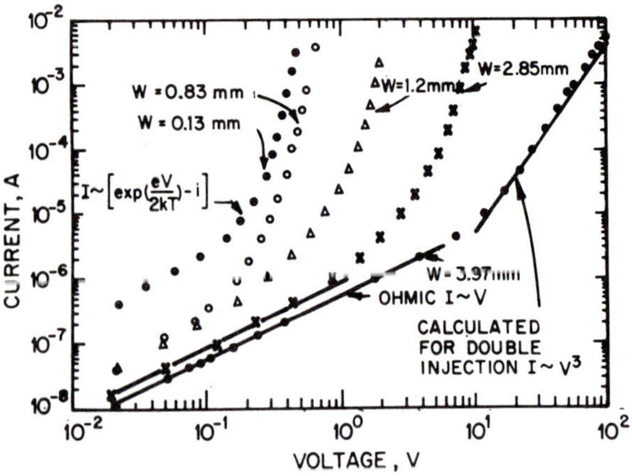

FIG. 13.10. The log current–log voltage characteristics of five diodes as a function of intrinsic length. The length of the intrinsic region was between 0.13 and 3.97 mm. The solid line shown for double injection was calculated on the basis of an effective length. The diodes were Si PIN structures prepared by Li ion drift. (From Mayer *et al.*, 1965a, Fig. 3.)

to the data, and it is seen that a quite good fit is obtained with these samples. Note, in the last column of figures, that the ratio of active sample length L to ambipolar diffusion length L_a is large in all of these samples. As discussed in Section 13.3.2, the corrections due to the carrier-accumulation effects at the injecting junctions are minimized under this circumstance. For Fig. 13.10, also taken from this work, we see that at smaller active sample lengths the dependence of current on voltage goes over into the exponential form, extreme left, characteristic of diffusion-dominated, junction-current injection and expected here for small L/L_a ratios. (The "W" of this figure becomes L in our notation.) Note again, on the extreme right, that for the larger sample, $W = 4$ mm, a good fit is obtained with a cube law. Figure 13.11, from the same authors, presents data pertinent to the thickness-dependence of the insulator injected-plasma current-voltage characteristic. The expected proportionality of $J/\tau\mu_n\mu_p$ to L^{-5} at fixed voltage $V = 10$ V, does not hold if the actual, uncorrected lengths L (solid circles) are used. If the corrected lengths $L_{\text{eff}} = L - 2.2 \times (2L_a)$ are used, then the open circles are obtained. This kind of correction is indicated by the theory (Section 13.3.2 and Eq. (13.92) in particular). The solid line is obtained from the insulator cube law (11.14). It is seen that the corrected, experimental points, namely the open circles, give a reasonably good fit to the theoretical line.

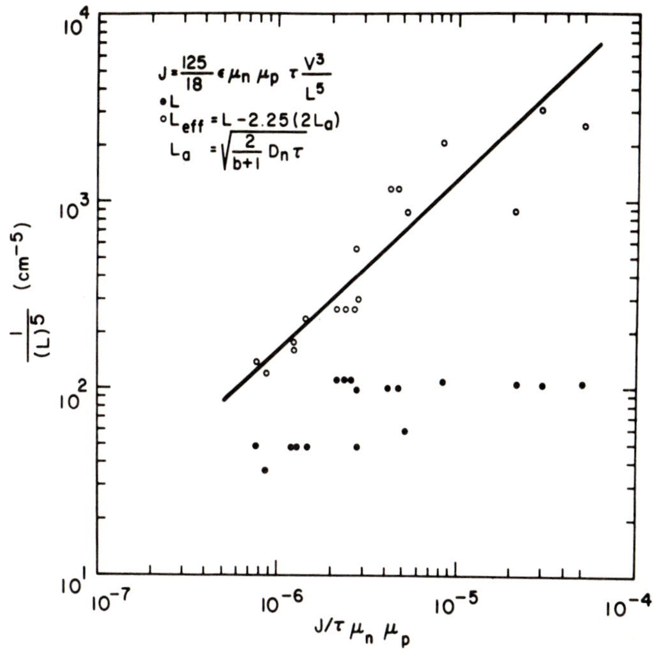

FIG. 13.11. Comparison of reciprocal of the fifth power of the sample length with the quantity $J/\tau\mu_n\mu_p$ for 15 samples. The current values used were taken from the currents in the V^3-portion extrapolated to 10 V as shown in Fig. 13.9. The closed points represent the measured length of the intrinsic region. The open points represent the effective length calculated from $L_{eff} = L - 2.2 \times (2L_a)$. The solid line represents the effective length required to fit the observed current. (From Mayer *et al.*, 1965a, Fig. 5.)

The insulator cube-law has also been verified by Okazaki and Hiromatsu (1967) working with p-type silicon. These authors likewise confirm the L^{-5} thickness dependence of the current at fixed voltage. The cube law has also been confirmed in much work by the Illinois group under Holonyak (see Moore, 1967, and Moore *et al.*, 1966, 1967a, b) working with silicon PIN diodes, the controlling dopant in the i-region being gold. In this work, the injected plasma concentration is about two orders of magnitude smaller than the Au concentration. The striking feature in this work, which is the subject of considerable study by this group, is the occurrence of spontaneous oscillations in the positive-resistance, cube-law regime of the I–V characteristic. Underlying these oscillations is presumably a dynamic negative resistance whose precise nature has not yet been delineated.

14

DOUBLE INJECTION WITH TRAPPING—THE NEGATIVE RESISTANCE CASE

In the discussion of double injection problems up to this point, a minimal role has been assigned to the localized defect states; they could be recombination centers, but then changes in their occupancy were assumed to be negligible. The realization of this condition requires ultra-pure material, such as Ge, Si, or InSb. In the more usual cases of double injection, the concentration of injected free carriers is smaller than that of the defect states. This is the situation connoted by "double injection with trapping."

The capture and immobilization of injected carriers may be expected to affect drastically the double-injection current-voltage characteristic, judging from the effects of like behavior (trapping) on the single-injection characteristic. One of the most interesting, and potentially useful, effects is a current-controlled negative-resistance regime in the characteristic, deriving from a free-carrier lifetime which increases substantially with injection level. Two cases leading to a negative resistance of this type are discussed in detail below.

The full range of possible structure in the current-voltage characteristic for double injection with trapping has simply not been delineated yet. This range is certainly larger than that for one-carrier injection, because a larger number of independent physical processes come into play. Concomitantly, the analytical difficulties are much greater than for one-carrier injection. Even if diffusion currents are neglected, the combination of space-charge and recombination effects produces complexities that are simply beyond the capabilities of rigorous analysis. On the other hand, deep inroads into the problem have been made via the Method of Regional Approximations. In the following, we review a few of the simple models that have been studied in some detail. Sections 14.1 and 14.2 are taken largely from the paper by Waxman and Lampert (1969).

14.1 Recombination Centers Fully Occupied

We first study the problem illustrated in Fig. 14.1, namely, where there is a single set of recombination centers lying well below the Fermi level

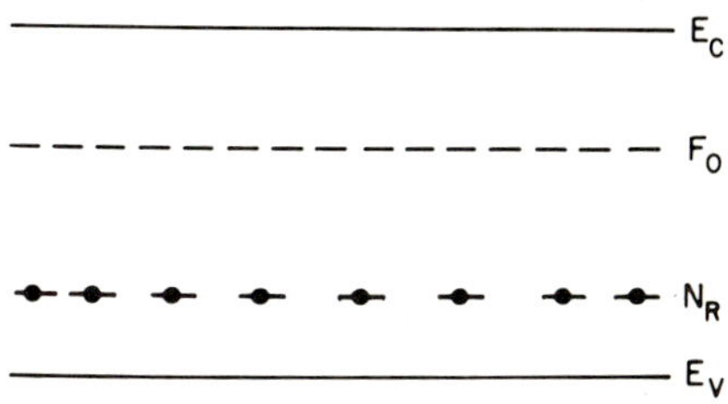

FIG. 14.1. Schematic, thermal-equilibrium, energy-band diagram for problem of double-injection with trapping by filled recombination centers.

and therefore completely occupied by electrons in thermal equilibrium. It is assumed that there are a negligible number of free carriers in the insulator at the outset. Further, it is assumed that $\sigma_p \gg \sigma_n$, as would be the case if the states N_R were acceptor-like, that is, negatively charged when electron occupied. Local charge neutrality, in thermal equilibrium, may be imagined maintained by exact compensation by shallow donors which play no further role in the electrical behavior of the insulator. (The opposite case, $\sigma_n \gg \sigma_p$, has already been studied in Section 11.1. Because the centers will remain filled at all injection levels, the hole lifetime is a constant in the problem and the solution corresponds to a one-carrier, electron SCL current at low voltages going over into an injected plasma at higher voltages.)

14.1.1 *Physical Arguments*

What main features of the double injection current-voltage characteristic may be anticipated by physical arguments? At very low levels of injection, where the occupancy of the states N_R is negligibly perturbed, the electron lifetime $\tau_{n,low}$ is essentially infinite since there are no empty recombination centers to capture injected electrons. On the other hand, the hole lifetime $\tau_{p,low} = (N_R\langle v\sigma_p\rangle)^{-1}$ is relatively short. Thus, in a manner of speaking, there is a "recombination barrier" to the passage of holes and none whatever to the passage of electrons. The injected current will be essentially a trap-free, electron SCL current which recombines with injected holes within a diffusion length, approximately, of the anode. When the applied voltage is high enough to drive holes appreciably across the crystal, that is, when a voltage V_{TH} is reached such that $t_{p,TH} = L^2/\mu_p V_{TH} \approx \tau_{p,low}$, the situation is drastically changed. Because $\sigma_p \gg \sigma_n$, the recombination centers preferentially capture holes and will therefore fill up with holes. This, in turn, will cause the hole lifetime to increase. Since both electrons and holes are now injected into the body of the insulator, it will be largely neutral. At high injection levels, where the injected electron and hole concentrations exceed N_R, they are necessarily free and approximately equal. With $n \approx p$, the steady state requires that $\tau_{n,high} \approx \tau_{p,high}$ where these are the respective high-injection-level lifetimes. Since now the recombination centers are almost completely occupied by holes, $\tau_{n,high} \simeq (N_R\langle v\sigma_n\rangle)^{-1}$. Thus, between low and high injection levels, the hole lifetime increases by the factor σ_p/σ_n, or more precisely $\langle v\sigma_p\rangle/\langle v\sigma_n\rangle$—a factor which may easily be many powers of ten. The "recombination barrier" to the passage of holes decreases with increasing injection level: The more holes injected, the easier it is for them to traverse the crystal. In fact it is so much easier that the voltage required actually decreases as the current increases. An outstanding feature of this double injection problem is a region of negative-resistance of the so-called "current-controlled" type, i.e., with voltage decreasing with increasing current. (This mechanism for producing negative resistance—a free carrier lifetime which increases with injection level—was first pointed out by Stafeev (1959), who discussed it qualitatively and with reference to problems of diffusive current flow.) We should like to determine the minimum voltage V_M and corresponding current J_M reached on the negative-resistance excursion. We note that beyond J_M, high injection level conditions obtain, that is, injected-plasma conditions for which $\tau_n = \tau_{n,high} = \tau_p = \tau_{p,high}$. In this regime, the electrons initially in the recombination centers have been transferred to the conduction band and the insulator behaves exactly like an n-type semiconductor with a thermal-equilibrium concentration N_R of electrons, the current-voltage characteristic for which has been derived in Section 12.1. The voltage V_M is simply the lower limit of validity of the semiconductor square-law and is determined by

$$t_{p,M} = L^2/\mu_p V_M \simeq \tau_{p,high} = (N_R\langle v\sigma_n\rangle)^{-1}.$$

The corresponding current

$$J_M \simeq eN_R\tau_{p,high}\mu_n\mu_p V_M^2/L^3 \simeq eN_R Lb/\tau_{p,high}\,.$$

Note that $V_{\text{TH}}/V_{\text{M}} \simeq \langle v\sigma_{\text{p}}\rangle/\langle v\sigma_{\text{n}}\rangle = \tau_{\text{p,high}}/\tau_{\text{p,low}}$. All of the features we have just discussed are seen on the complete current-voltage characteristic shown in Fig. 14.3.

We proceed to analytical discussion of the problem, first assuming local electrical neutrality, then dropping this assumption.

14.1.2 *Neutrality Solution*

In the simplified, neutrality theory, the equations characterizing the problem are:

current-flow equation:

$$J = e\mu_{\text{n}} n\mathscr{E} + e\mu_{\text{p}} p\mathscr{E} = \text{constant}, \tag{14.1}$$

where, as usual, diffusion currents have been neglected,

local neutrality condition:

$$n = p + p_{\text{R}}, \tag{14.2}$$

where p_{R} is the concentration of empty recombination centers,

particle-conservation equations:

$$\mu_{\text{n}}(d/dx)(n\mathscr{E}) = r = -\mu_{\text{p}}(d/dx)(p\mathscr{E}), \tag{14.3}$$

recombination-kinetic expressions:

$$r = \frac{n}{\tau_{\text{n}}} = \frac{p}{\tau_{\text{p}}}, \quad \frac{1}{\tau_{\text{n}}} = \langle v\sigma_{\text{n}}\rangle p_{\text{R}}, \quad \frac{1}{\tau_{\text{p}}} = \langle v\sigma_{\text{p}}\rangle n_{\text{R}}, \quad p_{\text{R}} + n_{\text{R}} = N_{\text{R}}. \tag{14.4}$$

The two equations in (14.3) are readily combined to give

$$\frac{d}{dx}[(n-p)\mathscr{E}] = \frac{(a+1)n}{\mu_{\text{p}}\tau_{\text{n}}} = \frac{(a+1)p}{\mu_{\text{p}}\tau_{\text{p}}}, \qquad a = \frac{\mu_{\text{p}}}{\mu_{\text{n}}} = \frac{1}{b}. \tag{14.5}$$

If we had used the full Poisson equation (14.29), in place of (14.2), we would have obtained a second-order differential equation characterizing the problem. In neglecting space-charge, the order of the final differential equation is lowered to first-order and only one boundary condition can be imposed. This is quite analogous to the situation obtaining in the semiconductor injected-plasma problem when space charge is ignored, and here, as in that problem, we employ the boundary condition: $\mathscr{E} = 0$ at the "minority-carrier" injecting contact. In this problem, the equivalent "minority-carriers" are the holes, and the proper boundary condition is then

$$\mathscr{E} = 0 \quad \text{at} \quad x = 0. \tag{14.6}$$

The system of equations (14.1)–(14.5), subject to boundary condition (14.6), has been solved without further approximation (Lampert, 1962). Here we prefer to approximate further, thereby obtaining a very much simpler, and more illuminating, solution. Using the familiar Regional Approximation Method, the insulator is divided into two regions: Region I extending out from the anode up to $x = x_1$, and Region II extending from x_1 to the cathode. In Region I, characterized by $n(x) > N_R$, the recombination centers are largely depopulated, their electrons having been transferred to the conduction band. In Region II, the recombination centers are still largely filled. We now study the separate regions in some detail.

Region I ($0 \leqslant x < x_1$): $n > N_R$, $p_R \simeq N_R$

Taking $p_R = N_R$ in (14.2), Eqs. (14.5) and (14.1) become, respectively,

$$N_R(d\mathscr{E}/dx) = [(a+1)n/\mu_p\tau_h], \qquad (1/\tau_h) = \langle v\sigma_n\rangle N_R, \tag{14.7}$$

where subscript "h" on τ denotes the common, high-injection-level lifetime, and

$$J = e\mu_n\mathscr{E}(a+1)\,n - e\mu_p N_R\mathscr{E}. \tag{14.8}$$

Substituting for $(a+1)\,n$ from (14.8) into (14.7) gives the differential equation

$$[\mathscr{E}/(\mathscr{E}+S)]\,d\mathscr{E} = T\,dx; \qquad S = J/e\mu_p N_R, \quad T = 1/\mu_n\tau_h, \tag{14.9}$$

with solution

$$\mathscr{E} - S\ln[(\mathscr{E}+S)/S] = Tx, \tag{14.10}$$

satisfying boundary condition (14.6). The potential drop from 0 to x is

$$V(x) - \int_0^x \mathscr{E}\,dx = \int_0^{\mathscr{E}} \mathscr{E}(dx/d\mathscr{E})\,d\mathscr{E} = (1/T)\{\tfrac{1}{2}\mathscr{E}^2 - S\mathscr{E} + S^2\ln[(\mathscr{E}+S)/S]\}. \tag{14.11}$$

At the right-hand end of Region I, we have

$x = x_1$:

$$n_1 = n(x_1) = N_R, \qquad \mathscr{E}_1 = \mathscr{E}(x_1) = (J/e\mu_n N_R) = aS, \tag{14.12}$$

from (14.8). Substituting (14.12) into (14.10) and (14.11) respectively gives

$$x_1 = (S/T)\{a - \ln(1 + a)\}, \qquad V_{a,1} = V(x_1) = (S^2/T)\{\tfrac{1}{2}a^2 - a + \ln(a + 1)\}. \tag{14.13}$$

Let J_M, V_M correspond to $x_1 = L$, i.e., $n = N_R$ at the cathode. This is the low voltage end of the negative-resistance regime in Fig. 14.3. Then the two equations (14.13) give, using (14.9),

$$\begin{aligned} J_M &= h(a)\frac{eN_R L}{\tau_h}, & h(a) &= \frac{a}{a - \ln(1 + a)}, \\ V_M &= \frac{1}{g(a)}\frac{L^2}{\mu_p \tau_h}, & g(a) &= \frac{\{a - \ln(1 + a)\}^2}{a\{\tfrac{1}{2}a^2 - a + \ln(1 + a)\}}. \end{aligned} \tag{14.14}$$

Remarkably, these expressions are precisely the same as obtained from the rigorous neutrality theory [Lampert, 1962, Eqs. (B9) and (B10)].

It is convenient to rewrite the equations in (14.13) as

$J \leqslant J_M$:

$$x_1 = L(J/J_M), \qquad V_{a,1} = V_M(J/J_M)^2. \tag{14.15}$$

Region II ($x_1 < x \leqslant L$): $n < N_R$, $p \ll n$, $p_R \simeq n$, $n_R \simeq N_R$

Taking $p_R = n$ and $n_R = N_R$ in (14.4), we obtain

$$p \simeq \frac{n^2\langle v\sigma_n\rangle}{N_R\langle v\sigma_p\rangle}. \tag{14.16}$$

Rewriting (14.1) as

$$n\mathscr{E} = (J/e\mu_n)(1 + \zeta)^{-1} \qquad \text{with} \quad \zeta = ap/n \simeq a\langle v\sigma_n\rangle n/N_R\langle v\sigma_p\rangle \ll 1$$

from (14.16), Eq. (14.3) now gives

$$-\frac{dn}{n^2} = \frac{eN_R\langle v\sigma_p\rangle}{aJ}\,dx, \tag{14.17}$$

with solution:

$$\frac{1}{n} = \frac{1}{N_R} + \frac{eN_R\langle v\sigma_p\rangle}{aJ}(x - x_1), \tag{14.18}$$

satisfying the boundary condition: $n = N_R$ at $x = x_1$. The field intensity $\mathscr{E}$ is given by:

$$\mathscr{E} \simeq \frac{J}{e\mu_n}\frac{1}{n} = \frac{J}{e\mu_n N_R} + \frac{N_R\langle v\sigma_p\rangle}{\mu_p}(x - x_1). \tag{14.19}$$

The potential drop from x_1 to L is given by

$$V_{1,c} = \int_{x_1}^{L} \mathscr{E}\, dx = \frac{J}{e\mu_n N_R}(L - x_1) + \frac{N_R\langle v\sigma_p\rangle}{2\mu_p}(L - x_1)^2. \tag{14.20}$$

For the full applied voltage, $V = V_{a,1} + V_{1,c}$, we have, using (14.14), (14.15) and (14.20),

$$V = V_M\left\{\left(\frac{J}{J_M}\right)^2 + f(a)\left(\frac{J}{J_M}\right)\left(1 - \frac{J}{J_M}\right)\right\} + V_{TH}\left(1 - \frac{J}{J_M}\right)^2,$$

$$f(a) = \frac{a\{a - \ln(1 + a)\}}{\frac{1}{2}a^2 - a + \ln(1 + a)}, \qquad V_{TH} = \frac{\langle v\sigma_p\rangle L^2 N_R}{2\mu_p} = \frac{L^2}{2\mu_p\tau_{p,\text{low}}}. \tag{14.21}$$

The basic features of the neutrality solution derive directly from (14.21). As J increases from 0 to J_M, V decreases monotonically from a threshold value V_{TH} to its minimum value V_M. The threshold voltage corresponds to the condition:

$$\bar{t}_{p,TH} = (L^2/\mu_p V_{TH}) = 2\tau_{p,\text{low}}, \tag{14.22}$$

where $\bar{t}_{p,TH}$ is, of course, approximately, the average hole transit time at the threshold voltage. Note that $f(a)$ is almost a constant, of order unity; it varies from $\frac{3}{2}$ at $a \ll 1$ to 2 at $a \gg 1$. Comparing V_{TH} and V_M, we have from (14.21) and (14.14)

$$\frac{V_{TH}}{V_M} = \frac{g(a)}{2}\frac{\langle v\sigma_p\rangle}{\langle v\sigma_n\rangle}, \tag{14.23}$$

Note, from (14.14), that $g(a)$ decreases monotonically from $\frac{3}{8}$ at $a \ll 1$ to $1/a$ at $a \gg 1$. Thus, unless a is very large, $V_{TH} \gg V_M$. In this case, Eq. (14.21) can be further simplified to

$$V \simeq V_M(J/J_M)^2 + V_{TH}[1 - (J/J_M)]^2. \tag{14.24}$$

On a log–log plot, such as in Fig. 14.3, the current rises very steeply with voltage near threshold, reaching $0.3J_M$ at only a 50% decrease in voltage, $V = V_{TH}/2$. Correspondingly, the negative resistance region is very flat, the current increasing from $0.3J_M$ to J_M over the large voltage decrease from $V_{TH}/2$ to V_M.

The negative-resistance regime, $J < J_M$, presents us with our first example of a planar injection problem in which the electric field distribution is not convex and, concomitantly, does not vary relatively gently over the crystal. Indeed since Region I has high conductivity, $n > N_R$, and Region II low conductivity, $n < N_R$, the applied voltage is largely piled up in Region II. Over Region I, $\mathscr{E}$ is convex and increases monotonically from 0 at $x = 0$ to $\mathscr{E}_1 = J/e\mu_n N_R$ at $x = x_1$, the slope $d\mathscr{E}/dx$ decreasing monotonically from ∞ at $x = 0$ to $(a + 1)/\mu_p\tau_h$ at $x = x_1$. Over Region II, $\mathscr{E}$ increases monotonically from $\mathscr{E}_1$ at $x = x_1$ to $\mathscr{E}_1 + N_R\langle v\sigma_p\rangle(L - x_1)/\mu_p$ at $x = L$, the slope being constant at $d\mathscr{E}/dx = N_R\langle v\sigma_p\rangle/\mu_p = 2V_{TH}/L^2$. Due to our simplifications, there is actually a discontinuity in slope at $x = x_1$: $(d\mathscr{E}/dx)_{II\ x_1}/(d\mathscr{E}/dx)_{I,x_1} = \langle v\sigma_p\rangle/(a + 1)\langle v\sigma_n\rangle \gg 1$. In the rigorous neutrality theory, the discontinuity in slope is, of course, smoothed out.

The electric field intensity in Region II, Eq. (14.19), can be rewritten, using (14.14), (14.15), and (14.21), as

$$\mathscr{E} = f(a)(J/J_M)(V_M/L) + (2V_{TH}/L^2)(x - x_1), \tag{14.25a}$$

from which

$$\mathscr{E}_c = \mathscr{E}(L) = f(a)(J/J_M)(V_M/L) + (2V_{TH}/L)[1 - (J/J_M)]. \tag{14.25b}$$

A convenient measure of the field pileup toward the cathode is the ratio $\mathscr{E}_c/\mathscr{E}_\Omega$, which can be written, using (14.25b) and (14.24), as

$$\frac{\mathscr{E}_c}{\mathscr{E}_\Omega} = \frac{\mathscr{E}_c L}{V} = \frac{f(a)\,V_M(J/J_M) + 2V_{TH}[1 - (J/J_M)]}{V_M(J/J_M)^2 + V_{TH}[1 - (J/J_M)]^2};$$
$$\left(\frac{\mathscr{E}_c}{\mathscr{E}_\Omega}\right)_{\max} \simeq \left(\frac{V_{TH}}{V_M}\right)^{1/2}, \tag{14.26}$$

where the maximum value of $\mathscr{E}_c/\mathscr{E}_\Omega$, obtained by setting its derivative with respect to J/J_M equal to 0, occurs at

$$V \simeq 2V_M, \qquad J/J_M \simeq 1 - (V_M/V_{TH})^{1/2}.$$

Beyond the negative-resistance regime, i.e., for $J > J_M$, there is only the Region I in the insulator and the current-voltage characteristic is determined implicitly through the equations

$$\mathscr{E}_c - S\ln[(\mathscr{E}_c + S)/S] = TL, \tag{14.27}$$

$$V = (1/T)\{\tfrac{1}{2}\mathscr{E}_c^2 - S\mathscr{E}_c + S^2\ln[(\mathscr{E}_c + S)/S], \tag{14.28}$$

obtained from (14.10) and (14.11), respectively, by taking $x = L$ and $\mathscr{E} = \mathscr{E}_c$, with S and T given by (14.9). For $J/J_M \gg 1$, $\mathscr{E}_c/S \ll 1$ and expansion of the logarithm gives $\mathscr{E}_c^2/2S \simeq TL$ and $V \simeq \mathscr{E}_c^3/3TS$, which combine to give the expected semiconductor square law:

$$J = 9eN_R\mu_n\mu_p\tau_h V^2/8L^3.$$

In order to discuss the self-consistency of the neutrality approximation, we first note that the current equation, in place of the neutrality condition (14.2), is the Poisson equation

$$(\epsilon/e)(d\mathscr{E}/dx) = p + p_R - n. \tag{14.29}$$

From the discussion following (14.24), we know that the neglected space charge is much larger in Region II than in Region I. In Region II, $p \ll p_R$, n so that (14.29) can be written there as

Region II:

$$n = p_R - (\epsilon/e)(d\mathscr{E}/dx). \tag{14.30}$$

Since, from (14.19), $d\mathscr{E}/dx$ is a constant in Region II, whereas, from (14.18), n decreases monotonically as x increases, it suffices to compare the neglected space charge to the free carrier density n_c at the cathode to establish self-consistency of the neutrality solution. From (14.14), (14.15), (14.18), and (14.19), we can write:

Region II:

$$\frac{(\epsilon/e)(d\mathscr{E}/dx)_c}{n_c} = \frac{t_{R,\Omega}\tau_h}{ah(a)\tau^2_{p,low}}\left(\frac{J_M}{J} - 1\right), \qquad \text{with} \quad t_{R,\Omega} = \frac{\epsilon}{e\mu_p N_R}. \tag{14.31}$$

Thus, we conclude that if $t_{R,\Omega}\tau_h \ll \tau^2_{p,tow}$, there is a substantial range of currents J over which the neutrality solution is self-consistent. However for J small enough, the neutrality solution is always inconsistent. This result was already anticipated in the physical arguments of the preceding section where we noted that the low current regime in the J–V characteristic would be the one-carrier, electron, trap-free SCL square law. For $t_{R,\Omega}\tau_h \gtrsim \tau^2_{p,low}$, there is a negligible range of currents J, just below J_M, over which self-consistency obtains.

In the following section we show how space charge can be self-consistently included in the analysis by extending the Regional Approximation analysis.

14.1.3 *Inclusion of Space Charge*

The inclusion of space charge in the problem, that is, the replacement of (14.2) by (14.29), requires the use of four regions in the analysis, as illustrated in Fig. 14.2. Regions I and II are the same as before, namely regions in which local neutrality, Eq. (14.2), obtains. However, now Region II may terminate inside the insulator, at a plane x_2 where, finally,

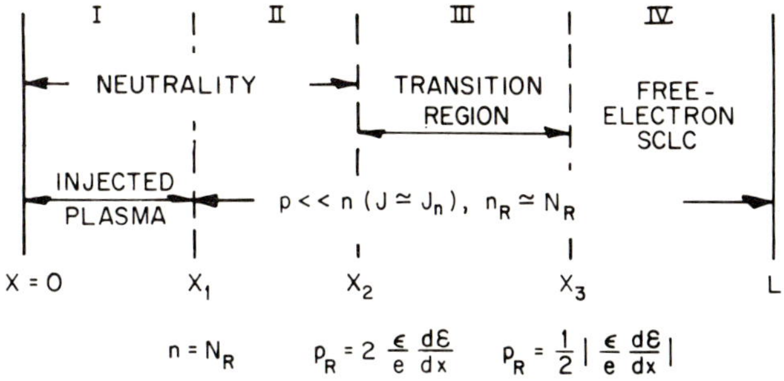

FIG. 14.2. Schematic diagram for Regional Approximation applied to problem of double-injection with trapping by filled recombination centers.

space charge can no longer be self-consistently neglected because it has become comparable to the terms in the Poisson equation that are retained: $(\epsilon/e)(d\mathscr{E}/dx)_2 \simeq p_{R,2} \simeq n_2$, where n_2 denotes $n(x_2)$, and so on. Regions III and IV are regions in which the role of space charge is no longer negligible; it is in these two regions that (14.2) is replaced by (14.29) or equivalently (14.30), since p is negligible in these two regions.

Region IV, adjacent to the cathode, is a region into which there has been negligible hole penetration: $p \ll p_R \ll n$, and therefore it is a region dominated by negative space charge. Since the recombination centers are filled at the outset, this negative space charge is free (mobile); therefore the current in Region IV is just the one-carrier, electron, trap-free SCL current injected from the cathode. The left-hand terminus of Region IV is that plane x_3 where the neglected, trapped holes can no longer be self-consistently neglected, that is, from (14.30), where

$$p_{R,3} \simeq n_3 \simeq |(\epsilon/e)(d\mathscr{E}/dx)_3|.$$

Region III is a "buffer" region needed to join smoothly the neutrality-dominated Region II to the negative-space-charge dominated Region IV. Since p_R and n are comparable at each end of Region III, as seen in the above discussion, they are comparable throughout this region and may be interchanged except where their difference is involved. In this respect, Region III is similar to Region II; the two regions differ in that (14.30) is needed to describe Region III, whereas (14.2) is adequate in Region II. The space charge in Region III is needed to "turn the field $\mathscr{E}$ around," since $\mathscr{E}$ is increasing throughout Region II, according to (14.19), and decreasing throughout Region IV, according to (14.37). In other words, $\mathscr{E}$ reaches its maximum, $d\mathscr{E}/dx = 0$, somewhere inside Region III.

The mathematical characterization of Regions III and IV is as follows:

Region III ($x_2 \leqslant x \leqslant x_3$): $p \ll n \simeq p_R \ll N_R$, $n_R \simeq N_R$

The differential equation for p_R in this region is directly obtained from (14.17), simply by replacing n by p_R :

$$-dp_R/p_R^2 = (eN_R\langle v\sigma_p\rangle/aJ)\,dx. \tag{14.32}$$

[This procedure is valid because the steps leading to (14.17) or (14.32) nowhere involve the difference, $n - p_R$.]

Equation (14.32) is readily integrated, yielding

$$\frac{1}{p_R(x)} - \frac{1}{p_{R,2}} = \frac{eN_R\langle v\sigma_p\rangle}{aJ}(x - x_2). \tag{14.33}$$

The spatial dependence of $\mathscr{E}$ in Region III is given by the solution to the differential equation (14.30), here rewritten as

$$(\epsilon/e)(d\mathscr{E}/dx) + (J/e\mu_n)(1/\mathscr{E}) - p_R = 0,$$

with p_R given by (14.33). Actually we shall not solve this differential equation since our primary interest here is the potential drop across Region III, $V_{2,3} = \int_{x_2}^{x_3} \mathscr{E}\,dx$; it is adequate, for this purpose, to use a simple interpolation scheme to approximate $\mathscr{E}(x)$, say, with a quadratic expression.

Region IV ($x_3 \leqslant x \leqslant L$): $p \ll p_R \ll n$, $n_R \simeq N_R$

The equations characterizing the current flow in this region are

$$J = e\mu_n n\mathscr{E} \tag{14.34}$$

and

$$n = -(\epsilon/e)(d\mathscr{E}/dx), \tag{14.35}$$

which is (14.30) with p_R neglected.

The equations, of course, describe one-carrier, electron, trap-free SCL current injection from the cathode at $x = L$. From Section 4.1, their solution is

$$\mathscr{E}(x) = (2J/\epsilon\mu_n)^{1/2}(L - x)^{1/2} \qquad \text{and} \qquad n(x) = (\epsilon J/2e^2\mu_n)^{1/2}[1/(L - x)^{1/2}]. \tag{14.36}$$

For further discussion of the problem, it is convenient to switch to dimensionless variables:

$$x = x^{**}w, \qquad \mathscr{E} = \mathscr{E}^{**}u, \qquad V = V^{**}v,$$
$$x^{**} = \frac{2a\mu_p J}{\epsilon N_R^2\langle v\sigma_p\rangle^2}, \quad \mathscr{E}^{**} = \frac{2aJ}{\epsilon N_R\langle v\sigma_p\rangle}, \quad V^{**} = \frac{4a^2\mu_p J^2}{\epsilon^2 N_R^3\langle v\sigma_p\rangle^3} = \mathscr{E}^{**}x^{**}. \tag{14.37}$$

The separate regions are now characterized as follows.

Region I ($0 \leqslant w \leqslant w_1$)

Equation (14.9), characterizing this region, becomes

$$\frac{u\,du}{u+A} = B\,dw, \qquad A = \frac{\epsilon\langle v\sigma_{\rm p}\rangle}{2ea\mu_{\rm p}} = \frac{S}{\mathscr{E}^{**}} \quad \text{and} \quad B = a\frac{\langle v\sigma_{\rm n}\rangle}{\langle v\sigma_{\rm p}\rangle} = \frac{Tx^{**}}{\mathscr{E}^{**}}, \tag{14.38}$$

with S and T given by (14.9). Equation (14.38) has the solution

$$u - A\ln[(u+A)/A] = Bw, \tag{14.39}$$

satisfying the boundary condition (14.6), namely $u = 0$ at $w = 0$.

The right-hand end of Region I is at w_1 corresponding to x_1 given by (14.12). Using (14.38), Eqs. (14.12) and (14.13) become

$$w_1 = (A/B)\{a - \ln(1+a)\}, \qquad u_1 = u(w_1) = aA. \tag{14.40}$$

From (14.13) and (14.38), the dimensionless voltage $v_{{\rm a},1}$ across Region I is given by

$$v_{{\rm a},1} = v(w_1) = (A^2/B)\{\tfrac{1}{2}a^2 - a + \ln(a+1)\}. \tag{14.41}$$

Region II ($w_1 \leqslant w \leqslant w_2$)

The Eq. (14.17) characterizing this region becomes

$$du/dw = 1 \tag{14.42}$$

with solution

$$u - u_1 = w - w_1, \tag{14.43}$$

where u_1 and w_1 are given by (14.40).

From the previous discussion, the right-hand end x_2 of Region II is chosen such that $(\epsilon/e)(d\mathscr{E}/dx)_2 \simeq p_{{\rm R},2} \simeq n_{{\rm R},2}$. From (14.37), it is seen that this choice corresponds to

$$u_2 = (Jx^{**}/\epsilon\mu_{\rm n}\mathscr{E}^{**2}) = \tfrac{1}{2}, \qquad w_2 = \tfrac{1}{2} - aA + (A/B)\{a - \ln(1+a)\}, \tag{14.44}$$

where (14.40) and (14.43) have also been used. [If we had used the criterion $(\epsilon/e)(d\mathscr{E}/dx)_2 = p_{{\rm R},2}/2$ to fix the right-hand of Region II, we would have obtained $u_2 = \frac{1}{4}$. The final result, namely, the current voltage characteristic, is insignificantly changed by this relatively small change in the sizes of a couple of adjacent regions. Indeed, the Regional Approximation Method is most reliable precisely when the current-voltage characteristic is fairly insensitive to the precise criteria fixing the boundaries of the separate regions.]

The voltage $v_{1,2}$ across Region II is given by

$$v_{1,2} = \int_{w_1}^{w_2} u\, dw = \int_{u_1}^{u_2} u\, du = \tfrac{1}{2}(u_2^2 - u_1^2) = \tfrac{1}{2}(\tfrac{1}{4} - a^2A^2), \qquad (14.45)$$

where (14.42), (14.40), and (14.44) have been used.

Region III ($w_2 \leqslant w \leqslant w_3$)

We are interested in Eq. (14.33) evaluated at $x = x_3$. Taking as the criterion to fix $x_3 : p_{R,3} = p_R(x_3) = n(x_3)/2 = -(\epsilon/2e)(d\mathscr{E}/dx)_3$ [above we have already used $p_{R,2} = p_R(x_2) = (\epsilon/e)(d\mathscr{E}/dx)$ to fix x_2], at $x = x_3$, Eq. (14.33) becomes, in the dimensionless units (14.37):

$$\frac{1}{-\frac{1}{2}(du/dw)_3} - \frac{1}{(du/dw)_2} = 2(w_3 - w_2) \qquad \text{or} \qquad 4(w_c - w_3)^{1/2} = 1 + 2(w_3 - w_2), \qquad (14.46)$$

where we have used (14.42) and (14.53).

As already remarked above, we shall simply approximate $u(\mathscr{E})$ in Region III by an interpolation scheme. Any number of such schemes are possible and useful. Here we proceed as follows: Let $u = U_2(w)$ be the equation for the tangent to the curve $u = u(w)$ in Region II at $w = w_2$; correspondingly, $u = U_3(w)$ is the tangent to the curve $u = u(w)$ in Region IV at $w = w_3$. Then

$$\begin{aligned} U_2(w) &= u_2 + (w - w_2)(du/dw)_2 = u_2 + (w - w_2), \\ U_3(w) &= u_3 + (w - w_3)(du/dw)_3 = u_3 - [(w - w_3)/2(w_c - w_3)^{1/2}], \end{aligned} \qquad (14.47)$$

where we have used (14.42) and (14.53).

Furthermore, let $\alpha(w)$ and $\beta(w)$ be two linear weighting functions:

$$\begin{aligned} \alpha(w) &= (w_3 - w)/(w_3 - w_2), \qquad \alpha(w_3) = 0, \qquad \text{and} \qquad \alpha(w_2) = 1, \\ \beta(w) &= (w - w_2)/(w_3 - w_2), \qquad \beta(w_3) = 1, \qquad \text{and} \qquad \beta(w_2) = 0, \end{aligned} \qquad (14.48)$$

where $\alpha + \beta = 1$.

The our interpolation approximation is

$$u(w) = U_2(w)\,\alpha(w) + U_3(w)\,\beta(w) \qquad (14.49)$$

or, substituting from (14.47) and (14.48),

$$u(w) = (u_2 + w - w_2)\frac{(w_3 - w)}{(w_3 - w_2)} + \left(u_3 - \frac{w - w_3}{2(w_c - w_3)^{1/2}}\right)\frac{(w - w_2)}{(w_3 - w_2)}. \qquad (14.50)$$

Note that $u(w_2) = u_2$ and $u(w_3) = u_3$, assuring continuity of the $\mathscr{E}$ field across the transition planes $x = x_2$ and $x = x_3$.

The voltage $v_{2,3}$ across Region III is given by

$$v_{2,3} = \int_{w_2}^{w_3} u\,dw = \tfrac{1}{2}(u_2 + u_3)(w_3 - w_2) + \tfrac{1}{6}\{1 + [1/2(w_c - w_3)^{1/2}]\}(w_3 - w_2)^2. \tag{14.51}$$

Region IV ($w_3 \leqslant w \leqslant w_c$)

The equations (14.34) and (14.35) combine to give, in dimensionless form,

$$du^2/dw = -1, \tag{14.52}$$

with solution

$$u = (w_c - w)^{1/2}, \qquad du/dw = -1/2(w_c - w)^{1/2}, \tag{14.53}$$

satisfying $u_c = u(w_c) = 0$, that is $\mathscr{E}(L) = 0$.

The voltage across Region IV is

$$v_{3,c} = \int_{w_3}^{w_c} u\,dw = \tfrac{2}{3}(w_c - w_3)^{3/2}. \tag{14.54}$$

We are ready now to collect all of the results. The current-voltage characteristic is given by a plot of $1/w_c$ vs v_c/w_c^2, since

$$\frac{1}{w_c} = \frac{2a\mu_p}{\epsilon N_R^2\langle v\sigma_p\rangle^2 L} J, \qquad \frac{v_c}{w_c^2} = \frac{\mu_p}{N_R\langle v\sigma_p\rangle L^2} V. \tag{14.55}$$

Writing $w_c = w_2 + (w_3 - w_2) + (w_c - w_3)$, and using (14.40) and (14.44), we have

$$\frac{1}{w_c} = \frac{1}{-1 - aA + (A/B)\{a - \ln(1 + a)\} + (y + 1)^2}, \qquad y = (w_c - w_3)^{1/2}. \tag{14.56}$$

From (14.54), $v_{3,c} = 2y^3/3$. From (14.46), (14.51), and (14.53), $v_{2,3} = 5y^2/3 + y/4 - 1/4 + 1/48y$. Inserting these into the voltage relation

$$v_c = v_{a,1} + v_{1,2} + v_{2,3} + v_{3,c},$$

we have, using (14.45) and (14.41),

$$\frac{v_c}{w_c^2} = \frac{\frac{1}{2}(\frac{1}{4} - a^2A^2) + (A^2/B)\{\frac{1}{2}a^2 - a + \ln(a+1)\} + \frac{2}{3}y^3 + \frac{5}{3}y^2 + \frac{1}{4}y - \frac{1}{4} + (1/48y)}{[-1 - aA + (A/B)\{a - \ln(a + 1)\} + (y + 1)^2]^2} \tag{14.57}$$

The relations (14.56) and (14.57) give a parametric representation of the current-voltage characteristic in terms of the single parameter y and the material constants, a, A, and B.

In terms of the dimensionless variables (14.37), the critical point J_M, V_M on the neutrality-based current-voltage characteristic becomes

$$\left(\frac{1}{w_c}\right)_M = \frac{h(a)B}{aA}, \qquad \left(\frac{v_c}{w_c^2}\right)_M = \frac{B}{ag(a)}. \tag{14.58}$$

The neutrality-based current-voltage characteristic (14.21) can be written as

$$\frac{V}{V_M} = \left(\frac{J}{J_M}\right)^2 + f(a)\frac{J}{J_M}\left(1 - \frac{J}{J_M}\right) + \frac{ag(a)}{2B}\left(1 - \frac{J}{J_M}\right)^2, \tag{14.59}$$

with

$$\frac{V}{V_M} = \frac{(v_c/w_c^2)}{(v_c/w_c^2)_M} \quad \text{and} \quad \frac{J}{J_M} = \frac{(1/w_c)}{(1/w_c)_M}, \tag{14.60}$$

and where (14.23) has been used.

The voltage threshold for the neutrality based characteristic is, from (14.21) and (14.55),

$$(v_c/w_c^2)_{TH} = \tfrac{1}{2}. \tag{14.61}$$

The electron, trap-free square law, $J = 9\epsilon\mu_n V^2/8L^3$, becomes

$$(1/w_c) = (9/4)(v_c/w_c^2)^2. \tag{14.62}$$

The space-charge-based J–V characteristic (14.56) and (14.57) is compared to the neutrality-based characteristic (14.59) and (14.60) for two cases in Fig. 14.3. The solid straight line, of slope 2, is the electron trap-free square law (14.62). The two solid curves are neutrality-based characteristics and the two dashed curves are the corresponding space-charge-based characteristics. The material parameters chosen for the lower pair of curves were $\epsilon/\epsilon_0 = 12$, $\mu_n = \mu_p = 10^4$ cm²/V sec

$$N_R = 10^{15}\ \text{cm}^{-3}, \qquad \langle v\sigma_n\rangle = 9 \times 10^{-10}\ \text{cm}^3/\text{sec},$$

and

$$\langle v\sigma_p\rangle = 3 \times 10^{-6}\ \text{cm}^3/\text{sec},$$

as might pertain to a silicon experiment at liquid nitrogen temperature. The corresponding dimensionless parameters are $a = 1$, $A = 10^{-3}$, and $B = 3 \times 10^{-4}$. For the upper pair of curves, the changes in the material

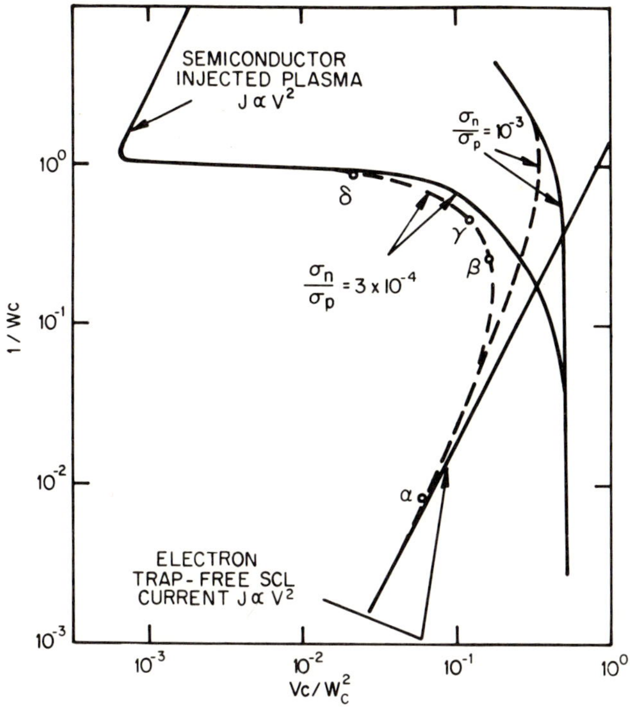

FIG. 14.3. Prototype universal current-voltage characteristics for double injection into an insulator with a single set of recombination centers completely filled in thermal equilibrium. $J \propto 1/w_c$ and $V \propto v_c/w_c^2$. The two cases calculated are $\sigma_n/\sigma_p = 10^{-3}$ and $\sigma_n/\sigma_p = 3 \times 10^{-4}$. The heavy solid lines are the characteristics obtained assuming neutrality; the dashed lines are those obtained including space charge.

parameters are $\langle v\sigma_n \rangle = 10^{-9}$ cm³/sec and $\langle v\sigma_p \rangle = 10^{-6}$; correspondingly, $A = 3 \times 10^{-4}$ and $B = 10^{-3}$.

Some further particulars are of interest. On the lower dashed curve, at point α, the electron space-charge Region IV occupies 83% of the crystal and absorbs 81% of the applied voltage. The Transition Region III takes up the rest of the crystal and the applied voltage, both. Regions I and II are negligible. Going to point β, the neutrality Regions I and II are now in the picture. They take 32% of the crystal width, though only 2% of the voltage. Region III takes 43% of the width and 72% of the voltage, Region IV 25% of the width and 26% of the voltage. At point γ, Regions I and II now occupy 55% of the width and take 5% of the voltage, Region III 34% of the width and 80% of the voltage, Region IV 11% of the width and 5% of the voltage. At this point, going up on the curve, Region IV is shrinking fast. At point δ, the two neutrality Regions I and II occupy 99% of the crystal width and 98% of the voltage, the residual width and voltage being in the Transition Region III.

14.2 Recombination Centers Partially Occupied

We next study the problem illustrated in Fig. 14.4, namely where there is a single set of recombination centers lying not too far from the Fermi

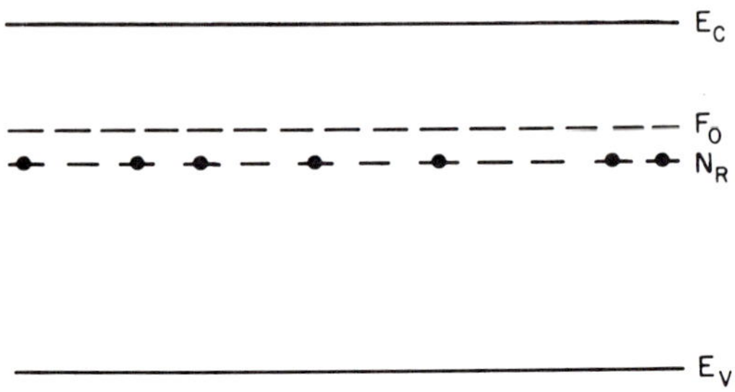

Fig. 14.4. Schematic, thermal-equilibrium, energy-band diagram for problem of double-injection with trapping by partially filled recombination centers.

level and therefore only partially occupied by electrons in thermal equilibrium. Again we assume that there are a negligible number of free carriers in the insulator at the outset, and we further assume that $\sigma_p \gg \sigma_n$. In fact the problem we shall now study is obtained from the previous problem, defined by Fig. 14.1, simply by moving the recombination centers much closer to the Fermi level (and on either side of it).

The double-injection, current-voltage characteristic for this problem has several features in common with that for the previous problem. Again, because $\sigma_p \gg \sigma_n$, the hole lifetime will increase with increasing injection level, allowing the voltage across the insulator to drop, that is, producing a current-controlled negative resistance. Again, above the low-voltage turn-around point the insulator behaves exactly like an n-type semiconductor with a thermal-equilibrium concentration $n_{R,0}$ of electrons, and injected-plasma conditions obtain. The striking difference from the previous problem appears at low injection levels. In this problem, since there are a substantial number of empty centers, as well as filled centers, at the outset, there is a "recombination barrier" to the transit of electrons as well as holes. In consequence there is a voltage threshold for current flow. This threshold, which we calculate below, is determined by trapped space charge, that is, space charge held in recombination centers. Therefore the neutrality approximation is a poor starting point for the study of this problem.

The equations characterizing the problem are the same as those for the previous problem: (14.1) and (14.3)–(14.5), except that the Poisson equation (14.21) is here replaced by

$$(\epsilon/e)(d\mathscr{E}/dx) = p + (p_R - p_{R,0}) - n, \tag{14.63}$$

thereby taking into account the initially empty recombination centers $p_{R,0}$, and the boundary condition (14.6) is replaced by

$$\mathscr{E} = 0 \quad \text{at} \quad x = 0 \quad \text{and} \quad x = L \tag{14.64}$$

as were also used in the previous problem when space charge was included (the four-region approximation).

This set of equations is studied below via the Regional Approximation Method. We first study them at the voltage threshold, that is, in the limit of vanishingly small current.

14.2.1 *Voltage Threshold*

At low injection levels, the dominant contribution to the total current is the electron current, since the injected holes are preferentially captured by the filled centers ($\sigma_p \gg \sigma_n$). Therefore, in place of (14.1), we can write

$$J \simeq en\mu_n\mathscr{E} = \text{constant}. \tag{14.65}$$

Further, since most of the injected carriers are captured by the recombination centers, we can neglect p and n in (14.63):

$$(\epsilon/e)(d\mathscr{E}/dx) \simeq p_R - p_{R,0}\,. \tag{14.66}$$

Multiplying both sides of the recombination-kinetic relation (14.4), namely $p\langle v\sigma_p\rangle\, n_R = n\langle v\sigma_n\rangle\, p_R$, by $e\mu_p\mathscr{E}$, and substituting from (14.65), we obtain

$$J_p \simeq aJ(\langle v\sigma_n\rangle/\langle v\sigma_p\rangle)(p_R/n_R). \tag{14.67}$$

Since

$$(d/dx)(p_R/n_R) = (N_R/n_R{}^2)(dp_R/dx) \simeq (\epsilon N_R/en_R{}^2)(d^2\mathscr{E}/dx^2),$$

Eq. (14.67) gives

$$dJ_p/dx \simeq (\epsilon aJ\langle v\sigma_n\rangle N_R/e\langle v\sigma_p\rangle n_R{}^2)\, d^2\mathscr{E}/dx^2. \tag{14.68}$$

From (14.3), $dJ_p/dx = -er = -en\langle v\sigma_n\rangle p_R$ or, using (14.65),

$$dJ_p/dx \simeq -(J\langle v\sigma_n\rangle p_R/\mu_n\mathscr{E}). \tag{14.69}$$

Equations (14.68) and (14.69) together yield the differential equation

$$\mathscr{E}\frac{d^2\mathscr{E}}{dx^2} \simeq -\frac{e\langle v\sigma_p\rangle n_R^{\ 2}p_R}{\epsilon\mu_p N_R} \simeq -\frac{e\langle v\sigma_p\rangle n_{R,0}^2 p_{R,0}}{\epsilon\mu_p N_R}. \tag{14.70}$$

The appearance of J in both (14.68) and (14.69) leads to its absence from (14.70) and this, mathematically speaking, corresponds to the existence of a threshold voltage, independent of current. The replacement of n_R, p_R by $n_{R,0}$, $p_{R,0}$, respectively, in (14.70) must always be valid at sufficiently low injection level.

The differential equation (14.70) is solved below. Here we note that a simple dimensional analysis, namely replacing $\mathscr{E}\,d^2\mathscr{E}/dx^2$ by $-V_{TH}^2/L^4$, gives a good approximation for the threshold voltage V_{TH}:

$$V_{TH} \simeq \{e\langle v\sigma_p\rangle L^4 n_{R,0}^2 p_{R,0}/\epsilon\mu_p N_R\}^{1/2}. \tag{14.71}$$

The exact result, (14.106a), namely the solution of (14.70) subject to boundary conditions (14.64), differs from (14.71) only through the numerical factor $1/(4\pi)^{1/2}$. Note that if a parabolic approximation is used for $\mathscr{E}$, namely $\mathscr{E} = 4\mathscr{E}_{max}x(L-x)/L^2$, then $d^2\mathscr{E}/dx^2 = -8\mathscr{E}_{max}/L^2$ and $\bar{\mathscr{E}} = 2\mathscr{E}_{max}/3$, so that $\overline{\mathscr{E}\,d^2\mathscr{E}/dx^2} = -12V_{TH}^2/L^4$. Using this in (14.70) multiplies the right-hand side of (14.71) by $1/\sqrt{12}$, which is very close to the correct factor $1/(4\pi)^{1/2}$. The parabolic approximation for $\mathscr{E}$ is very close to the correct distribution.

14.2.2 *Finite Current Flow*

The system of equations (14.1), (14.3)–(14.5), (14.63) cannot be solved analytically. Once again progress can be made by use of the Method of Regional Approximations. Three regions are necessary to the analysis, as illustrated in Fig. 14.5. Regions I and II are very similar to the same regions of the preceding problem (Fig. 14.2). They are both regions in which local neutrality obtains. Region I, extending from the anode out to $x = x_1$, is characterized by $n(x) > n_{R,0}$ and by depopulation of the recombination centers, their electrons having been transferred to the conduction band. In Region II, characterized by $n(x) < n_{R,0}$, the electron population of the recombination centers is not drastically changed from its thermal equilibrium value. Region II terminates at a plane x_2 where space charge can no longer be self-consistently neglected. Region III, extending from x_2 to the cathode, is a region dominated by

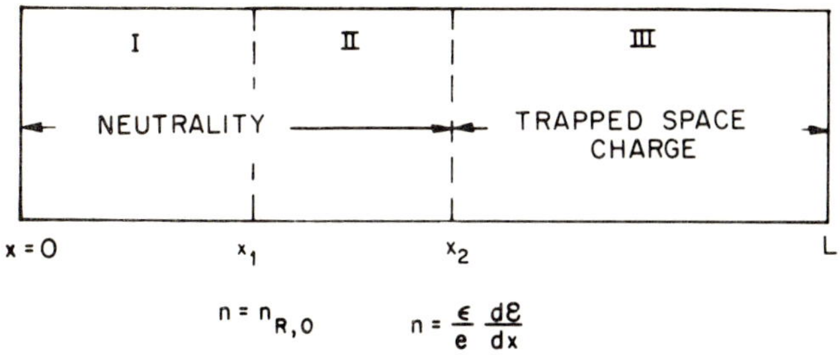

FIG. 14.5. Schematic diagram for Regional Approximation applied to problem of double-injection with trapping by partially filled recombination centers.

space charge trapped in the recombination centers; it is the region already described in Section 14.2.1 except that here it no longer fills the insulator.

We now consider the separate regions in some detail.

Region I ($0 \leqslant x \leqslant x_1$): $n > n_{R,0}$, $p_R \simeq N_R$

In place of (14.63) we have local neutrality, which for $p_R = N_R$, is specified by

$$n - p = N_R - p_{R,0} = n_{R,0}. \tag{14.72}$$

Using (14.72) in (14.5), we have

$$n_{R,0}(d\mathscr{E}/dx) = [(a+1)n/\mu_p\tau_h], \qquad (1/\tau_h) = \langle v\sigma_n \rangle N_R. \tag{14.73}$$

Using (14.72) in (14.1), we obtain

$$J = e\mu_n\mathscr{E}(a+1)\,n - e\mu_p n_{R,0}\mathscr{E}. \tag{14.74}$$

Except for the replacement of N_R by $n_{R,0}$, Eqs. (14.72) and (14.73) are identical to (14.7) and (14.8), respectively. Thus we have the following results, (14.75)–(14.81), paralleling (14.9)–(14.15), respectively,

$$[\mathscr{E}/(\mathscr{E} + S_0)]\,d\mathscr{E} = T\,dx; \qquad S_0 = (J/e\mu_p n_{R,0}), \quad T = (1/\mu_n\tau_h), \tag{14.75}$$

with solution

$$\mathscr{E} - S_0 \ln[(\mathscr{E} + S_0)/S_0] = Tx, \tag{14.76}$$

satisfying $\mathscr{E} = 0$ at $x = 0$.

$$V(x) = \int_0^x \mathscr{E}\,dx = (1/T)\{\tfrac{1}{2}\mathscr{E}^2 - S_0\mathscr{E} + S_0^2 \ln[(\mathscr{E} + S_0)/S_0]\}. \tag{14.77}$$

At the right-hand of Region I,

$x = x_1$:

$$n_1 = n(x_1) = n_{R,0}\,, \qquad \mathscr{E}_1 = \mathscr{E}(x_1) = (J/e\mu_n n_{R,0}) = aS_0\,, \tag{14.78}$$

$$x_1 = (S_0/T)\{a - \ln(1+a)\}, \qquad V_{a,1} = (S_0{}^2/T)\{\tfrac{1}{2}a^2 - a + \ln(a+1)\}. \tag{14.79}$$

At the low-voltage end of the negative-resistance regime,

$$J_M = h(a)(en_{R,0}L/\tau_h), \qquad V_M = (1/g(a))(L^2/\mu_p\tau_h), \tag{14.80}$$

with $h(a)$ and $g(a)$ given by (14.14).

$J \leqslant J_M$:

$$x_1 = L(J/J_M), \qquad V_{a,1} = V_M(J/J_M)^2. \tag{14.81}$$

Region II ($x_1 \leqslant x \leqslant x_2$): $n < n_{R,0}\,, p \ll n, p_R - p_{R,0} \simeq n, n_R \simeq n_{R,0}$

Here there is not quite the close parallelism with Region II of the former problem that obtained with Region I for the two problems. The neutrality condition in this region, replacing (14.63) and dropping p, is

$$n = p_R - p_{R,0} = n_{R,0} - n_R\,. \tag{14.82}$$

In place of (14.74) we now have the simpler

$$J = e\mu_n n\mathscr{E}. \tag{14.83}$$

Because of (14.83), it is readily seen that (14.67) is also valid here. Now using $(d/dx)(p_R/n_R) = (N_R/n_R{}^2)(dp_R/dx) = (N_R/n_R{}^2)(dn/dx)$ from (14.82), it follows that

$$\frac{dJ_p}{dx} = \frac{aJ\langle v\sigma_n\rangle N_R}{\langle v\sigma_p\rangle n_R{}^2}\frac{dn}{dx}. \tag{14.84}$$

Since, from (14.3), $dJ_p/dx = -er = -en\langle v\sigma_n\rangle\, p_R\,,$ we have, using (14.82)

$$-\frac{dn}{n(n + p_{R,0})} = \frac{e\langle v\sigma_p\rangle n_R{}^2}{aJN_R}\,dx \simeq \frac{e\langle v\sigma_p\rangle n_{R,0}^2}{aJN_R}\,dx. \tag{14.85}$$

Note that we have replaced n_R by $n_{R,0}$ in (14.85). This is completely analogous to our replacement of n_R by N_R in Region II of the preceding problem, and is valid for the same reason: $n < n_{R,0}$ over all of Region II and $n < n_{R,0}/2$ over most of Region II (everywhere except near $x = x_1$). (It is possible to deal with n_R exactly in (14.85), namely writing $n_R = n_{R,0} - n$ from (14.82), but the amount of algebraic manipulation then required is very large and there is no significant gain in accuracy.) On the other hand, we do not replace $n + p_{R,0} = p_R$ by $p_{R,0}$ on the left-hand side of (14.85), precisely because we have made no assumption about the relative magnitudes of $p_{R,0}$ and n in Region II. Note that if $p_{R,0} = 0$, $n_{R,0} = N_R$, then (14.85) reduces to (14.17), as indeed it must.

From (14.83), $dn = -J\, d\mathscr{E}/e\mu_n\mathscr{E}^2$, from which (14.85) may be rewritten as

$$\frac{d\mathscr{E}}{\mathscr{E} + (J/e\mu_n p_{R,0})} = \frac{e\langle v\sigma_p\rangle n_{R,0}^2 p_{R,0}}{aJN_R}\, dx. \tag{14.86}$$

The right-hand terminus x_2 of Region II is taken where the neglected space charge overtakes the retained terms in the Poisson equation, that is, where

$$n_2 = n(x_2) = (\epsilon/e)(d\mathscr{E}/dx)_2\,. \tag{14.87}$$

For further discussion of the problem, it is convenient to switch to dimensionless variables:

$$x = x^\ddagger w, \qquad \mathscr{E} = \mathscr{E}^\ddagger u, \qquad V = V^\ddagger v,$$

$$x^\ddagger = \frac{aN_R J}{e\langle v\sigma_p\rangle n_{R,0}^2 p_{R,0}}, \qquad \mathscr{E}^\ddagger = \frac{aN_R J}{\{e\langle v\sigma_p\rangle n_{R,0}^2 p_{R,0}\epsilon\mu_p N_R\}^{1/2}}, \tag{14.88}$$

$$V^\ddagger = \frac{(aN_R J)^2}{n_{R,0}^3\{e\langle v\sigma_p\rangle p_{R,0}\}^{3/2}\{\epsilon\mu_p N_R\}^{1/2}} = \mathscr{E}^\ddagger x^\ddagger.$$

The separate regions are now characterized as follows.

Region I ($0 \leqslant w \leqslant w_1$)

The equation (14.75), characterizing this region, becomes

$$\frac{u\, du}{u + C} = D\, dw, \qquad C = \left\{\frac{\epsilon\langle v\sigma_p\rangle p_{R,0}}{ea^2\mu_p N_R}\right\}^{1/2},$$

$$D = \langle v\sigma_n\rangle N_R \left\{\frac{a^2\epsilon N_R}{e\langle v\sigma_p\rangle n_{R,0}^2 p_{R,0}\mu_p}\right\}^{1/2}, \tag{14.89}$$

which has the solution

$$u - C\ln[(u + C)/C] = Dw, \tag{14.90}$$

satisfying the boundary condition (14.6), namely $u = 0$ at $w = 0$.

The right-hand end of Region I is at w_1 corresponding to x_1 given by (14.79), namely

$$w_1 = (C/D)\{a - \ln(1 + a)\}, \qquad u_1 = aC, \tag{14.91}$$

where (14.78) has also been used.

The dimensionless voltage $v_{a,1}$ across Region I is, from (14.79),

$$v_{a,1} = (C^2/D)\{\tfrac{1}{2}a^2 - a + \ln(1 + a)\}. \tag{14.92}$$

For $J > J_M$, $V > V_M$ with J_M, V_M given by (14.80), there is only the Region I in the insulator terminating at $w_c < w_1$, where $w_c = w(L)$. In place of (14.91) and (14.92), we now have

$$w_c = (C/D)\{a(u_c/u_1) - \ln[a(u_c/u_1) + 1]\} \tag{14.93}$$

and

$$v_c = (C^2/D)\{\tfrac{1}{2}a^2(u_c/u_1)^2 - a(u_c/u_1) + \ln[a(u_c/u_1) + 1]\}, \tag{14.94}$$

with u_1 given by (14.91).

At large injection levels, $u_c \ll u_1$, $w_c \simeq u_c{}^2/2CD$, and $v_c \simeq u_c{}^3/3CD$. The corresponding current-voltage characteristic is

$u_c \ll u_1$:

$$(v_c/w_c{}^2)^2 \simeq (8/9)\, CD(1/w_c) \qquad \text{or} \qquad J \simeq (9/8)\, e\mu_p\mu_n n_{R,0}\tau_h(V^2/L^3), \tag{14.95}$$

which is the semiconductor, injected-plasma characteristic corresponding to $n_0 - p_0 = n_{R,0}$, that is, to transfer of the electrons from the filled recombination centers to the conduction band.

Region II ($w_1 \leqslant w \leqslant w_2$)

The equation (14.86), characterizing this region, becomes

$$du/(u + E) = dw, \qquad E = \{\epsilon\langle v\sigma_p\rangle n_{R,0}^2/e\mu_p p_{R,0}N_R\}^{1/2} \tag{14.96}$$

with solution

$$w - w_1 = \ln[(u + E)/(u_1 + E)]. \tag{14.97}$$

The right-hand end of Region II is specified by (14.87) which, using (14.96), can be written

$$u_2(u_2 + E) = 1 \qquad \text{with} \quad u_2 = u(w_2). \tag{14.98}$$

Using (14.98) in (14.97), we get

$$w_2 - w_1 = \ln[(u_2 + E)/(u_1 + E)] = -\ln u_2(u_1 + E). \tag{14.99}$$

For the voltage $v_{1,2}$ across Region II we have, using (14.96),

$$v_{1,2} = \int_{w_1}^{w_2} u\,dw = \int_{u_1}^{u_2} u/(u+E)\,du = u_2 - u_1 - E\ln[(u_2+E)/(u_1+E)]. \tag{14.100}$$

Region III ($w_2 \leqslant w \leqslant w_c$)

This is the region discussed at length in Section 14.2.1; whereas in that discussion this region was taken to fill the entire insulator (a reasonable approximation at very low injection levels), here it is one of three regions we must deal with, namely the region adjacent to the cathode.

The equation (14.70) characterizing this region, becomes

$$u\,d^2u/dw^2 = -1. \tag{14.101}$$

We have encountered this equation previously, namely as (11.31), characterizing the injected-plasma limit of the Parmenter–Ruppel problem, Section 11.2.1. Adopting the solution (11.32) to our current problem, we obtain the solution to (14.100):

$$u = u_m \exp(-s^2), \qquad w - w_2 = \sqrt{2}u_m \int_{s_2}^{s} ds \exp(-s^2), \tag{14.102}$$

where u_m is obviously the maximum value reached by u, namely at $s = 0$. The differences between (14.102) and (11.32) stem from the fact that (14.101) is an equation for u while (11.31) is an equation for u^2.

From continuity of u at $x = x_2$ ($w = w_2$), we have, from (14.102),

$$u_2 = u_m \exp(-s_2^2),$$

with u_2 given by (14.98). Further, since u increases monotonically with w in Region I and II, from (14.90) and (14.97), it follows that u must reach its maximum, hence $s = 0$, inside Region III. Thus, s_2 must be negative and can be written as $-|s_2|$. Further, the boundary condition $\mathscr{E} = 0$ at $x = L$, i.e., $u = 0$ at $w = w_c$, corresponds to $s_c = s(w_c) = \infty$ in (14.102). Using this in (14.102), we obtain for $w_c = w(L)$:

$$w_c - w_2 = \sqrt{2}u_2(\exp s_2^2) \int_{-|s_2|}^{\infty} ds \exp(-s^2). \tag{14.103}$$

The voltage across Region III is

$$v_{2,c} = \int_{w_2}^{w_c} u\,dw = \int_{-|s_2|}^{\infty} u \frac{dw}{ds}\,ds = u_2^2(\exp 2s_2^2) \int_{-\sqrt{2}|s_2|}^{\infty} ds \exp(-s^2), \tag{14.104}$$

where we have used (14.102) and the identity:

$$\sqrt{2}\int_{-|s_2|}^{\infty} ds\,\exp(-2s^2) = \int_{-\sqrt{2}|s_2|}^{\infty} ds\,\exp(-s^2).$$

We are now ready to collect all of the results. Once again, the current-voltage characteristic is given by a plot of $1/w_c$ vs $v_c/w_c{}^2$:

$$\frac{1}{w_c} = \frac{aN_R}{e\langle v\sigma_p\rangle n_{R,0}^2 p_{R,0} L}\, J, \qquad \frac{v_c}{w_c{}^2} = \frac{1}{n_{R,0}L^2}\left\{\frac{\epsilon\mu_p N_R}{ep_{R,0}\langle v\sigma_p\rangle}\right\}^{1/2} V. \tag{14.105}$$

Note that if Region III occupies the entire insulator (which, for practical purposes, it does at low currents), then $w_2 = 0$, $|s_2| = \infty$, and $v_{2,c} = v_c$ in (14.103) and (14.104). From (14.103) and (14.104), the corresponding, dimensionless, threshold voltage $(v_c/w_c{}^2)_{TH}$ is given by $(v_c/w_c{}^2)_{TH} = [2\int_{-\infty}^{\infty} ds\,\exp(-s^2)]^{-1} = 1/(4\pi)^{1/2}$. Using this in (14.105), we obtain for the threshold voltage

$$V_{TH} = \{e\langle v\sigma_p\rangle L^4 n_{R,0}^2 p_{R,0}/4\pi\epsilon\mu_p N_R\}^{1/2}. \tag{14.106a}$$

Ashley's (1963) more general result for V_{TH} is

$$V_{TH} = F\frac{L^2}{2\mu_p\tau_{p,\text{low}}}, \quad F = \frac{\Gamma(2\alpha+1)\,\Gamma(2\beta+1)/\Gamma(2\alpha+2\beta+1)}{[\Gamma(\alpha+1)\,\Gamma(\beta+1)/\Gamma(\alpha+\beta+1)]^2},$$

$$\alpha = \frac{\tau_h}{t'_{n,\Omega}}, \quad \beta = \frac{\tau_{p,\text{low}}}{t'_{p,\Omega}}\frac{n_{R,0}}{N_R}, \quad t'_{n,\Omega} = \frac{\epsilon}{en_{R,0}\mu_n}, \quad t'_{p,\Omega} = \frac{\epsilon}{ep_{R,0}\mu_p}, \tag{14.106b}$$

where $\Gamma(x)$ is the well-known gamma function. For $\alpha \gg 1$, $\beta \gg 1$, the use of Stirling's approximation gives

$\alpha \gg 1, \beta \gg 1$:

$$F \simeq \{(\beta/\pi)[1 + (\beta/\alpha)]\}^{1/2}. \tag{14.106c}$$

For $\langle v\sigma_p\rangle \gg \langle v\sigma_n\rangle$, $\tau_h \gg n_{R,0}\tau_{p,\text{low}}/N_R$, and for $n_{R,0} \gtrsim N_R/2$ and $\mu_p \lesssim \mu_n$, $t'_{n,\Omega} < t'_{p,\Omega}$ so that $\alpha \gg \beta$. If further, $p_{R,0}$ is large enough that $\beta \gg 1$, then $F \simeq (\beta/\pi)^{1/2}$ which, inserted into (14.106b), gives (14.106a).

Returning to the finite-current problem, writing

$$w_c = (w_c - w_2) + (w_2 - w_1) + w_1$$

and using (14.91), (14.98), and (14.103), we obtain

$$w_c = \sqrt{2}u_2(\exp s_2{}^2)\int_{-|s_2|}^{\infty} ds\,\exp(-s^2) - \ln u_2(aC + E) + (C/D)\{a - \ln(1 + a)\}, \tag{14.107}$$

where, from (14.98),

$$u_2 = (E/2)\{-1 + [1 + (4/E^2)]^{1/2}\}. \tag{14.108}$$

Since $v_c = v_{a,1} + v_{1,2} + v_{2,c}$, we have, from (14.92), (14.100), and (14.104),

$$\begin{aligned} v_c = {} & (C^2/D)\{\tfrac{1}{2}a^2 - a + \ln(1+a)\} + (u_2 - aC) \\ & - E\ln[(u_2+E)/(aC+E)] + u_2^2 \exp(2s_2^2)\int_{-\sqrt{2}|s_2|}^{\infty} ds \exp(-s^2). \end{aligned} \tag{14.109}$$

A parametric representation of $1/w_c$ vs v_c/w_c^2 is given by (14.107)–(14.109) with $|s_2|$ as the variable parameter. Further detailed results require numerical computation. The current-voltage characteristic for one choice of parameters is shown in Fig. 14.6; the corresponding parameter values are

$$\epsilon/\epsilon_0 = 12, \qquad \mu_n = \mu_p = 10^4 \text{ cm}^2/\text{V sec}, \qquad n_{R,0} = p_{R,0} = 5 \times 10^{14} \text{ cm}^{-3},$$

$\langle v\sigma_n\rangle = 10^{-9}$, and $\langle v\sigma_p\rangle = 10^{-7}$ cm^3/sec, corresponding to $a = 1$, $C = E = 5.8 \times 10^{-3}$ and $D = 2.3 \times 10^{-4}$. The threshold voltage is $(v_c/w_c^2)_{TH} = 0.28$ and the turn-around voltage is $(v_c/w_c^2)_M = 2.1 \times 10^{-5}$, the turn-around current being $(1/w_c)_M = 0.13$. At the point labeled α, Region I occupies almost 2% of the insulator, Region II 1%, and Region III 97%; essentially 100% of the applied voltage is across Region III. At point β, Regions I and II occupy 14 and 7%, respectively, of the insulator but still absorb negligible voltage; Region III occupies 79% of the insulator and absorbs 100% of the voltage. At point γ, Region I occupies 48% of the insulator and absorbs 4% of the voltage, Region II occupies 26% of the insulator and absorbs 14% of the voltage, and Region III occupies 26% of the insulator and absorbs 82% of the voltage. At point δ, Region I occupies 56% of the insulator and absorbs 8% of the voltage, Region II occupies 29% of the insulator and absorbs 33% of the voltage, and Region III occupies 15% of the insulator and absorbs 59% of the voltage. The dashed curve in Fig. 14.6 is the neutrality-determined solution (Regions I and II assumed to fill the insulator), which yields an infinite threshold voltage (Keating, 1964).

Because of the empty recombination centers there will also be, in this problem, a threshold voltage for one-carrier, electron SCL current flow, namely the trap-filled-limit voltage $V_{TFL} = ep_{R,0}L^2/2\epsilon = L^2/2\mu_p t'_{p,\Omega}$, in the notation of (14.106). For $\alpha \gg \beta \gg 1$ and $n_{R,0} > N_R/2$,

$$V_{TH} \simeq (1/\pi\beta)^{1/2} V_{TFL} < V_{TFL}.$$

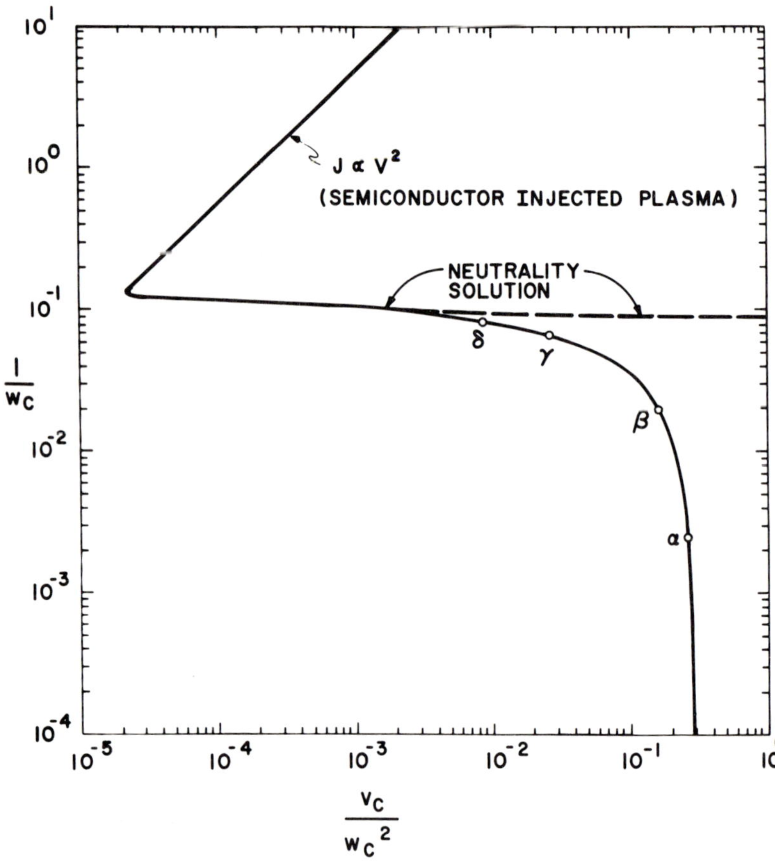

FIG. 14.6. Prototype universal current-voltage characteristic (solid line) for double injection into an insulator with a single set of recombination centers partially filled in thermal equilibrium.

In this, there will be no electron SCL current preceding the double-injection current. Where $V_{TFL} < V_{TH}$, the trap-filled-limit, electron SCL current-voltage characteristic will be observed prior to the onset of double injection. (This initial regime in the current–voltage characteristic did not emerge from our analysis, nor from Ashley's (1963), because of our neglect of thermal reemission of electrons from the recombination centers back into the conduction band. That is, with empty centers $p_{R,0}$ present initially, and the thermal reemission of electrons neglected, electron injection is suppressed until holes are also injected.) Since the TFL electron flow prior to double injection completely fills the recombination centers, we would expect the double-injection portion of the current-voltage characteristic to be described precisely by the analysis of Section 14.1. Where this latter characteristic intersects the electron TFL characteristic, there will naturally be some rounding-off, in order that the transition from the dominant one-carrier SCL flow to the double-injection flow be smooth.

14.3 Thermal Free Carriers Present

In the problems studied thus far in this chapter, we have ignored any effects due to the presence of thermally generated free carriers on the insulator. We turn our attention now to this problem. In the problem of Section 14.2, illustrated in Fig. 14.4, we now assume that the insulator is initially p-type, with a thermal concentration p_0 of holes.

The equations characterizing this problem are the current-flow equation (14.1) rewritten here for convenience,

$$J = e\mu_n n\mathscr{E} + e\mu_p p\mathscr{E} = \text{constant}, \tag{14.110}$$

the Poisson equation

$$(\epsilon/e)(d\mathscr{E}/dx) = (p - p_0) + (p_R - p_{R,0}) - n, \tag{14.111}$$

which is (14.63) generalized by including p_0 but still taking $n_0 = 0$, and the particle-conservation equations

$$\begin{gathered}\mu_n(d/dx)(n\mathscr{E}) = -\mu_p(d/dx)(p\mathscr{E}) = r - g;\\ r - g = r_n = r_p - g_p\,,\end{gathered} \tag{14.112}$$

$$\begin{gathered} r_n = \frac{n}{\tau_n}\,, \qquad r_p = \frac{p}{\tau_p}\,, \qquad g_p = \frac{p_R}{\tau_R}\,;\\ \frac{1}{\tau_n} = \langle v\sigma_n\rangle p_R\,, \qquad \frac{1}{\tau_p} = \langle v\sigma_p\rangle n_R\,, \qquad \frac{1}{\tau_R} = \langle e_p\rangle N_v\,,\end{gathered} \tag{14.113}$$

where we are clearly ignoring the thermal reemission of electrons, consistent with our taking $n_0 = 0$. Rewriting (14.110) as

$$J = e\mu_n n\mathscr{E}[1 + a(\delta p/n)] + e\mu_p p_0\mathscr{E}, \qquad \delta p = p - p_0\,, \tag{14.114}$$

we now assume that $\delta p/n$ is a constant, independent of injection level, hence of position. This will be true if, in (14.112), (14.113), $r_n \simeq n/\tau_{n,\mathrm{low}}$ and $r_p - g_p \simeq \delta p/\tau_{p,\mathrm{low}}$, with

$$\tau_{p,\mathrm{low}}^{-1} = \langle v\sigma_n\rangle p_{R,0} \qquad \text{and} \qquad \tau_{p,\mathrm{low}}^{-1} = \langle v\sigma_p\rangle n_{R,0}\,,$$

as we indeed expect at not-too-high injection levels. Then $\delta p/n = \tau_{\mathrm{p,low}}/\tau_{\mathrm{n,low}}$. Equation (14.114) now gives

$$\mu_{\mathrm{n}}\, d(n\mathscr{E})/dx = -\mu_{\mathrm{p}} p_0 \{1 + a\,\delta p/n\}^{-1}\, d\mathscr{E}/dx.$$

We make the further assumption, appropriate to the negative resistance problem, that $a\,\delta p/n \ll 1$, that is, that $\tau_{\mathrm{p,low}} \ll \tau_{\mathrm{n,low}}$. Finally, we have then

$$-\mu_{\mathrm{p}} p_0 \frac{d\mathscr{E}}{dx} \simeq \frac{n}{\tau_{\mathrm{n,low}}}\left(1 + \frac{\delta p_{\mathrm{R}}}{p_{\mathrm{r},0}}\right) \simeq \frac{n}{\tau_{\mathrm{n,low}}}, \qquad \delta p_{\mathrm{R}} = p_{\mathrm{R}} - p_{\mathrm{R},0} \tag{14.115}$$

where it is assumed that $\delta p_{\mathrm{R}}/p_{\mathrm{R},0} \ll 1$, consistent with the above. [This equation could have been written down, out of hand, taking $p = p_0$ in the hole conservation equation (14.112). The above derivation is more useful for exploring the limitations on the validity of (14.115).] It is convenient to rewrite (14.115) in the form

$$\frac{\epsilon}{e}\frac{d\mathscr{E}}{dx} = -\frac{t_{\mathrm{p},\Omega}}{\tau_{\mathrm{n,low}}}\, n; \qquad t_{\mathrm{p},\Omega} = \frac{\epsilon}{e p_0 \mu_{\mathrm{p}}}, \quad \tau_{\mathrm{n,low}} = \frac{1}{\langle v\sigma_{\mathrm{n}}\rangle p_{\mathrm{R},0}}. \tag{14.116}$$

Consistent with our approximations, Eq. (14.114) can be rewritten as

$$J \simeq e\mu_{\mathrm{n}} n\mathscr{E} + e\mu_{\mathrm{p}} p_0 \mathscr{E} = \text{constant.} \tag{14.117}$$

It will be recognized that Eqs. (14.116) and (14.117) have basically the same form as Eqs. (4.44) and (4.1) of Section 4.3 characterizing the one-carrier SCL shallow-trap problem, the equivalent θ-factor for this problem being $\tau_{\mathrm{n,low}}/t_{\mathrm{p},\Omega}$. [This is obvious if, for example, we set $n + ap_0 = n'$. The negative sign in the equivalent or pseudo-Poisson equation (14.116) derives from the placement of the cathode at $x = L$, the electrons then flowing from right to left.] The resulting regimes of the current-voltage characteristic can now be written out of hand.

At low voltages Ohm's law obtains:

$$J = e p_0 \mu_{\mathrm{p}} (V/L). \tag{14.118}$$

At the crossover voltage $V_x \simeq L^2/\mu_{\mathrm{n}}\tau_{\mathrm{n,low}}$, corresponding to $t_{\mathrm{n},x} \simeq L^2/\mu_{\mathrm{n}} V_x \simeq \tau_{\mathrm{n,low}}$, the current goes over into the square law

$$J = (9/8)(\tau_{\mathrm{n,low}}/t_{\mathrm{p},\Omega})\, \epsilon\mu_{\mathrm{n}} (V^2/L^3). \tag{14.119}$$

Equation (14.119) is, of course, the analog of (4.45). An important difference is that here the equivalent θ-factor, $\tau_{\mathrm{n,low}}/t_{\mathrm{p},\Omega}$ can be larger than unity, although it need not be.

The square law (14.119) was first derived in Ashley's Ph.D. thesis (1963) [see also Ashley and Milnes (1964)]. Note that it is a purely recombination-determined law; the Poisson equation plays no role in its derivation. The explicit appearance of ϵ in (14.119) is misleading; it is, of course, canceled by the ϵ contained in $t_{p,\Omega}$, as defined in (14.116).

Although the Poisson equation (14.111) has not entered the above discussion it is, as always, a valid equation and yields some interesting information. Substituting (14.116) into (14.111), and neglecting $p - p_0$, we obtain, using (14.115),

$$\delta p_R = \alpha n, \qquad \alpha = 1 - (t_{p,\Omega}/\tau_{n,low}). \tag{14.120}$$

There are two cases to consider:

I $(t_{p,\Omega} < \tau_{n,low})$:
Then $\delta p_R \simeq n$; that is, the injected electrons "come from" the recombination centers, much as in Region I of the problems of Sections 14.1 and 14.2.

II $(t_{p,\Omega} > \tau_{n,low})$:
Then $(\epsilon/e)(d\mathscr{E}/dx) \simeq \delta p_R$; that is, the injected negative charge lies mostly in the recombination centers, consistent with the equivalent θ-factor $\tau_{n,low}/t_{p,\Omega}$ being smaller than unity.

A problem that remains to be explored in detail is the transition from the square-law (14.119) to a negative-resistance regime. Ashley and Milnes' (1964) conjecture that this transition is adequately described by following the regime (14.119) up to its intersection with, and thereafter switching over to, the J–V characteristic obtained (say via the procedures of Section 14.2) neglecting p_0, and therefore g_p, in the defining equations (14.110)–(14.113). Although this is likely true under some circumstances, it is not obvious that this is a valid prescription under all circumstances of physical interest. Only a detailed examination of the problem can clarify this problem.

14.4 Filament Formation

Uniform current flow in a regime of current-controlled, bulk-distributed, negative resistance is unstable against filament formation, as pointed out by Ridley (1963). A pioneering study of filamentary conduction associated with current-controlled negative resistance was made by Barnett (1966) in his Ph.D. thesis at the Carnegie Institute of Technology. [See also Barnett and Milnes (1966).] Barnett's studies were made on semi-insulating p-type silicon at liquid nitrogen temperature, 77°K (sample resistivity was $\simeq 10^4\ \Omega$ cm, corresponding to $p_0 \simeq 2 \times 10^{10}$ cm^{-3}). The silicon was doped with indium, in a concentration of 6×10^{16} cm^{-3}, providing an acceptor level 0.16 eV above the valence band. The sample also contained a concentration $N_D - N_A \simeq 4 \times 10^{13}$ cm^{-3}, with N_D, N_A shallow donors and shallow acceptors, respectively. The filaments were studied photographically via the recombination radiation (at peak wavelengths of 1.06 and 1.2 μ) of the injected electrons and holes. A current-voltage characteristic obtained with a diode of cathode–anode spacing $L = 165\ \mu$ and equivalent cross-sectional radius $R = (A/\pi)^{1/2} = 440\ \mu$, with A the (rectangular) cross-section area, is shown in Fig. 14.7. The pre-"breakdown" Ohm's law and square-law are in accord with theory, Eqs. (14.118) and (14.119), respectively. Filament structure was seen at a current of 10 mA and

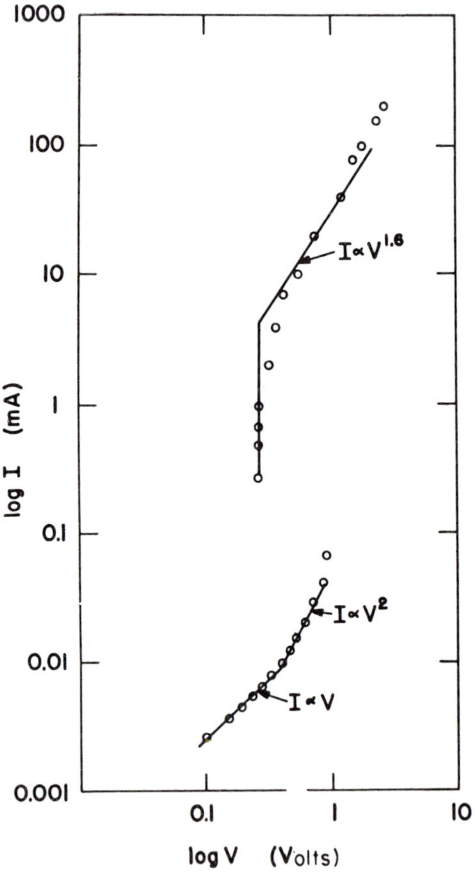

FIG. 14.7. Current-voltage characteristic for device #25-1 (length 1.65×10^{-2} cm, radius 4.4×10^{-2} cm). The smooth curves are based on theoretical considerations. (From Barnett and Milnes, 1966, Fig. 8.)

larger, exposure times on the order of half an hour being required because of the low efficiency of radiative recombination in silicon. Note then that all filament studies were made in the post-"breakdown" positive resistance regime. Sensitivity limitations prevented the study of filaments in the vertical-current regime immediately following "breakdown." Up to about 50 mA, a single filament was seen. The profile of the current density across the filament is exhibited in Fig. 14.8 at three different current values. Also shown are corresponding theoretical curves. The 10-mA curve was fitted to the experimental data at two points; no further adjustments were made in fitting the two upper curves. Above about 50 mA, a second filament is formed, as seen in the current-density profiles at 80 and 160 mA, shown in Fig. 14.9. The theoretical curves in

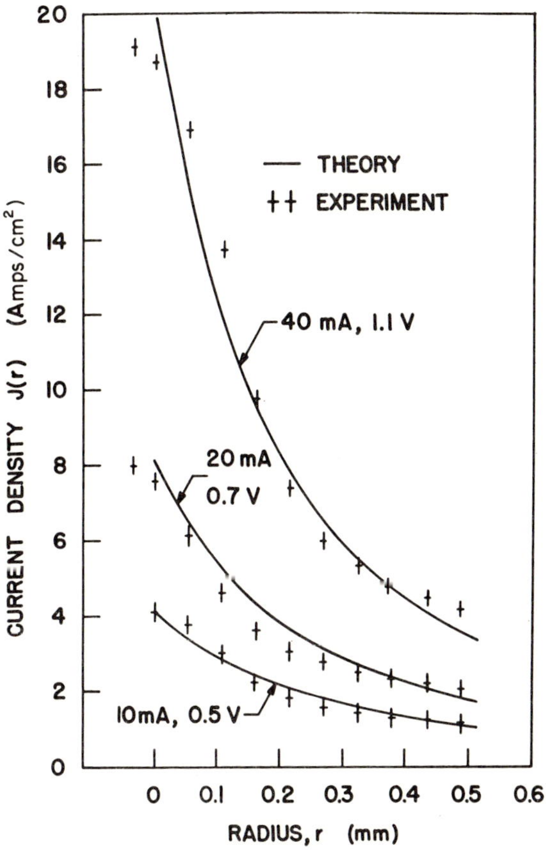

FIG. 14.8. Current density versus distance across the 10-, 20-, and 40-mA filaments of device #25-1. (From Barnett and Milnes, 1966, Fig. 5.)

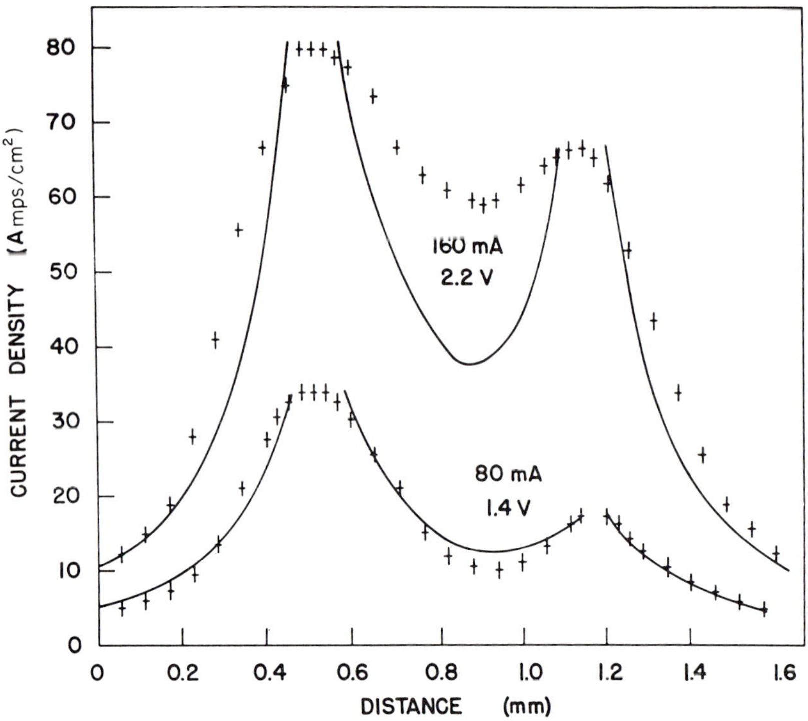

FIG. 14.9. Current density versus distance for the multiple filaments of device #25-1. (From Barnett and Milnes, 1966, Fig. 7.)

this figure are a linear superposition of single-filament density profiles. The discrepancy between theory and experiment at the higher current is likely due to an interaction between the filaments in their region of overlap. Generally, where filament formation was observed, the positive-resistance, post-"breakdown" regime in the current-voltage characteristic was of the form: $I \propto V^{\alpha}$ with $1.5 \leqslant \alpha \leqslant 1.9$.

Barnett's theoretical treatment of the current filament rests heavily on phenomenology. His starting point is the phenomenological equation (10.8) characterizing plane-parallel current flow in the insulator injected plasma, rewritten here as

$$\bar{n} = (8\epsilon\bar{\tau}\mu_{\rm n}\mu_{\rm p}V^2)/[e(\mu_{\rm n} + \mu_{\rm p})L^4]. \tag{14.121}$$

He further assumes that recombination everywhere throughout the filament is

$$\bar{\tau} = 1/F\bar{n}. \tag{14.122}$$

Inserting (14.122) into (14.121), we obtain

$$\bar{n} = \{(8\epsilon\mu_n\mu_p)/[eF(\mu_n + \mu_p)]\}^{1/2}(V/L^2). \tag{14.123}$$

Substitution of (14.123) into the current expression, $J = e\bar{n}(\mu_n + \mu_p)V/L$, yields

$$J = \{[8e\epsilon\mu_n\mu_p(\mu_n + \mu_p)]/F\}^{1/2}(V^2/L^3). \tag{14.124}$$

[Note that $\bar{n} \propto V/L^2$, $J \propto V^2/L^3$ also characterize the semiconductor injected plasma under monomolecular (constant) lifetime conditions, as seen in (10.19) and (10.20) or (12.7)].

Barnett now makes the bold assumption that the plane-parallel flow expressions (14.123) and (14.124) are valid at the center of the current filament, $\bar{n}$ now becoming $\bar{n}_0 = \bar{n}(0)$.

The radial carrier concentration is determined by the diffusion equation

$$D_a \nabla^2 n = n/\tau = Fn^2, \tag{14.125}$$

with $\nabla^2 = (d^2/dr^2) + (1/r)(d/dr)$ with D_a the ambipolar diffusion coefficient, given by (13.13). Barnett approximates the solution to this radial flow equation by

$$\bar{n}(r) \simeq \bar{n}_0/[1 + (\bar{n}_0 F/6D_a)^{1/2} r], \tag{14.126}$$

which is the *exact* solution to the identical *planar-flow*, diffusion equation, namely, taking $\nabla^2 = d^2/dr^2$. This approximation is convincingly argued in Appendix A of Barnett and Milnes (1966; see particularly Fig. 10).

Combining (14.126) and (14.123) with the relation,

$$J(r) = e\bar{n}(r)(\mu_n + \mu_p)\, V/L,$$

we obtain

$$J(r) = G/[(1 + dr)^2], \tag{14.127}$$

$$G = \left\{\frac{8e\epsilon\mu_n\mu_p(\mu_n + \mu_p)}{F}\right\}^{1/2} \frac{V^2}{L^3}; \qquad d = \left\{\frac{2\epsilon\mu_n\mu_p F}{e(\mu_n + \mu_p)}\right\}^{1/2} \left(\frac{V}{3D_a L^2}\right)^{1/2}.$$

Integration of (14.127) over the cross section of the sample yields for the total current I:

$$I = (2\pi G/d^2)\{\ln(1 + Rd) - [Rd/(1 + Rd)]\}. \tag{14.128}$$

Although (14.128) is obviously not an exact power law, it can be approximated by a power law over a limited range of voltages. The authors state that the power law approaches $I \propto V^{1.5}$ for $dR > 1.2$ and $I \propto V^{1.9}$ for $dR < 0.8$.

The vertical segment of the I–V characteristic immediately following "breakdown" is not explained by this work, although a successful estimate of the low current end I_M of this segment is made utilizing the measured V_M and estimating the hole density at the high-voltage breakdown point.

It is obvious that much more detailed theoretical work must be done to obtain insight into the filamentary conduction process. One might be optimistic about the type of approach followed in treating the two-dimensional semiconductor injected plasma, Section 12.4. The basic aim would be to decouple the longitudinal and transverse flows by suitable approximations. Presumably the equations then characterizing the (average) longitudinal flow would resemble those characterizing the corresponding planar-flow problems (Section 14.2 or 14.1) and could then be treated by the Regional Approximation Method. The equation characterizing the transverse flow would presumably be a pure diffusion equation. In a sense Barnett's theoretical treatment is an attempt to guess the outcome of such a decoupling theory.

Concomitantly with a more sophisticated theory, considerably more experimental work would be desirable. For example, the assumption of bimolecular recombination in the filament, Eq. (14.122), can be explored experimentally. In all probability, the filamentary conduction process is quite as complex as the Gunn domain propagation accompanying the voltage-controlled, bulk-distributed type of negative resistance observed, e.g., in n-GaAs [see Butcher's review article (1967)] and will require, in the long run, a comparable effort for its elucidation.

Interestingly enough, filamentary conduction was not observed in all samples exhibiting the negative resistance. In particular, it was not seen in diodes with L exceeding 220 μ. Also, in these larger diodes, the post-"breakdown" positive resistance regime followed closely the $I \propto V^2$ predicted by the planar-flow theory of Sections 14.1 and 14.2. [The latter was likewise the experience of Mayer *et al.* (1965b) in their study of long p-π-n Si structures ($L > 2$ mm) at $-40°$ C.] Where filamentary conduction is not present, the planar-flow theory of Sections 14.1 and 14.2 should prove directly useful.

14.5 Earlier Experimental Work

In the following discussion, we shall make no reference to double injection studies with forms of matter (for example, powders) other than single crystals because of the difficulties of interpretation associated with such studies. The reader interested in such studies is referred to the well-documented bibliographies of Ivey (1961) and Henisch (1962).

The earliest observations, known to the authors, that could be unambiguously associated with double injection with trapping are those of Tyler (1954) on iron-doped germanium at 77°K. The Ge was n-type, of high resistivity, with a fused In contact for hole injection and a fused Sn contact for electron injection. The cathode-anode separation was about 3 mm. The current-voltage characteristic was studied on an oscilloscope with 60 cps voltage applied to the sample. Below 50 V peak, forward injection currents on the order of microamperes were observed, corresponding to a resistivity of $\sim 10^{12}\ \Omega$ cm. At about 50 V peak, a "breakdown" effect was observed with the effective resistivity dropping to $\sim 10^{2}\ \Omega$ cm. This "breakdown" condition was maintained down to several volts peak. Irradiation of the diode with strongly absorbed light ($h\nu > E_G$) greatly reduced the "breakdown" initiation voltage. On the other hand, just above the minimum sustaining voltage, irradiation of the diode with longer-wavelength light ($h\nu$ in the range 0.35–0.7 eV) produced a quenching of the "breakdown."

Tyler's observations could all be accounted for qualitatively in terms of the simple negative resistance model discussed in Sections 14.1 and 14.2. Since the Ge was n-type, the Fermi level was high enough in the gap that the Fe centers were largely filled with electrons. Further, these centers being acceptor-like, their hole capture cross section at 77°K is large, $>10^{-14}$ cm². Thus the filled Fe centers present a recombination barrier to the passage of the injected holes. At about 50 V peak, $t_p = L^2/\mu_p V \simeq 10^{-7}$ sec, taking $\mu_p \approx 10^4$ cm²/V sec. Taking $\langle v\sigma_p \rangle \approx 10^{-7}$ cm³/sec, we note that if the density N_R of Fe centers (not given by Tyler) were 10^{14} cm^{-3}, then $\tau_{p,\mathrm{low}} \simeq 10^{-7}$ sec, the same magnitude as t_p. This suggests that the observed "breakdown" is the "overcoming" of the recombination barrier by the injected holes. Sustaining of the "breakdown" at lower voltages is no doubt due to a thermal release time for the holes at 77°K which is longer than 1/60 sec, so that the diode was unable to "recover" during the reverse-bias half-cycle. With the diode illuminated by light with $h\nu > E_G$, the free holes created by the light are captured by the Fe centers, thereby "lowering" the recombination barrier to injected holes, hence the lower "breakdown" voltage. On the other hand, irradiation in the band 0.35–0.7 eV excites captured holes out of the Fe centers back into the valence band, thereby quenching the breakdown when the voltage is just above the sustaining voltage. This same irradiation is known to quench the intrinsic photoconductivity of Fe-doped Ge, by precisely the same mechanism.

Lebedev *et al.* (1956), in experiments on Au-doped Ge at 77°K, observed spontaneous oscillations under dc applied voltage. These oscillations, no doubt, were associated with the negative resistance produced by the Au recombination centers under double injection. Other work with Ge is discussed below.

Cadmium sulfide is a material which has long been popular in the electroluminescence field. Further, double injection in single crystals of CdS has been studied by several groups. Unfortunately, a good hole-injecting contact is not available for this material; hole injection is a field-induced phenomenon. For this reason, the discussion of double injection in CdS is postponed until the following section, where it is more appropriate.

Holonyak and collaborators (1962a,b, 1963) opened up a considerable area for exploration in their studies, at room temperature, of double injection phenomena associated with deep-lying centers in GaAs, Si, and Ge. They observed the negative-resistance double-injection *I–V* characteristic associated with a hole lifetime which increases with injection level in high-resistivity (10^5–10^7 Ω cm) GaAs, in epitaxially grown GaAs p-si-n diodes with the si (semi-insulating) region Cu-doped, in high purity, high resistivity ($\sim 10^5$ Ω cm) Si doped with Au, in n-type Si counterdoped with Au, Zn, Cd, or Co, and in n-type Ge counterdoped with Cu, Fe, Ni, Co, or Mn. Their diodes were fabricated by alloying procedures, diffusion procedures, and combinations of the two. *I–V* characteristics exhibited on an oscilloscope screen are shown in Figs. 14.10 a–d, taken from Holonyak (1962a). Figure 14.10a shows the *I–V* characteristic for a GaAs diode under ambient light, curve 1, and under illumination with a microscope lamp, curve 2. The latter shows a slight lowering of the "breakdown" voltage, as in Tyler's Ge experiments. The highest V_{TH}/V_M value exhibited by their GaAs units was about 4. The pre-breakdown regime in curve 1 varied as V^2, as predicted by the Ashley theory, Section 14.3. The post-breakdown, high-current variation with voltage was not discussed. The *I–V* characteristic for another GaAs unit is shown in Fig. 14.10b. Here two negative-resistance regimes are clearly exhibited, indicating some complexity in the recombination processes. Figure 14.10c shows *I–V* characteristics for a Si unit under successively heavier illumination. Here the effect of illumination on lowering the "breakdown" threshold, *à la* Tyler, is very marked, the negative-resistance being almost eliminated under the heaviest irradiation, curve 4. The post-breakdwon current in this unit is independent of voltage, indicating that the high-injection-level diffusion length already exceeds the thickness of the active region. Their Si units also exhibited pre-breakdown currents varying as the square of the voltage. Finally, Fig. 14.10d shows *I–V* characteristics for a Ge diode under successively heavier illumination. The general features are similar to those of the Si unit shown in Fig. 14.10c except for a broader negative-

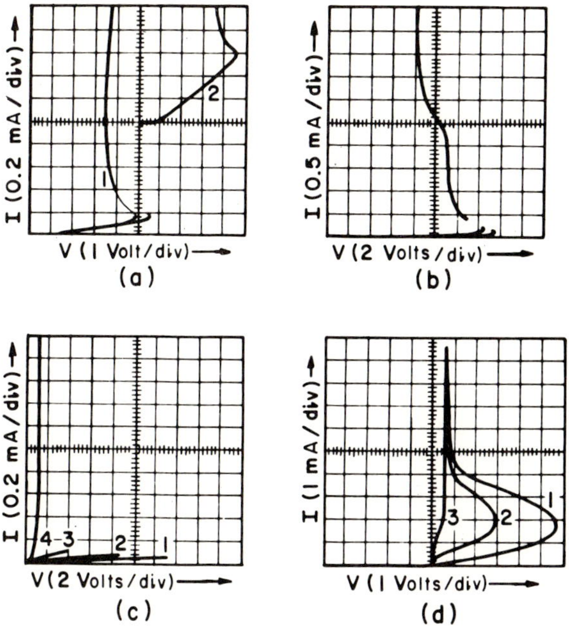

FIG. 14.10. Double-injection current-voltage characteristics (at room temperature) for PIN diodes fabricated out of several materials: (a) GaAs with Mn-diffused p^+ region, Sn-alloyed n^+ region: (1) diode in room light; (2) diode illuminated with microscope lamp; (b) GaAs with Zn-diffused p^+ region, Sn-alloyed n^+ region; (c) Si counterdoped with Co, Al–B alloyed p^+ region, Au–Sb alloyed n^+ region: (1) diode in room light; (2), (3), and (4) diode under successively heavier illumination; (d) Ge counterdoped with Cu: (1), (2), and (3) diode under successively heavier illumination. (From Holonyak, 1962a, Figs. 1, 3, 7, and 10.)

resistance transition. The possibility that some (for example, the GaAs results), and conceivably all, of these results are related to a light feedback mechanism is discussed in Section 14.6.2.

Ashley (1963) carried out an extensive study of double-injection negative resistance in Au-doped germanium. His diodes were fabricated with soldered Sn–In electron-injecting contacts and soldered In hole-injecting contacts. His sample cross section was 3×3 mm^2 and L varied from 0.1 to 3 mm for various diodes. The gold concentration was $N_R \simeq 5 \times 10^{14}$ cm^{-3} and there was a partially compensating, shallow-donor concentration $N_D \simeq 2 \times 10^{14}$ cm^{-3}. Thus $n_{R,0} \simeq 2 \times 10^{14}$ cm^{-3} and $p_{R,0} \simeq 3 \times 10^{14}$ cm^{-3} in his experiments. The $n_{R,0}$ centers were singly negatively charged, corresponding to the lowest gold acceptor level located 0.15 eV above the valence band, and correspondingly, the $p_{R,0}$ centers were neutral. The experiments were carried out at 77°K. The ambipolar diffusion length at high injection levels was $\simeq 0.1$ mm, corresponding to $\mu_n = \mu_p \simeq 2 \times 10^4$ cm^2/V sec and $\tau_h \simeq 10^{-6}$ sec (from $\langle v\sigma_n \rangle \simeq 2 \times 10^{-9}$ cm^3/sec and $N_R = 5 \times 10^{14}$ cm^{-3}). Ashley observed the Ohm's law, square-law, and negative resistance regimes in the I–V characteristic, as expected. (With some samples there was a supra-square-law transition region between the Ohm's law and square-law regimes which Ashley associated with a non-ideal electron-injecting contact and a relatively small-area hole-injecting contact.) No study was made of the V-dependence of I above the negative-resistance regime. The characteristic voltages V_{TH}, V_M, and V_{X_1} (the voltage of transition from Ohm's law to the pre-breakdown square law) all varied with L^2, as expected. From the observed V_{TH} and V_{X_1}, Ashley obtained $\sigma_p \simeq 0.5 \times 10^{-14}$ cm^2 and $\sigma_n \simeq 2 \times 10^{-16}$ cm^2 at 77°K. The latter value is consistent with the values in the literature, whereas the former value is somewhat low [although the values given in the literature show very great scatter, and the smallest, that cited by Vogl *et al.* (1961), is only a factor of about four larger].

A negative resistance observed by Ryuzan (1961) and Van Ruyven and Adriaens (1962) with plastically deformed germanium p–n junctions has been interpreted by the latter authors in terms of a single-recombination-center model as discussed in the previous sections. In the work of Van Ruyven and Adriaens, the p-type Ge was In-doped, 0.15 Ω cm resistivity, and the n-type Ge was Bi-doped, 0.8 Ω cm resistivity. The Ge bar, of dimensions $2 \times 3 \times 20$ mm^3 with the $\langle 111 \rangle$ direction perpendicular to the long axis, was twisted about its long axis in a hydrogen atmosphere at 600°C. In and In + 1% As contacts were provided after cooling. The intrinsic region in the resulting PIN structure has a width of about 0.2 mm. The resulting double-injection I–V characteristics at various temperatures are shown in Fig. 14.11. At very low currents Ohm's law is observed, going over into a square law and finally breaking over into a negative resistance. At the highest temperature, $T_1 = 100$°C, the negative resistance is absent because of thermal re-emission of captured carriers by the recombination centers.

A double-injection negative resistance was observed in InSb at 77°K by Melngailis and Rediker (1962a,b). They attribute the effect to an electron lifetime which increases from $\sim 10^{-9}$ sec at low injection levels to 10^{-6} to 10^{-7} sec at higher levels as electron trapping is overcome.

A very large negative-resistance (switching) effect associated with double injection was observed for Si at 4°K by Jonscher (1961). Here the double injection produces a transfer of electrons, initially in the acceptor states, back into donor states, so that thereafter neither acceptor nor donor states are effective as free-carrier traps.

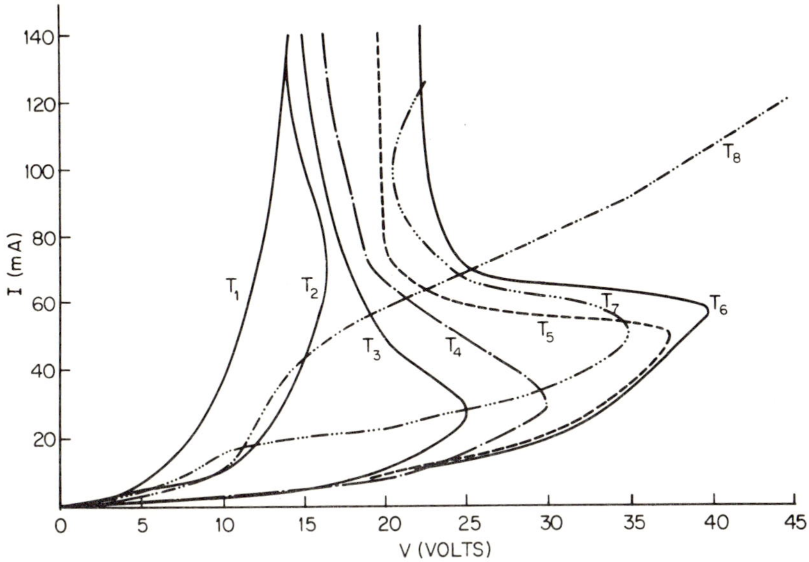

FIG. 14.11. Double-injection current-voltage characteristics for a Ge crystal with a plastically deformed p–n junction. The different curves correspond to different temperatures: $T_1 = +100°C$, $T_2 = +88°C$, $T_3 = +20°C$, $T_4 = 0°C$, $T_5 = -60°C$, $T_6 = -80°C$, $T_7 = -140°C$, and $T_8 = -196°C$. (From van Ruyven and Adriaens, 1962, Fig. 2.)

Double-injection current-voltage characteristics without negative resistance have been reported for SiC by Patrick (1957), for ZnS and ZnSe by Aven and Cusano (1964), and for $BaTiO_3$ by Cox and Tredgold (1963). Double injection electroluminescence in anthracene has been reported in an important series of papers by Helfrich and Schneider (1965, 1966) and by Dresner (1969).

14.6 Other Negative Resistance Mechanisms

The current-controlled negative resistance studied previously in this chapter is a pure bulk effect in that good injecting contacts are assumed for both signs of carrier. The origin of the negative resistance, in this case, is a free-carrier lifetime that increases markedly with injection level. We have seen that there are many known instances in which this mechanism is operative. However, this is not the only source of a current-controlled negative resistance associated with the presence of free carriers of both signs of charge. Another broad class of possibilities is that in which a contact, in one way or another, plays a key role in the negative resistance. We shall consider here two possibilities: first, that in which the injection of one of the carriers (the minority carrier, if the sample is a semiconductor) is initially suppressed by a noninjecting, or poorly injecting, contact and then promoted through a field-induced breakdown at, or in the vicinity of, the contact, and second, that in which recombination light emitted at a contact and reabsorbed in the bulk produces the negative resistance.

14.6.1 *Field-Induced Breakdown as a Source of Negative Resistance*

Suppose a sample of insulator or semiconductor has attached a pair of electrodes which are injecting for the majority carrier. Then the observed *I–V* characteristic, at "low" fields, will be a one-carrier characteristic, namely that for the majority carrier, as illustrated in Fig. 14.12, the dashed

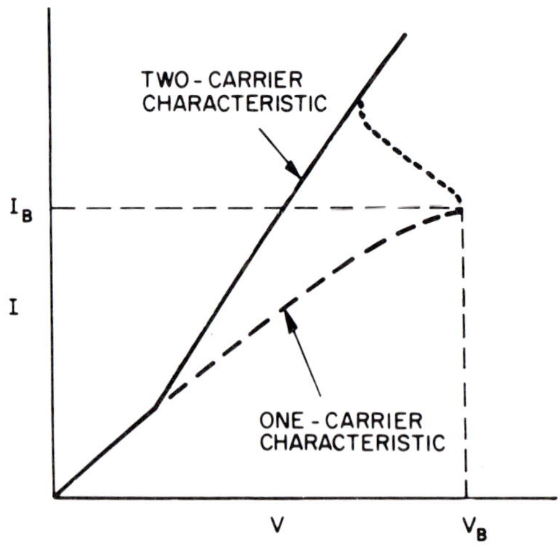

FIG. 14.12. Schematic current-voltage characteristics of a semiconductor for one-carrier injection (both contacts ohmic for the majority carrier) and for two-carrier injection (one contact injecting for electrons, the other for holes). I_B , V_B are the breakdown current and voltage, respectively, for the one-carrier characteristic.

line. In this particular example, the majority-carrier characteristic is Ohm's law, tending toward saturation at high fields, as might be found in Ge or Si at room temperature. The solid curve in Fig. 14.12 is the two-carrier characteristic that would be observed if injecting contacts were provided for both signs of carrier. In the particular example, the double-injection characteristic is that for the semiconductor injected plasma. In the present instance, this characteristic is suppressed, at "low" fields, by the unavailability of minority carriers. However, now assume that at the voltage V_B and current I_B there is a field-induced "breakdown" at, or in the vicinity of, the exiting electrode for the majority carrier, leading to the copious injection of minority carriers into the bulk. Such a "breakdown" might be the field-induced emission (tunneling) of minority carriers from the electrode into the bulk or it might be the field-induced avalanche breakdown across the bandgap in the neighborhood of the electrode. In either case, the bulk can now carry the current as a two-carrier current which requires less voltage than the same one-carrier current, a seen in Fig. 14.12. The drop in voltage across the bulk appears as a negative resistance. The actual path followed in going over from the one-carrier to the two-carrier characteristic, shown as a dotted curve in the figure, will depend on the external circuit.

An example of this kind of negative resistance is seen in Fig. 14.13, taken

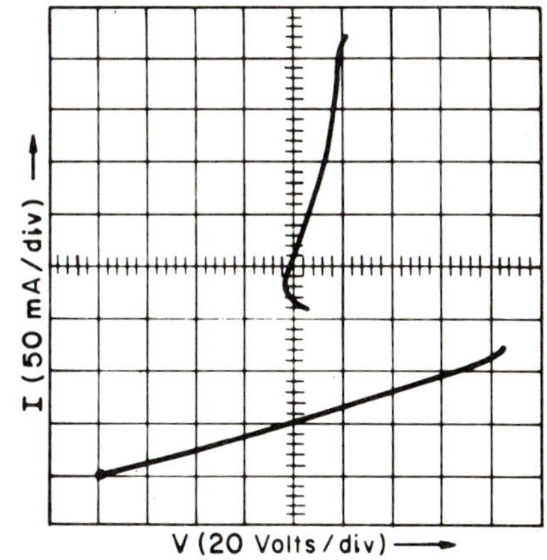

FIG. 14.13. Avalanche-breakdown, double-injection, negative-resistance current-voltage characteristic for a Ge crystal (20 Ω cm p-type). The cathode was a tungsten point contact (ohmic at low current levels) of diameter 0.03 cm; the anode was a broad area, ohmic contact. (From Steele *et al.*, 1962, Fig. 2.)

from Steele *et al.* (1962). The *I–V* characteristic exhibited here was obtained with a 20 Ω cm p-type Ge crystal with hole-injecting contacts. The crystal had a 1 mm × 1 mm square cross section and the cathode–anode spacing was 5 mm. The cathode contact was a tungsten point of 0.3 mm diameter. Because of near-saturation of the hole drift velocity, as the current increased, the crystal was finally driven into avalanche breakdown in the neighborhood of the point, in order to carry the current flowing in from the bulk. The avalanche breakdown provided a reservoir of electrons at the cathode contact which thereafter (at higher currents) functioned as a good electron emitter. That there was a drop in the voltage across the *bulk*, associated with the observed negative resistance, was directly established by potential measurements. The same effects were also studied by Ando *et al.* (1963) using Ge samples with a "bottleneck" geometry more accessible to potential probes. [Similar effects have been obtained with Ge and Si by Kikuchi (1963), who leaves open the interpretation, but argues that avalanche breakdown is not the explanation for his results.] In interpreting this type of negative resistance quantitatively, one must know not only the fraction of the total applied voltage needed to produce and sustain the "breakdown" but also whether the "breakdown" has produced the full equivalent of a reservoir-type, minority-carrier, injecting contact or only a partially good contact.

The avalanche breakdown in the above-described experiments was localized at, or near, an injecting contact through imposition of a geometric constraint, namely narrowing of the current flow path near that contact. This is by no means the only avalanche breakdown geometry leading to current-controlled negative resistance. Consider a rod of n-type InSb of *uniform* cross section and with injecting contacts for electrons. At fields on the order of several hundred volts per centimeter, the initially present electrons will "heat up" sufficiently to produce avalanching, an effect which the holes cannot produce because of the low fields. Since the applied field initially is uniformly distributed across the bulk, the avalanche breakdown will also be approximately uniform. However, the cathode, being injecting for electrons, is blocking for holes. Thus the holes created by the avalanche will pile up near the cathode, in a manner already studied at length in Sections 13.3.1 and 13.3.2. These holes will be neutralized by an equal pileup of electrons injected at the cathode. This plasma, created by the hole pileup, is a relatively high conductivity region and the field across it must be lower than the initial field in order that the total current have no divergence. If the extent of the plasma, namely a couple of ambipolar diffusion lengths, is a substantial fraction of the total cathode–anode spacing, there will be a significant drop in voltage across the semiconductor, i.e., a significant negative resistance. Effectively, the region of avalanching is driven toward the anode by the hole pileup at the cathode. Clearly the above argument presupposes a hole lifetime comparable to or larger than the hole transit time at the breakdown field.

Minority-carrier pileup at the majority-carrier injecting contact, together with a reasonable magnitude of ambipolar diffusion length, constitute a sufficient, but not necessary, condition for a negative resistance accompanying avalanche breakdown. (We are not considering here breakdowns geometrically constrained to occur at a contact, as discussed earlier.) For example, an avalanche-breakdown negative resistance has been seen by Tosima and Ando (1967) with a fork-shaped sample of n-InSb, illustrated in Fig. 14.14. The breakdown is initially

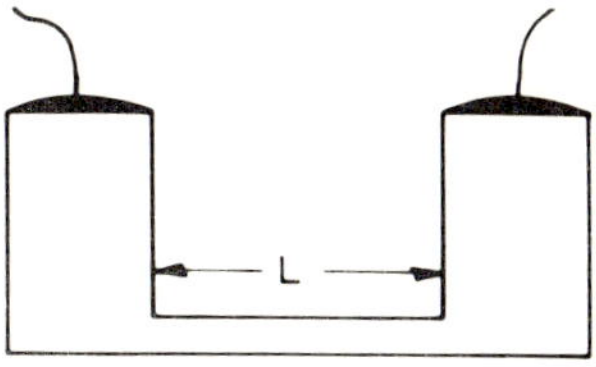

FIG. 14.14. Sample geometry which circumvents problem of free-carrier pile-up at blocking, exiting electrode. The "active" region of the sample is over the length L.

uniform over the active region L but then tends to be confined toward the anode end of this region concurrently with the negative resistance. Hole pileup obviously does not occur at the cathode end of the active region L and thus cannot explain the negative resistance. Tosima (1967) has explained the negative resistance in terms of the different field dependences of the electron and hole mobilities respectively, at high fields, the effect here being related to the reverse carrier-transport phenomenon of Prior (1960).

Negative resistance associated with avalanche breakdown in the depletion region of a back-biased p-n junction was first studied by Gunn (1956, 1957). More recent work is referenced by Hoefflinger (1966) who also presents models for negative resistance accompanying one-sided avalanche breakdown in the depletion layer, e.g., in $n^{++}–p^{+}–p^{++}$ structures and two-sided avalanche breakdown in the depletion layer, e.g., in $n^{++}–p^{+}–i–p^{++}$.

All of the above discussion refers solely to the dc negative resistance. There is a considerable literature on high-frequency, transit-time negative resistance, starting with Read (1958); the reader is referred to two special issues of the *IEEE Transactions on Electron Devices*, **ED-13**, No. 1, January 1966 and **ED-14**, No. 9, September 1967, for further studies in this area.

We turn now to a double-injection, negative resistance associated with a field-induced emission of holes from an anode, namely the $I–V$ characteristic shown in Fig. 14.15, taken from Smith's electroluminescence studies (1957) on CdS. This curve was obtained with a very pure, single crystal platelet with a cathode-anode spacing of 1 mm (current along the platelet) and with Ga electrodes. The characteristic exhibits both negative resistance and hysteresis. The lower dashed branch is a one-carrier, electron SCL characteristic corresponding to the use of electron-injecting contacts. That double injection is involved in the upper solid branch is confirmed by the observation of electroluminescence, with $h\nu \simeq E_G$ (the bandgap), distributed throughout the sample (Fig. 3 of Smith's paper). An interesting feature of the observed electroluminescent intensity I_L is its proportionality to the current. Because $h\nu$ is very close to the bandgap energy, we expect that $I_L \propto pn$, the product of the injected

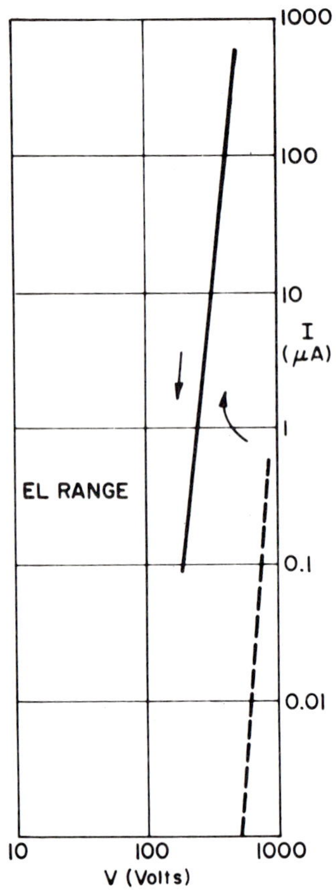

FIG. 14.15. Double-injection current-voltage characteristic for a CdS crystal with $L = 0.1$ cm. The dashed curve is obtained with current increasing, the solid curve with current decreasing. The overlap in current of the two branches constitutes a hysteresis. The transition from one branch to the other is discontinuous. Light emission is observed under operation on the upper branch of the characteristic, hence the designation EL (electroluminescent) Range. (From Smith, 1957, Fig. 2.)

free-carrier concentrations. Making the reasonable assumption that I (the electrical current) $\propto n$, then p cannot be proportional to n. A possible explanation for this lies in the assumption that the electrons are involved in a trap-filled-limit situation over the entire range of currents exhibited in Fig. 14.15, as indeed they appear to be over the one-carrier (dashed) branch. Since the corresponding range of voltage in Fig. 14.15 is only a factor of about three, the *total* number of injected electrons would then undergo the same relatively small change. The large change in n evidenced in the curve would then be due to the very fast-changing ratio of n to n_t, typical of the TFL problem as discussed in Sections 2.5 and 4.4. So long as the free holes are not involved in a similar TFL situation, p would vary by only the same small factor of about three and np would appear to vary with I over the limited voltage range. A theoretical model to explain all of the main features of Fig. 14.15 has not been given. For one thing, both the holes and electrons injected from the contacts are largely captured by defect states, the electrons by traps and the holes by recombination centers or traps or both. Theoretical models with that much complexity have not yet been studied. A second point is that it is not at all clear, and is perhaps unlikely, that the anode has been converted into the equivalent of a good reservoir contact for hole emission by the field-induced breakdown. Electroluminescent studies on CdS, similar to those of Smith, have been made by Litton and Reynolds (1964), Yamashita *et al.* (1963), and Keating (1963).

Another example in which field emission (tunneling) at a contact leads to negative resistance is the "thyristor" (Mueller and Hilibrand, 1958).

14.6.2 *Light Emission at a Contact as a Source of Negative Resistance*

In order to explain the negative-resistance *I–V* characteristic obtained by Weiser and Levitt (1964) [see also Weiser (1965)] with a Zn- and Mn-doped GaAs PIN diode, shown in Fig. 14.16, Dumke (1964) has pro-

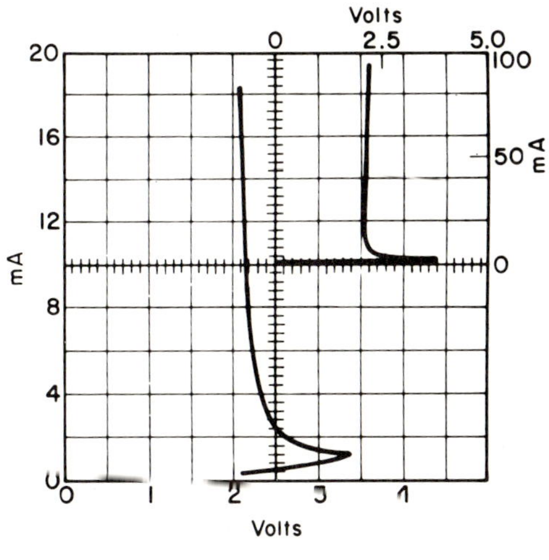

FIG. 14.16. Experimental current-voltage characteristic for Mn-doped GaAs PIN diode at 77°K. (From Weiser and Levitt, 1964, Fig. 1. See also, Weiser, 1965.)

posed a model involving light feedback. As applied to the GaAs experiment, this model has the light generated by recombination of electrons arriving at the *p*-region with the dominant, Zn-dopant atoms. Some of this light is absorbed by electron-filled Mn-dopant atoms in the *i*-region, exciting the electrons into the conduction band. If the light-generation rate were proportional to the current, this internal photoconductivity would appear as a steep rise in current at some critical voltage, i.e., a "breakdown." In the GaAs sample used, the light-generation rate is supralinear with current and consequently negative resistance accompanies the breakdown.

We present here the simple model developed by Dumke. In the pre-"breakdown" regime, the current, because of the presence of thermal holes, follows the square law (14.119) which we here write as

$$J = \beta V'^2; \qquad V' = (V - V_j), \quad \beta = (9/8)(\tau_{n,low}/t_{p,\Omega})\,\epsilon\mu_n\,, \tag{14.129}$$

where V_j is the voltage required to overcome the junction barrier. The electrons arriving at the p^+-region, and recombining there, produce photons with a quantum efficiency γ, i.e., at the rate $\gamma J/e$. A fraction ϕ of these photons are absorbed by the electron-filled, deep-lying acceptors in the *i*-region, the resulting electron excitation being to the conduction band. Assuming approximately uniform absorption of the light throughout the *i*-region, the rate of generation $g_{n,L}$ of photoelectrons is

$$g_{n,L} = n_L/\tau_{n,low} = \phi\gamma J/Le, \tag{14.130}$$

where n_L is the photoelectron concentration, $n_L/\tau_{n,low}$ is the rate of recombination, and we are assuming that the electron lifetime has its low-injection-level value. For the sake of simplicity, now assume that the photocurrent J_L is ohmic, $J_L = en_L\mu_n V'/L$, and can simply be added to the square-law injection current (14.129). The total current can then be written as

$$J = \beta V'^2 + [\tau_{n,low}\phi\gamma\mu_n(V'/L^2)]\,J. \tag{14.131}$$

Solving (14.131) for J, we get

$$J = \frac{\beta V'^2}{1 - \phi\gamma(\tau_{n,low}/t_n(V'))} \qquad \text{with} \quad t_n(V') = \frac{L^2}{\mu_n V'}\,. \tag{14.132}$$

Electrical "breakdown" occurs where $\phi\gamma\tau_{n,low}/t_n(V') \simeq 1$. For γ constant, independent of current, this condition leads to a steep current rise, but without negative resistance. However, if γ increases as the current increases, then the "breakdown" condition can be maintained at progressively lower voltages as it proceeds, and this corresponds to a negative resistance.

Now switch to dimensionless variables

$$j = J/\beta V_1'^2, \qquad v = V'/V_1', \qquad V_1' = L^2/(\phi\mu_n\tau_{n,low}), \tag{14.133}$$

where V_1' is the "breakdown" voltage for $\gamma = 1$. Equation (14.131) now becomes

$$j = v^2 + \gamma j v \tag{14.134}$$

with solution

$$v = -(\gamma j/2) + [(\gamma j/2)^2 + j]^{1/2}. \tag{14.135}$$

The j–v characteristics corresponding to (14.135) are plotted in Fig. 14.17 for quantum efficiences γ of the form: $\gamma = j^n/(j^n + \text{constant})$ for

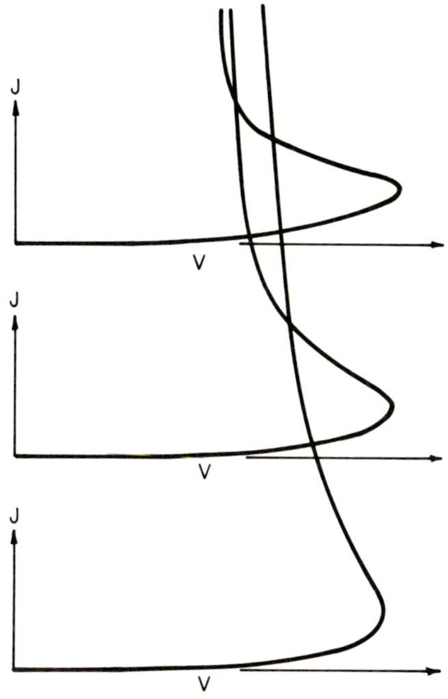

FIG. 14.17. Theoretical, dimensionless current-voltage characteristics for the PIN diode with radiative feedback. The radiative quantum efficience γ is expressed in the form $\gamma = j^n/(j^n + \text{const.})$; $n = 4, 2, 1$ (from top to bottom). (From Dumke, 1964, Fig. 3.)

$n = 1, 2$, and 4. It is seen that these theoretical characteristics strongly resemble the experimental characteristic exhibited in Fig. 14.16. That γ does indeed increase with current, as required by Dumke's theory, is seen in the experimental light output versus current plots taken with the same diodes (Fig. 2 of Dumke's paper). The effect of external illumination on the current-voltage characteristic has also been calculated by Dumke (his Fig. 4) and the results again strongly resemble the experimental data.

The question arises as to which of the other experimental observations of negative resistance noted in Section 14.5 might have been caused by Dumke's mechanism. Since formation of current filaments would be ruled out in this particular mechanism, observation of the filaments constitutes conclusive evidence against the mechanism. Likewise a post "breakdown" square law or near-square law is strong evidence against it. In other cases, the question is simply not settled, e.g., with Holonyak's (1962) data taken with Ge. As noted by Dumke, with long electron lifetimes, a very low quantum efficiency γ can still yield breakdown as a result of radiative feedback. An incisive choice of mechanism in these ambiguous situations must await the availability of quantitative measurements of the elusive quantities involved.

14.7 The Scaling Law

In Section 4.7 we derived the scaling law for the general, simplified theory of one-carrier, space-charge-limited currents. Here we derive the scaling law for the general, simplified theory of two-carrier injection currents. The equations characterizing this theory are:

current-flow equations:

$$J_\mathrm{n} = e\mu_\mathrm{n} n\mathscr{E}, \qquad J_\mathrm{p} = e\mu_\mathrm{p} p\mathscr{E}, \qquad J = J_\mathrm{n} + J_\mathrm{p} = \text{constant}, \tag{14.136}$$

Poisson equation:

$$(\epsilon/e)(d\mathscr{E}/dx) = (p - p_0) - \sum_j (n_{\mathrm{t}j} - n_{\mathrm{t}j,0}) - (n - n_0), \tag{14.137}$$

particle-conservation equations:

$$\mu_\mathrm{n}(d/dx)(n\mathscr{E}) = -\mu_\mathrm{p}(d/dx)(p\mathscr{E}) = r - g, \tag{14.138a}$$

$$r - g = \sum_j (r_j - g_j); \qquad r_j - g_j = r_{\mathrm{n},j} - g_{\mathrm{n},j} = r_{\mathrm{p},j} - g_{\mathrm{p},j}, \tag{14.138b}$$

$$\begin{aligned} r_{\mathrm{n},j} &= n\langle v\sigma_{\mathrm{n}j}\rangle p_{\mathrm{t}j}, \qquad g_{\mathrm{n},j} = n_{\mathrm{t}j}\langle e_{\mathrm{n}j}\rangle N_\mathrm{c}, \quad n_{\mathrm{t}j} + p_{\mathrm{t},j} = N_{\mathrm{t}j}, \\ r_{\mathrm{p},j} &= p\langle v\sigma_{\mathrm{p},j}\rangle n_{\mathrm{t}j}, \qquad g_{\mathrm{p},j} = p_{\mathrm{t}j}\langle e_{\mathrm{p}j}\rangle N_\mathrm{v}, \end{aligned} \tag{14.138c}$$

$$\langle e_{\mathrm{n}j}\rangle = \frac{1}{\bar{g}_{\mathrm{n}j}}\left(\exp\frac{E_{\mathrm{t}j} - E_\mathrm{c}}{kT}\right)\langle v\sigma_{\mathrm{n}j}\rangle, \qquad \langle e_{\mathrm{p}j}\rangle = \frac{1}{\bar{g}_{\mathrm{p}j}}\left(\exp\frac{E_v - E_{\mathrm{t}j}}{kT}\right)\langle v\sigma_{\mathrm{p}j}\rangle, \tag{14.138d}$$

with $\bar{g}_{\mathrm{n}j}$ and $\bar{g}_{\mathrm{p}j}$ the usual trap statistical weight factors and where we have neglected the direct, bimolecular recombination of n and p and correspondingly, the direct generation of pairs across the gap. (The scaling law is not changed if we include these terms.)

Equations (14.136)–(14.138) must be solved subject to the usual boundary conditions characterizing simplified theories:

$$\mathscr{E} = 0 \quad \text{at} \quad x = 0 \quad \text{and at} \quad x = L. \tag{14.139}$$

Equations (14.138) give directly

$$\frac{d}{dx}[(n - p)\mathscr{E}] = \frac{(n+1)(r-g)}{\mu_{\mathrm{p}}} = \frac{(n+1)}{\mu_{\mathrm{p}}} \sum_j \{n\langle v\sigma_{\mathrm{n}j}\rangle p_{\mathrm{t}j} - n_{\mathrm{t}j}\langle e_{\mathrm{n}j}\rangle N_{\mathrm{c}}\}. \tag{14.140}$$

Equation (14.136) can be written as

$$J = e\mu_{\mathrm{n}}\mathscr{E}(n + ap), \qquad a = \mu_{\mathrm{p}}/\mu_{\mathrm{n}}\,. \tag{14.141}$$

The $n_{\mathrm{t}j}$ and $p_{\mathrm{t}j}$ may be eliminated from the defining equations via the relations:

$$\begin{aligned} n_{\mathrm{t}j} &= \frac{(n\langle v\sigma_{\mathrm{n}j}\rangle + \langle e_{\mathrm{p}j}\rangle N_{\mathrm{v}})N_{\mathrm{t}j}}{p\langle v\sigma_{\mathrm{p}j}\rangle + n\langle v\sigma_{\mathrm{n}j}\rangle + \langle e_{\mathrm{n}j}\rangle N_{\mathrm{c}} + \langle e_{\mathrm{p}j}\rangle N_{\mathrm{v}}}, \\ p_{\mathrm{t}j} = N_{\mathrm{t}} - n_{\mathrm{t}j} &= \frac{(p\langle v\sigma_{\mathrm{p}j}\rangle + \langle e_{\mathrm{n}j}\rangle N_{\mathrm{c}})N_{\mathrm{t}j}}{p\langle v\sigma_{\mathrm{p}j}\rangle + n\langle v\sigma_{\mathrm{n}j}\rangle + \langle e_{\mathrm{n}j}\rangle N_{\mathrm{c}} + \langle e_{\mathrm{p}j}\rangle N_{\mathrm{v}}}. \end{aligned} \tag{14.142}$$

We now go over to dimensionless variables defined as follows:

$$\begin{aligned} &n = N_0 s, \qquad p = N_0 t, \qquad x = x_1 w, \qquad V(x) = V_1 v, \\ &x_1 = (\epsilon J/e^2 N_0^2 \mu_{\mathrm{n}}), \qquad V_1 = (\epsilon J^2/e^3 N_0^3 \mu_{\mathrm{n}}^2), \end{aligned} \tag{14.143}$$

where N_0 is an arbitrary concentration, chosen for convenience. (For example, N_0 might be n_0 or p_0 or some $N_{\mathrm{t}j}$, etc.)

With this choice of variables, Eq. (14.137) becomes, using (14.141),

$$-\frac{1}{(at+s)^2}\left(a\frac{dt}{dw} + \frac{ds}{dw}\right) = t - s + \frac{n_0 - p_0}{N_0} - \sum_j \left\{\frac{N_{\mathrm{t}j}}{N_0}\, \frac{sf_{\mathrm{nc},j} + f_{\mathrm{pe},j}}{sf_{\mathrm{nc},j} + tf_{\mathrm{pc},j} + f_{\mathrm{ne},j} + f_{\mathrm{pe},j}} - \frac{n_{\mathrm{t}j,0}}{N_0}\right\}, \tag{14.144}$$

and Eq. (14.140) becomes

$$\frac{d}{dw}\left[\frac{s-t}{s+at}\right] = \sum_j \frac{N_{\mathrm{t}j}}{N_0} \left\{\frac{s(tf_{\mathrm{pc},j} + f_{\mathrm{ne},j})(f_{\mathrm{nc},j}/f_{0,\mathrm{r}}) - (sf_{\mathrm{nc},j} + f_{\mathrm{pe},j})(f_{\mathrm{ne},j}/f_{0,\Omega})}{sf_{\mathrm{nc},j} + tf_{\mathrm{pc},j} + f_{\mathrm{ne},j} + f_{\mathrm{pe},j}}\right\}, \tag{14.145}$$

where the f's are all frequencies defined as follows:

$$f_{\mathrm{nc},j} = N_0\langle v\sigma_{\mathrm{n}j}\rangle, \qquad f_{\mathrm{pc},j} = N_0\langle v\sigma_{\mathrm{p}j}\rangle, \qquad f_{\mathrm{ne},j} = \langle e_{\mathrm{n}j}\rangle N_{\mathrm{c}}\,,$$
$$f_{\mathrm{pe},j} = \langle e_{\mathrm{p}j}\rangle N_{\mathrm{v}}\,, \qquad f_{0,\Omega} = eN_0\mu_{\mathrm{p}}/(a+1)\epsilon. \tag{14.146}$$

The potential integral, $V(x) = \int_0^x \mathscr{E}\, dx$, becomes

$$v = \int_0^w dw/(at+s) \qquad \text{and} \qquad v_{\mathrm{c}} = \int_0^{w_{\mathrm{c}}} dw/(at+s). \tag{14.147}$$

Equations (14.144) and (14.145) are a pair of coupled differential equations for s and t as functions of w. The important observation for the scaling law is that the current J no longer appears in these equations [likewise for (14.147)]; indeed, the dimensionless variables (14.143) were chosen precisely so as to bring about the elimination of J. With $t(w)$ and $s(w)$ determined in principle, Eq. (14.147) gives v_{c} as a function of w_{c} (in principle).

Since

$$\frac{1}{w_{\mathrm{c}}} = \frac{\epsilon}{e^2N_0{}^2\mu_{\mathrm{n}}}\frac{J}{L} \qquad \text{and} \qquad \frac{v_{\mathrm{c}}}{w_{\mathrm{c}}{}^2} = \frac{\epsilon}{eN_0}\frac{V}{L^2}\,, \tag{14.148}$$

the scaling law is, symbolically

$$J/L = f(V/L^2). \tag{14.149}$$

This is the identical scaling law as for the one-carrier case (4.158). Note that the one-carrier theory of Section 4.8 is contained in the above theory; we need only take $\mu_{\mathrm{p}} = 0$, $p = p_0 = 0$, and $r_{\mathrm{n},j} = g_{\mathrm{n},j}$ in the defining equations above, to recover the one-carrier theory.

APPENDIX A

NOTATION

- C Capacitance per unit area
- C_0 Geometric capacitance per unit area
- D Diffusion coefficient
- D_n Electron diffusion coefficient
- D_p Hole diffusion coefficient
- D_a Ambipolar diffusion coefficient
- E Energy or energy level
- E_c Lowest conduction-band energy level
- E_v Highest valence-band energy level
- $E_t(E_{tj}, E_{t,n}, E_{t,n,j})$ Trap energy level
- F Electron quasi-Fermi level
- F_0 Thermodynamic Fermi level
- I Current
- J Current density
- J_{di} Diffusion current density
- J_{dr} Drift current density
- J_{th} Temperature-limited, saturation current density
- L Cathode-anode spacing in a planar geometry
- L_D Ambipolar diffusion length
- N $N_c \exp\{(E_t - E_c)/kT\}$
- N_c Effective density of states in the conduction band
- N_v Effective density of states in the valence band
- $N_t(N_{tj}, N_{t,n}, N_{t,n,j})$ Trap concentrations
- N_j $N_c \exp\{(E_{tj} - E_c)/kT\}$
- N_D Donor concentration
- N_A Acceptor concentration
- N_R Recombination center concentration
- Q Total charge per unit area in planar-flow problems (also, total charge in radial-flow problems)
- Q_0 $\epsilon V/L$
- R_c Electron capture rate by traps
- R_e Electron emission rate from traps
- T Temperature
- T_t Characteristic temperature of exponential trap distribution
- V Voltage or potential
- V_T kT/e
- a μ_p/μ_n
- b μ_n/μ_p
- e Magnitude of electronic charge (1.602×10^{-19} C)
- $\langle e \rangle$ Band average of probability per unit time of thermal ejection of trapped electron into the conduction band
- e_0 Base for natural logarithms (2.718...)
- f, f_j Fractional change in electron trap occupancy when F moves from F_0 to $F_0 + kT$ (also, f is illumination intensity)
- f^* Attempt-to-escape frequency: $f^* = \langle v\sigma \rangle_0 N_c/g$
- g, g_j Degeneracy factor of trap state
- g_{th} Thermal ionization rate per unit volume
- h Plank's constant (6.62×10^{-34} J sec)
- j_{cond} Time-dependent, position-dependent conduction current density
- j_{displ} Time-dependent, position-dependent displacement current density
- k Boltzmann's constant (8.62×10^{-5} eV/deg, K)
- l T_t/T
- m Effective mass of electron or hole in solid
- m_0 Vacuum electron mass (9.105×10^{-31} kg)
- n Total free-electron concentration: $n = n_0 + n_i$

n_0	Thermally generated free-electron concentration
n_i	Injected free-electron concentration
n_t	Total trapped-electron concentration: $n_t = n_{t,0} + n_{t,i}$
$n_{t,0}$	Thermally generated trapped-electron concentration
$n_{t,i}$	Injected trapped-electron concentration
n_R	Occupied recombination center concentration
p	Total free-hole concentration
$p_{t,0}$	Thermally generated trapped-hole concentration
p_R	Empty recombination center concentration
r	Trapping or recombination rate per unit volume (also, radial position coordinate)
r_c	Cathode radius in spherical radial-flow problems
r_a	Anode radius in spherical radial-flow problems
s	Surface recombination velocity
t	Time
t_Ω	Ohmic relaxation time: $t_\Omega = \epsilon/en_0\mu$
t_0	Space-charge-free, interelectrode transit time: $t_0 = L^2/\mu V$
t_1	$0.786t_0$
$\bar{t}_a$	Ambipolar transit time: $\bar{t}_a = L^2/\lvert\mu_a\rvert V$
u	Dimensionless electric field intensity variable
v	Free carrier drift velocity (also, microscopic drift velocity when used inside angular brackets $\langle\ \rangle$ which denote averaging operation; also, dimensionless voltage or potential variable)
v_{th}	Thermal velocity of electrons: $v_{th} = (8kT/\pi m)^{1/2}$
w	Dimensionless position variable
x	Longitudinal position coordinate
y	Dimensionless dummy variable (also, transverse position coordinate
z	Dimensionless radial position variable: $z = (r/r_c)^3$
$\mathscr{E}$	Electric field intensity
$\mathscr{E}_\Omega$	Ohmic electric field intensity: V/L in planar-flow problems
$\mathscr{N}_t(E)$	Trap concentration per unit energy range
$\mathscr{N}_0$	Defined by $\mathscr{N}_t(E) = \mathscr{N}_0 \exp(E - E_c)/kT_t$
$\mathscr{N}_n$	$\mathscr{N}_0 \exp(F_0 - E_c)/kT_t$
Γ	$(N_t/n_0)(n_0/N_c)^{1/l}$ (used with exponential distribution of traps)
$\varPsi$	Metal-vacuum barrier (i.e., work function)
Ψ	Metal-insulator barrier
α	$\epsilon/e\mathscr{N}_n k$ (used with uniform distribution of traps)
δ	$Q/e\bar{n}L$
ϵ	Static dielectric constant
ϵ_0	Vacuum dielectric constant (8.854×10^{-12} F/m)
η	Fraction of injected charge remaining free: $\eta = n_i/(n_i + n_{t,i})$
θ	Ratio of free to shallow-trapped charge
κ	Relative dielectric constant: $\kappa = \epsilon/\epsilon_0$
λ_s	Mean range (i.e., Schubweg)
μ	Free-carrier drift mobility
μ_e	Effective drift mobility: $\mu_e = \theta\mu$
μ_R	Recombination mobility: $\mu_R = \epsilon\langle v\sigma_R\rangle/2e$
μ_a	Ambipolar drift mobility: $\mu_a = (n - p)/[(n/\mu_p) + (p/\mu_n)]$
ν	Frequency ($\sec^{-1}$)
ν_n	μ_n/μ_R
ν_p	μ_p/μ_R
ρ	Free charge concentration
ρ_t	Trapped charge concentration
σ	Capture cross section
σ_R	Recombination cross section
τ	Free-carrier lifetime
τ_0	Photoconductivity response time
τ_c	Electron capture time
τ_e	Electron emission time
τ_s	Mean time against electron capture (used in Schubweg analysis)
τ_n	Free-electron lifetime
τ_p	Free-hole lifetime
$\bar{\tau}$	Common average lifetime for injected electrons and holes
τ_h	Common lifetime of injected electrons and holes at high injection levels

$\langle\ \rangle$	Denotes that quantity inside the brackets is to be averaged over the free-carrier velocity distribution

Subscripts

m	extremum (maximum or minimum) of a quantity
t	quantity referring to traps
n	quantity referring to electrons
p	quantity referring to holes
G	quantity referring to band-gap
TFL	quantity evaluated at trap-filled-limit voltage
M	quantity evaluated at the low-voltage turn-around point in the double-injection negative resistance current-voltage characteristic
TH	quantity evaluated at the double-injection threshold voltage or the high-voltage turn-around point in the double-injection negative resistance current-voltage characteristic
X	quantity evaluated at the cross-over of two adjacent regimes
LF	"low field"
HF	"high field"
s	quantity evaluated under conditions of current saturation
a	position-dependent quantity evaluated at anode
c	position-dependent quantity evaluated at cathode
0	quantity evaluated at thermal equilibrium
cr, or cr, i	quantity (current or voltage) evaluated at the transition between two regimes in the current-voltage characteristic
l, or low	quantity evaluated at low injection levels
h, or high	quantity evaluated at high injection levels

Superscripts

—	average of a position-dependent quantity

APPENDIX B

CONVERSION TABLE FOR FORMULAS

	mks Equation	Practical Units
(2.7):	$J \simeq \epsilon\mu(V^2/L^3)$	$J \simeq (8.85\times 10^{-14})\kappa\mu(V^2/L^3)$ (A/cm²)
(2.10):	$J = en_0\mu(V/L)$	$J = (1.60\times 10^{-19})n_0\mu(V/L)$ (A/cm²)
(2.12):	$V_X \simeq en_0L^2/\epsilon$	$V_X \simeq (1.81\times 10^{-6})(n_0L^2/\kappa)$ (V)
(2.12):	$t_\Omega = \epsilon/en_0\mu$	$t_\Omega = (5.52\times 10^5)(\kappa/n_0\mu)$ (sec)
(2.13):	$Q = (\epsilon/L)V$	$Q = (8.85\times 10^{-14})(\kappa V/L)$ (C/cm²)
(2.27):	$V_X = ekT_t\mathscr{N}_nL^2/\epsilon$	$V_X = 1.56\times 10^{-10}(\mathscr{N}_nL^2T_t/\kappa)$ (V)
(2.29):	$J \simeq e\mu N_c(\epsilon/e\mathscr{N}_0kT_t)^l(V^{l+1}/L^{2l+1})$	$J \simeq 1.6\times 10^{-19}N_c\mu(6.4\times 10^9\kappa/\mathscr{N}_0T_t)^l$ $\times(V^{l+1}/L^{2l+1})$ (A/cm²)
(2.31):	$J \simeq e\mu n_0(V/L)\exp\alpha(V/TL^2)$ with $\alpha = \epsilon/e\mathscr{N}_nk$	$J \simeq 1.6\times 10^{-19}n_0\mu(V/L)\exp\alpha(V/TL^2)$ with $\alpha = 6.4\times 10^9\kappa/\mathscr{N}_n$ (A/cm²)
(2.37):	$J \simeq \epsilon\mu\mathscr{E}_1^{1/2}(V^{3/2}/L^{5/2})$	$J \simeq (8.85\times 10^{-14})\kappa\mu\mathscr{E}_1^{1/2}(V^{3/2}/L^{5/2})$ (A/cm²)
(2.39):	$J \simeq \epsilon v_s(V/L^2)$	$J \simeq (8.85\times 10^{-14})\kappa v_s(V/L^2)$ (A/cm²)
(3.25):	$Q = \epsilon\mathscr{E}_a$	$Q = (8.85\times 10^{-14})\kappa\mathscr{E}_a$ (C/cm²)
(3.28):	$\bar{n} = Jt/eL$	$\bar{n} = (6.23\times 10^{18})(Jt/L)$ (cm⁻³)
(4.11):	$\bar{n} = 3\epsilon V/2eL^2$	$\bar{n} = (8.29\times 10^5)(\kappa V/L^2)$ (cm⁻³)
(8.8):	$I = (3\pi/2)\epsilon\mu(V^2/r_a)$	$I = (4.17\times 10^{-13})(\kappa\mu V^2/r_a)$ (A)
(8.24):	$I = 4\pi e\mu n_0[3\epsilon/c_2e(n_0+n_{t,0})]^{1/2}$ $\times[V/(1+c_1)]^{3/2}$	$I = (2.59\times 10^{-15})\mu n_0[\kappa/c_2(n_0+n_{t,0})]^{1/2}$ $\times[V/(1+c_1)]^{3/2}$ (A)
(8.26):	$V_{cr} = r_c^2(1+c_1)^3c_2e(n_0+n_{t,0})/3\epsilon$	$V_{cr} = (6.04\times 10^{-7})$ $\times[r_c^2(1+c_1)^3c_2(n_0+n_{t,0})/\kappa$ (V)
(8.27):	$I_{cr} = [4\pi e^2\mu n_0c_2(1+c_1)^3$ $\times(n_0+n_{t,0})r_c^3]/3\epsilon$	$I_{cr} = (1.21\times 10^{-24})$ $\times\{[\mu n_0c_2(1+c_1)^3(n_0+n_{t,0})r_c^3]\}/\kappa$ (A)
(8.62):	$I' = 2\pi\epsilon\mu(V^2/r_a^2)$	$I' = (5.57\times 10^{-13})(\kappa\mu V^2/r_a^2)$ (A)
(11.14):	$J = (125/18)\epsilon\tau\mu_p\mu_n(V^3/L^5)$	$J = (6.15\times 10^{-13})(\kappa\tau\mu_p\mu_nV^3/L^5)$ (A/cm²)

	mks Equation	Practical Units
(11.29):	$\mu_R = \epsilon\langle v\sigma_R\rangle/2e$	$\mu_R = (2.76\times 10^5)\kappa\langle v\sigma_R\rangle$ (cm²/V sec)
(11.35):	$J = (9\pi/8)^{1/2}\,\epsilon\{\mu_p\mu_n(\mu_p+\mu_n)/\mu_R\}^{1/2} \times(V^2/L^3)$	$J = (1.66\times 10^{-13})\,\kappa[\mu_p\mu_n(\mu_p+\mu_n)/\mu_R]^{1/2} \times(V^2/L^3)$ (A/cm²)
(12.7):	$J = (9/8)\,e(n_0-p_0)\mu_n\mu_p\tau(V^2/L^3)$	$J = (1.80\times 10^{-19})(n_0-p_0)\mu_n\mu_p\tau(V^2/L^3)$ (A/cm²)
(12.19):	$J = (8/9)e\mu_p \times\{(b+1)\mu_n(n_0-p_0)/\langle v\sigma_R\rangle\}^{1/2} \times(V^{3/2}/L^2)$	$J = (1.42\times 10^{-19})\mu_p \times[(b+1)\mu_n(n_0-p_0)/\langle v\sigma_R\rangle]^{1/2} \times(V^{3/2}/L^2)$ (A/cm²)
(14.38):	$A = \epsilon\langle v\sigma_p\rangle/2ea\mu_p$	$A = (2.76\times 10^5)(\kappa\langle v\sigma_p\rangle/a\mu_p)$
(14.89):	$C = \{\epsilon\langle v\sigma_p\rangle p_{R,0}/ea^2\mu_p N_R\}^{1/2}$	$C = (7.44\times 10^2)[\kappa\langle v\sigma_p\rangle p_{R,0}/a^2\mu_p N_r]^{1/2}$
(14.89):	$D = \langle v\sigma_n\rangle \times N_R\{a^2\epsilon N_R/e\langle v\sigma_p\rangle n_{R,0}^2\,p_{R,0}\mu_p\}^{1/2}$	$D = (7.44\times 10^2)\langle v\sigma_n\rangle \times N_R[\kappa a^2 N_R/\langle v\sigma_p\rangle n_{R,0}^2\,p_{R,0}\mu_p]^{1/2}$
(14.96):	$E = \{\epsilon\langle v\sigma_p\rangle n_{R,0}^2/e\mu_p\,p_{R,0}N_R\}^{1/2}$	$E = (7.44\times 10^2) \times[\kappa\langle v\sigma_p\rangle n_{R,0}^2/\mu_p\,p_{R,0}N_R]^{1/2}$
(14.106a):	$V_{TH} = \{e\langle v\sigma_p\rangle L^4 n_{R,0}^2\,p_{R,0}/4\pi\epsilon\mu_p N_R\}^{1/2}$	$V_{TH} = (3.8\times 10^{-4}) \times[\langle v\sigma_p\rangle L^4 n_{R,0}^2 p_{R,0}/\mu_p N_R]^{1/2}$ (V)
(14.121):	$\bar{n} = 8\epsilon\bar{\tau}\mu_n\mu_p V^2/e(\mu_n+\mu_p)L^4$	$\bar{n} = (4.42\times 10^6)(\kappa\bar{\tau}\mu_n\mu_p V^2/(\mu_n+\mu_p)L^4)$ (cm⁻³)

In the above, right-hand formulas, the units of $\mathcal{N}_0$ and $\mathcal{N}_n$ are $\text{cm}^{-3}(\text{eV})^{-1}$ and the units of T_t and T are °K.

REFERENCES

Adams, A. R., and Spear, W. E. (1964). *J. Phys. Chem. Solids* **25**, 1113.

Adams, A. R., Gibson, D. J., and Spear, W. E. (1964). *Solid State Comm.* **2**, 387.

Adirovich, E. I. (1961). *Soviet Phys.—Solid. State* **2**, 1282.

Adolph, J., Baldinger, E., Czaja, W., and Gränacher, I. (1963). *Phys. Letters* **6**, 137.

Aiken, J. G., and Jordan, A. G. (1965). *J. Appl. Phys.* **36**, 3943.

Alfrey, G. F., and Cooke, I. (1957). *Proc. Phys. Soc. (London)* **B70**, 1096.

Allen, J. W., and Cherry, R. J. (1961). *Nature* **189**, 297.

Ancker-Johnson, B. (1966). *In* "Semiconductors and Semimetals" (R. K. Willardson and A. Beer, eds.), Vol. 1, p. 379. Academic Press, New York.

Anderson, W. J., and Meltzer, B. (1961). *J. Electron. Control.* **11**, 111.

Ando, K., Steele, M. C., and Lampert, M. A. (1963). *J. Phys. Soc. Japan* **18**, 591.

Arizumi, T., Umeno, M., and Horiba, Y. (1965). *J. Phys. Soc. Japan* **20**, 1901.

Ashley, K. L. (1963). PhD. thesis: Investigation of the Effects of Space Charge on the Conduction Mechanism of Double Injection in Semi-Insulators, Carnegie Inst. Technol., Dept. of Elec. Eng.

Ashley, K. L., and Milnes, A. G. (1964). *J. Appl. Phys.* **35**, 369.

Aven, M., and Cusano, D. A. (1964). *J. Appl. Phys.* **35**, 606.

Barnett, A. M. (1966). PhD thesis, Filamentary Injection Currents in Semi-Insulating Silicon, Carnegie Inst. of Technol., Dept. of Elec. Eng.

Barnett, A. M., and Milnes, A. G. (1966). *J. Appl. Phys.* **37**, 4215.

Baron, R. (1965). *Phys. Rev.* **137**, A272.

Baron, R. (1968). *J. Appl. Phys.* **39**, 1435.

Baron, R., Mayer, J. W., and Marsh, O. J. (1964). *Appl. Phys. Letters* **4**, 65.

Baron, R., Nicolet, M. A., and Rodriguez, V. (1966). *J. Appl. Phys.* **37**, 4156.

Bässler, H., Becker, G., and Riehl, N. (1966). *Phys. Status Solidi* **15**, 347.

Blakemore, J. S. (1962). "Semiconductor Statistics." Pergamon Press, Oxford.

Bloch, F. (1928). *Z. Phys.* **52**, 555.

Böer, K. W., and Kümmel, U. (1952). *Z. Phys. Chem.* **200**, 180.

Böer, K. W., and Kümmel, U. (1954a). *Z. Naturforsch.* **9a**, 117.

Böer, K. W., and Kümmel, U. (1954b). *Ann. Phys. (Leipzig)* **14**, 341.

Böer, K. W., and Kümmel, U. (1957). *Ann. Phys. (Leipzig)* **20**, 303.

Böer, K. W., and Kümmel, U. (1958). *Z. Naturforsch.* **16a**, 678.

Bok, J., Sohn, J. C., Zylbersztein, A. (1961). *Proc. Intern. Conf. Semiconductor Phys. 5th, Prague*, p. 138.

Bowlt, C. (1962). *Proc. Phys. Soc. (London)* **B80**, 810.

Branwood, A., and Tredgold, R. H. (1960). *Proc. Phys. Soc. (London)* **B76**, 93.

Branwood, A., Hughes, O. H., Hurd, J. D., and Tredgold, R. H. (1962). *Proc. Phys. Soc. (London)* **B79**, 1161.

Bräunlich, P. (1967). *Phys. Status Solidi* **21**, 383.

Broser, I., and Broser-Warminsky, R. (1958). *J. Phys. Chem. Solids* **6**, 386.

Brown, F. C. (1955). *Phys. Rev.* **97**, 355.

Bube, R. H. (1960). "Photoconductivity of Solids." Wiley, New York.

Bube, R. H. (1961). *J. Appl. Phys.* **32**, 1621.
Bube, R. H. (1962). *J. Appl. Phys.* **33**, 1733.
Büchner, W., and Mehl, W. (1965). *Z. Phys. Chem.* **69**, 376.
Butcher, P. N. (1967). *Rept. Progr. Phys.* **30**, Part 1, 97.
Cardon, F. (1961). *Physica* **27**, 841.
Chynoweth, A. G. (1968). *In* "Semiconductors and Semimetals" (R. K. Willardson and A. Beer, eds.), Vol. 4, p. 263. Academic Press, New York.
Conwell, E. M. (1967). "High Field Transport in Semiconductors." Academic Press, New York.
Cox, G. A., and Tredgold, R. H. (1963). *Phys. Letters* **4**, 199.
Dacey, G. C. (1953). *Phys. Rev.* **90**, 759.
Delacote, G., Filliard, J. P., and Schott, M. (1964). *Solid State Commun.* **2**, 373.
Denda, S., and Nicolet, M. A. (1966). *J. Appl. Phys.* **37**, 2412.
Dresner, J. (1969). *RCA Rev.* **30**, 322.
Dresner, J., and Shallcross, F. V. (1962). *Solid State Electron.* **5**, 305.
Driedonks, F., and Zijlstra, R. J. J. (1966). *Phys. Letters* **23**, 527.
Dumke, W. P. (1964). *Proc. Intern. Conf. Semiconductors, 7th* p. 611. Dumod, Paris.
Edwards, J., Kawabe, K., Stevens, G., and Tredgold, R. H. (1965). *Solid State Commun.* **3**, 99.
Engelhardt, H., and Riehl, N. (1965). *Phys. Letters* **14**, 20.
English, A. C., and Drews, R. E. (1963). *Sci. Electron.* **9**, 1.
Fan, H. Y. (1948). *Phys. Rev.* **74**, 1505.
Fassett, J. R., and Van Vliet, K. M. (1962). *Proc. Intern. Conf. Semiconductor Phys., 6th, Exeter* p. 886.
Fischer, A. G. (1961). *Solid State Electron.* **2**, 232.
Forgue, S. V., Goodrich, R. R., and Cope, A. D. (1951). *RCA Rev.* **12**, 335.
Franks, J., and Keating, P. N. (1961). *J. Phys. Chem. Solids* **22**, 25.
Franz, W. (1956). *In* "Handbuch der Physik." Vol. 17, p. 1550. Springer, Berlin.
Gilleo, M. A. (1953). *Phys. Rev.* **91**, 534.
Good, R. H. Jr., and Müller, W. (1956). *In* "Handbuch der Physik," Vol. 21. Springer, Berlin.
Goodman, A. (1966). *Phys. Rev.* **144**, 588; **152**, 780, 785.
Gränacher, I. (1965). *Solid State Commun.* **3**, 331.
Greener, E. H., and Whitmore, D. M. (1961). *J. Appl. Phys.* **32**, 1320.
Gregory, B. L., and Jordan, A. G. (1964). *Phys. Rev.* **134**, A1378.
Guerst, J. A. (1966). *Phys. Status Solidi* **15**, 107.
Gunn, J. B. (1956). *Proc. Phys. Soc. (London).* **B69**, 781.
Gunn, J. B. (1957). *Progr. Semiconductors* **2**, 213.
Harnik, E. (1965). *J. Appl. Phys.* **36**, 3850.
Hartke, J. L. (1962). *Phys. Rev.* **125**, 1177.
Haynes, J. R., and Shockley, W. (1949). *Phys. Rev.* **75**, 691.
Hecht, K. (1932). *Z. Phys.* **77**, 235.
Heilmeier, G. H., and Warfield, G. (1963). *J. Chem. Phys.* **38**, 163.
Heilmeier, G. H., and Heyman, P. M. (1967). *Phys. Rev. Letters* **18**, 583.
Helfrich, W. (1964). *Phys. Status Solidi* **7**, 863.
Helfrich, W. (1968). Space-Charge-Limited and Volume Controlled Currents in Organic Solids, *In* "Physics and Chemistry of the Organic Solid State." Vol. 3. Wiley, New York.
Helfrich, W., and Mark, P. (1962a). *Z. Phys.* **166**, 370.
Helfrich, W., and Mark, P. (1962b). *Z. Physik* **168**, 495.
Helfrich, W., and Mark, P. (1963). *Z. Phys.* **171**, 527.

Helfrich, W., and Lipsett, F. R. (1965). *J. Chem. Phys.* **43**, 4368.
Helfrich, W., and Schneider, W. G. (1965). *Phys. Rev. Letters* **14**, 229.
Helfrich, W., and Schneider, W. G. (1966). *J. Chem. Phys.* **44**, 2902.
Henisch, H. K. (1962). "Electroluminescence." Macmillan, New York.
Hinotani, K., and Sugigami, M. (1965). *Japan. J. Appl. Phys.* **4**, 731.
Hirota, R., Tosima, S. and Lampert, M. A. (1963). *J. Phys. Soc. Japan* **18**, 535.
Hoefflinger, B. (1966). *IEEE Trans.* **ED-13**, 151.
Hoestery, D. C., and Letson, G. M. (1963). *J. Phys. Chem. Solids* **24**, 1609.
Hofstein, S. R. (1967a). *Appl. Phys. Letters* **10**, 291.
Hofstein, S. R. (1967b). *Appl. Phys. Letters* **11**, 95.
Holonyak, N. Jr. (1962a). *Proc. IRE* **50**, 2421.
Holonyak, N. Jr., Ing, S. W. Jr., Thomas, R. C., and Bevacqua, S. F. (1962b). *Phys. Rev. Letters.* **8**, 426.
Holonyak, N. Jr., and Bevacqua, S. F. (1963). *Appl. Phys. Letters* **2**, 71.
Ioffe, A. F. (1960). "Physics of Semiconductors." Academic Press, New York.
Ivey, H. F. (1961). *J. Electrochem. Soc.* **180**, 590.
Jansen, P., Helfrich, W., and Riehl, N. (1964). *Phys. Status Solidi* **7**, 851.
Jonscher, A. K. (1961). *Brit. J. Appl. Phys.* **12**, 363.
Kallman, H. P., and Pope, M. (1960a). *Nature* **186**, 31.
Kallman, H. P., and Pope, M. (1960b). *J. Chem. Phys.* **32**, 300.
Känzig, W. (1955). *Phys. Rev.* **98**, 549.
Keating, P. N. (1963). *J. Phys. Chem. Solids* **24**, 1101.
Keating, P. N. (1964). *Phys. Rev.* **135**, A1407.
Kepler, R. G. (1960). *Phys. Rev.* **119**, 1226.
Kepler, R. G. (1962). *In* "Organic Semiconductors." Macmillan, New York.
Kikuchi, M. (1963). *Japan. J. Appl. Phys.* **1**, 31.
Koenig, S. H., and Gunther-Mohr, G. R. (1957). *J. Phys. Chem. Solids* **2**, 268.
Lampert, M. A. (1956). *Phys. Rev.* **103**, 1648.
Lampert, M. A. (1958). *J. Appl. Phys.* **29**, 1082.
Lampert, M. A. (1959). *RCA Rev.* **20**, 682.
Lampert, M. A. (1962). *Phys. Rev.* **125**, 126.
Lampert, M. A. (1964). *Rept. Progr. Phys.* **27**, 329.
Lampert, M. A. and Rose, A. (1961). *Phys. Rev.* **121**, 26.
Lampert, M. A., and Edelman, F. (1964). *J. Appl. Phys.* **35**, 2971.
Lampert, M. A., and Schilling, R. B. (1970). Current injection in solids: the Regional Approximation Method, *In* "Semiconductors and Semimetals" (R. K. Willardson and A. Beer, eds.), Vol. 6. Academic Press, New York.
Lampert, M. A., Rose, A., and Smith, R. W. (1959). *J. Phys. Chem. Solids* **8**, 464.
Lampert, M. A., Many, A., and Mark, P. (1964). *Phys. Rev.* **135**, A1444.
Landsberg, P. T. (1951). *Z. Phys. Chem.* **198**, 75.
Langmuir, I. (1929). *Phys. Rev.* **33**, 954.
Lanyon, H. P. D. (1963). *Phys. Rev.* **130**, 134.
Lanyon, H. P. D., and Spear, W. E. (1961). *Proc. Phys. Soc. (London)* **B77**, 1157.
Larrabee, R. D. (1961). *Phys. Rev.* **121**, 317.
Lax, B., and Mavroides, J. G. (1955). *Phys. Rev.* **100**, 1650.
Lebedev, A. A., Stafeev, V. I., and Tuchkevich, V. M. 1956, *Soviet Phys.-Tech. Phys.* **1**, 2071.
Lefkowitz, I. (1963). *Nature* **198**, 657.
Lemke, H. (1963). *Phys. Status Solidi* **3**, 2083.
Lemke, H., 1966. *Phys. Status Solidi* **16**, 427.

Lemke, H., and Müller, G. O. (1967). *Phys. Status Solidi* **24**, 127.

Lindmayer, J., Reynolds, J., and Wrigley, C. (1963). *J. Appl. Phys.* **34**, 809.

Litton, C. W., and Reynolds, D. C. (1964). *Phys. Rev.* **133**, A536.

MacDonald, J. R. (1962). *Solid State Electron.* **5**, 11.

Many, A. (1965). *J. Phys. Chem. Solids* **26**, 575.

Many, A., and Rakavy, G. (1962). *Phys. Rev.* **126**, 1980.

Many, A., Simhony, M., Weisz, S. Z., and Levinson, J. (1961). *J. Phys. Chem. Solids* **22**, 285.

Many, A., Weisz, S. Z., and Simhony, M. (1962). *Phys. Rev.* **126**, 1989.

Many, A., Simhony, M., Weisz, S. Z., and Toucher, Y. (1964). *J. Phys. Chem. Solids* **25**, 721.

Many, A., Simhony, M., and Grushvitz, Y. (1965). *J. Phys. Chem. Solids* **26**, 1925.

Mark, P. (1961). *Z. Naturforsch.* **169**, 950.

Mark, P. (1965). *Phys. Rev.* **137**, A203.

Mark, P., and Helfrich, W. (1962). *J. Appl. Phys.* **33**, 205.

Marlor, G. A., and Woods, J. (1965). *Brit. J. Appl. Phys.* **16**, 1449.

Marsh, O. J., Mayer, J. W. and Baron, R. (1965). *Appl. Phys. Letters* **5**, 74.

Mayer, J. W., Baron, R., and Marsh, O. J. (1965a). *Phys. Rev.* **137**, A286.

Mayer, J. W., Baron, R., and Marsh, O. J. (1965b). *Appl. Phys. Letters* **6**, 38.

McWhorter, A. L., and Rediker, R. H. (1959). *Proc. IRE* **47**, 1207.

McWhorter, A. L., and Rediker, R. H. (1961). *Proc. Intern. Conf. Semiconductor Phys. 5th, Prague* p. 134.

Mehl, W., and Funk, B. (1967). *Phys. Letters* **25A**, 364.

Mehl, W., and Hale, J. M. (1967). *Advan. Electrochem. Electrochem. Eng.* **6**, 399.

Melngailis, I., and Rediker, R. H. (1962a). *J. Appl. Phys.* **33**, 1892.

Melngailis, I., and Rediker, R. H. (1962b). *Proc. IRE* **50**, 2428.

Meltzer, B. (1960). *J. Electron. Control* **8**, 671.

Moore, J. S. (1967). PhD thesis. Some Properties of Silicon PIN Diodes Compensated with Gold. Elec. Eng. Dept., Univ. of Illinois, Urbana, Illinois.

Moore, J. S., Penchina, C. M., Holonyak, N., Sirkis, M. D., and Yamada, T. (1966). *J. Appl. Phys.* **37**, 2009.

Moore, J. S., Holonyak, N., Sirkis, M. D., and Blouke, M. M. (1967a). *Appl. Phys. Letters* **10**, 58.

Moore, J. S., Holonyak, N., and Sirkis, M. D. (1967b). *Solid State Electron.* **10**, 823.

Mort, J., and Spear, W. E. (1962). *Phys. Rev. Letters* **8**, 314.

Mott, N. F., and Gurney, R. W. (1940). "Electronic Processes in Ionic Crystals" Oxford Univ. Press (Clarendon), London and New York.

Mueller, C. W. and Hilibrand, J. (1958). *IRE Trans. Electron. Dev.* **ED-5**, 2.

Müller, G. O. (1962). *J. Phys. Chem. Solids* **23**, 1503.

Niekisch, E. A. (1961). *Z. Phys. Chem.* **217**, 110.

Okazaki, S., and Hiromatsu, M. (1967). *Solid State Commun.* **5**, 475.

Ozarow, V., and Hysell, R. E. (1962). *J. Appl. Phys.* **33**, 3013.

Page, D. J. (1966). *Solid State Electron.* **9**, 255.

Page, D. J., Kayali, A. A., and Wright, G. T. (1962). *Proc. Phys. Soc. (London)* **B80**, 1133.

Papadakis, A. C. (1967). *J. Phys. Chem. Solids* **28**, 641.

Parmenter, R. H., and Ruppel, W. (1959). *J. Appl. Phys.* **30**, 1548.

Patrick, L. (1957). *J. Appl. Phys.* **28**, 765.

Pilkuhn, M. H. (1968). *Phys. Status Solidi* **25**, 9.

Polke, M. (1964). *Phys. Status Solidi.* **5**, 279.

Polke, M., Stuke, J., and Vinaricky, E. (1963). *Phys. Status Solidi* **3**, 1885.
Pratt, R. G., and Ridley, B. K. (1963). *Proc. Phys. Soc. (London)* **81**, 996.
Pratt, R. G., and Ridley, B. K. (1965a). *Proc. Phys. Soc. (London)* **85**, 293.
Pratt, R. G., and Ridley, B. K. (1965b). *J. Phys. Chem. Solids* **26**, 11.
Prior, A. C. (1960). *Proc. Phys. Soc. (London)* **76**, 465.
Rashba, E. I., and Tolpygo, K. B. (1955). *J. Tech. Phys. USSR* **25**, 1335.
Read, W. T. Jr. (1958). *Bell System Tech. J.* **37**, 401.
Redington, R. W. (1958). *J. Appl. Phys.* **29**, 189.
Ridley, B. K. (1963). *Proc. Phys. Soc. (London)* **82**, 954.
Ridley, B. K., and Pratt, R. G. (1963). *Phys. Letters* **4**, 300.
Ridley, B. K. and Pratt, R. G. (1965). *J. Phys. Chem. Solids* **26**, 21.
Roberts, G. G. and Tredgold, R. H. (1962). *Phys. Letters* **2**, 6.
Robinson, B. (1969). (To be published).
Rose, A. (1951). *RCA Rev.* **12**, 362.
Rose, A. (1955). *Phys. Rev.* **97**, 1538.
Rose, A. (1957a). *Helv. Phys. Acta* **30**, 242.
Rose, A. (1957b). *Progr. Semiconductors* **2**, 111.
Rose, A. (1963). "Concepts in Photoconductivity and Allied Problems." Wiley, New York.
Rose, A., and Lampert, M. A. (1959a). *Phys. Rev.* **113**, 1227.
Rose, A., and Lampert, M. A. (1959b). *RCA Rev.* **20**, 57.
Rosenberg, L., and Lampert, M. A. (1969). *J. Appl. Phys.* (to be published).
Ruppel, W. (1958). *Helv. Phys. Acta* **31**, 311.
Ruppel, W. (1961a). *Proc. Intern. Conf. Semiconductor Phys., 5th, Prague* p. 1057.
Ruppel, W. (1961b). *J. Phys. Chem. Solids* **22**, 199.
Ruppel, W., and Smith, R. W. (1959). *RCA Rev.* **20**, 702.
Russ, M. J. (1963). *J. Appl. Phys.* **34**, 1831.
Ryuzan, O. (1961). *J. Phys. Soc. Japan.* **16**, 2177.
Schilling, R. B., and Lampert, M. A. (1970). *J. Appl. Phys.* (to be published).
Schilling, R. B., and Schachter, H. (1967a). *J. Appl. Phys.* **38**, 841.
Schilling, R. B., and Schachter, H. (1967b). *Solid State Electron.* **10**, 689.
Schilling, R. B., and Schachter, H. (1967c). *J. Appl. Phys.* **38**, 1643.
Schumann, W. O. (1933). *Z. Tech. Phys.* **14**, 23.
Schwartz, L. M., and Hornig, J. F. (1965). *J. Phys. Chem. Solids* **26**, 1821.
Sergiescu, V. (1965). *Brit. J. Appl. Phys.* **16**, 1435.
Shao, J., and Wright, G. T. (1961). *Solid State Electron.* **3**, 291.
Shockley, W. (1950). "Electrons and Holes in Semiconductors." van Nostrand, Princeton, New Jersey.
Shockley, W. (1951). *Bell System Tech. J.* **30**, 990.
Shockley, W. (1952). *Proc. IRE* **40**, 1289.
Shockley, W., and Prim, R. C. (1953). *Phys. Rev.* **90**, 753.
Shulman, C. (1955). *Phys. Rev.* **98**, 384.
Silver, M., Mark, P., Olness, D., Helfrich, W., and Jarnagin, R. C. (1962a). *J. Appl. Phys.* **33**, 2988.
Silver, M., Swicord, M., Jarnagin, R. C., Many, A., Weisz, S. Z., and Simhony, M. (1962b). *J. Phys. Chem. Solids* **23**, 419.
Simhony, M. (1963). *J. Phys. Chem. Solids* **24**, 1297.
Simhony, M., and Gorelik, J. (1965). *J. Phys. Chem. Solids* **26**, 1133.
Skinner, S. M. (1955). *J. Appl. Phys.* **26**, 509.
Smith, R. W. (1955). *Phys. Rev.* **97**, 1525.

Smith, R. W. (1957). *Phys. Rev.* **105**, 900.
Smith, R. W. (1959). *RCA Rev.* **20**, 69.
Smith, R. W., and Rose, A. (1955). *Phys. Rev.* **97**, 1531.
Spangenberg, K. R. (1948). "Vacuum Tubes," Chapter 8. McGraw-Hill, New York.
Spear, W. E. (1957). *Proc. Phys. Soc. (London)* **B70**, 669.
Spear, W. E. (1960). *Proc. Phys. Soc. (London)* **B76**, 826.
Spear, W. E. (1961). *J. Phys. Chem. Solids* **21**, 110.
Spear, W. E., and Lanyon, H. P. D. (1960). *Proc. Intern. Conf. Semiconductor Phys., 5th, Prague* p. 987.
Spear, W. E., and Mort, J. (1963). *Proc. Phys. Soc. (London)* **81**, 130.
Stafeev, V. I. (1958). *Soviet Phys. Tech. Phys.* **3**, 1502.
Stafeev, V. I. (1959). *Soviet Phys. Solid State* **1**, 763.
Steele, M. C., Ando, K., and Lampert, M. A. (1962). *J. Phys. Soc. Japan* **17**, 1729.
Steinberger, I. T. (1960). *J. Phys. Chem. Solids* **15**, 354.
Stöckmann, F. (1961). "Halbleiter Probleme," Vol. 2, p. 279. Vieweg, Braunschweig.
Stratton, R. (1961). *Progr. Dielectrics* **3**, 235.
Suits, G. H. (1957). *J. Appl. Phys.* **28**, 454.
Sussman, A. (1967). *J. Appl. Phys.* **38**, 2738, 2748.
Thomas, D. G. (1968). *Phys. Today* **21**, 42.
Tödheide-Haupt, U., Gyster, A., and Ruppel, W. (1966). *Phys. Status Solidi* **15**, 567.
Tosima, S. (1967). *J. Phys. Soc. Japan* **22**, 1025.
Tosima, S., and Ando, K. (1967). *J. Phys. Soc. Japan* **23**, 812.
Tredgold, R. H. (1965). "Space Charge Conduction in Solids," pp. 74–82. Elsevier, Amsterdam.
Trester, S., Cobas, A., and Weisz, S. Z. (1966). *J. Phys. Chem. Solids* **27**, 1701.
Tyler, W. W. (1954). *Phys. Rev.* **96**, 226.
van der Ziel, A. (1966). *Solid State Electron.* **9**, 123.
van Ruyven, L. J., and Adriaens, W. H. Th. (1962). *Phys. Letters* **3**, 109.
Vogl, T. P., Hansen, I. R., and Garbuny, M. (1961). *J. Opt. Soc. Am.* **51**, 70.
von Hippel, A., Kalnajs, J., and Westphal, W. B. (1962). *J. Phys. Chem. Solids* **23**, 779.
Warter, P. (1967). Private communication.
Waxman, A., and Lampert, M. A. (1969). *Phys. Rev.* (to be published).
Webb, P. W., and Wright, G. T. (1964). *Brit. J. Appl. Phys.* **15**, 385.
Weimer, P. K. (1961). *IRE Trans. Electron. Dev.* **8**, 421.
Weimer, P. K. (1962). *Proc. IRE* **50**, 1462.
Weimer, P. K., and Cope, A. D. (1951). *RCA Rev.* **12**, 314.
Weiser, K. (1965). *IBM J. Res. Develop.* **9**, 315.
Weiser, K., and Levitt, R. S. (1964). *J. Appl. Phys.* **35**, 2431.
Weisz, S. Z., Jarnagin, R. C., Silver, M., Simhony, M., and Balberg, J., (1964). *J. Chem. Phys.* **40**, 3365.
Weisz, S. Z., Cobas, A., Richardson, P. E., Szmant, H. H., and Trester, S., (1966). *J. Chem. Phys.* **44**, 1364.
Weisz, S. Z., Cobas, A., Trester, S., and Many, A. (1968). *J. Appl. Phys* **39**, 2296.
Williams, R. (1961). *J. Phys. Chem. Solids* **22**, 129.
Williams, R. (1962). *Phys. Rev.* **125**, 850.
Williams, R. (1965). *Phys. Rev.* **140**, A569.
Williams, R. (1966). *J. Appl. Phys.* **37**, 1491.
Williams, R., (1970). Injection by internal photoemission, *In* "Semiconductors and Semimetals" (R. K. Willardson and A. Beer, eds.) Vol. 6. Academic Press, New York. (in press).

Wright, G. T. (1959). *Proc. IEEE* **106B**, Suppl. No. 17, 915.
Wright, G. T. (1961). *Solid State Electron.* **2**, 165.
Wright, G. T. (1962). *Solid State Electron.* **5**, 117.
Wright, G. T. (1964). *Solid State Electron.* **7**, 167.
Yamashita, H., Ibuki, S., and Komiya, H. (1963). Mitsubishi Denki Lab. Rept. No. 4. p. 117.
Zijlstra, R. J. J., and van der Ziel, A. (1963). *Physica* **29**, 78.
Zuleeg, R. (1963). *Solid State Electron.* **6**, 193.
Zuleeg, R., and Muller, R. S. (1964). *Solid State Electron.* **7**, 575.

AUTHOR INDEX

Numbers in italics refer to the pages on which the complete references are listed.

SUBJECT INDEX

A

B

C

D

E

T

U

V

W

Z

ELECTRICAL SCIENCE

A Series of Monographs and Texts

Editors

Joseph E. Rowe. Nonlinear Electron-Wave Interaction Phenomena. 1965

Max J. O. Strutt. Semiconductor Devices: Volume I. Semiconductors and Semiconductor Diodes. 1966

Austin Blaquiere. Nonlinear System Analysis. 1966

Victor Rumsey. Frequency Independent Antennas. 1966

Charles K. Birdsall and William B. Bridges. Electron Dynamics of Diode Regions. 1966

A. D. Kuz'min and A. E. Salomonovich. Radioastronomical Methods of Antenna Measurements. 1966

Charles Cook and Marvin Bernfeld. Radar Signals: An Introduction to Theory and Application. 1967

J. W. Crispin, Jr., and K. M. Siegel (eds.). Methods of Radar Cross Section Analysis. 1968

Giuseppe Biorci (ed.). Network and Switching Theory. 1968

Ernest C. Okress (ed.). Microwave Power Engineering:
Volume 1. Generation, Transmission, Rectification. 1968
Volume 2. Applications. 1968

T. R. Bashkow (ed.). Engineering Applications of Digital Computers. 1968

Julius T. Tou (ed.). Applied Automata Theory. 1968

Robert Lyon-Caen. Diodes, Transistors, and Integrated Circuits for Switching Systems. 1969

M. Ronald Wohlers. Lumped and Distributed Passive Networks. 1969

Michel Cuenod and Allen E. Durling. A Discrete-Time Approach for System Analysis. 1969

K. Kurokawa. An Introduction to the Theory of Microwave Circuits. 1969

H. K. Messerle. Energy Conversion Statics. 1969

George Tyras. Radiation and Propagation of Electromagnetic Waves. 1969

Georges Metzger and Jean-Paul Vabre. Transmission Lines with Pulse Excitation. 1969

C. L. Sheng. Threshold Logic. 1969

Dale M. Grimes. Electromagnetism and Quantum Theory. 1969

Robert O. Harger. Synthetic Aperture Radar Systems: Theory and Design. 1970

M. A. Lampert and P. Mark. Current Injection in Solids. 1970

W. V. T. Rusch and P. D. Potter. Analysis of Reflector Antennas. 1970